MW00573966

Praise for *The Community-Scale Perm*

"I haven't been to D Acres, but *The Community-Scale Permaculture Farm* made me want to go. And as any good permaculture project should, this book stacks functions. It's at once a chronicle of the reinvention of an old family property as a twenty-first-century enterprise, a first-hand guidebook for developing a successful community, and a useful how-to for ecological homesteading and farming. If you are doing any of those—or thinking about it—this book should be in your hands."

—Toby Hemenway, author of *Gaia's Garden: A Guide to Home-Scale Permaculture* and *The Permaculture City*

"Homesteading on a hamlet scale, teaching ecological design, building for the future, and creating new family and tribal bonds in a time of dissolution, the D Acres vision sheds light on the uses of adversity to counter the centrifugal forces of our culture. With many rich vignettes and inventive solutions on offer, this book will reward careful reading."

—Peter Bane, author of *The Permaculture Handbook* and publisher of *Permaculture Activist* magazine

"One cannot discuss sustainable agriculture without considering its community context, or vice versa. Trought has given a lot of thought to both, and his experience and observations are well worth sharing. I've visited D Acres and been very impressed by the depth of its vision, which is clearly expressed throughout the book."

—Will Bonsall, author of *Will Bonsall's Essential Guide to Radical, Self-Reliant Gardening* and *Through the Eyes of a Stranger*

"If we want people to live ecologically, get along, and heal the Earth, we need to build a local, ecological economy. Josh Trought's detailed guide points the way to localizing where we get our basic needs and at the same time stays flexible and practical, not utopian-egalitarian. D Acres started with almost nothing, but through permaculture and perseverance, and a lot of hard knocks, it's getting to where it's doing everything right. *The Community-Scale Permaculture Farm* is not merely a model for farmers or for an education center, but for a great kind of life."

—Albert Bates, author of *The Post-Petroleum Survival Guide* and *The Biochar Solution*

"What a broad-minded book and thorough record of achievement for a visionary farm enterprise that takes practical steps to counter the capitalist disaster—an inspiration for the next generation of growers."

—Peter Schumann, founder of Bread and Puppet Theater

"Every community should be so blessed to have a permaculture learning center and demonstration farm in its midst. Josh Trought and company provide pure inspiration for learning essential earth skills. Pick up this book and you too can help people connect with the land and a sustainable way of living."

—Michael Phillips, author of *The Holistic Orchard*

"Josh's manual offers a wealth of practical advice that will be very useful to those adventuring in the direction that D Acres has developed over twenty years in intentional living and permaculture farming. Truly inspiring!"

—Andrew Faust, founder of Center for Bioregional Living

"At a crucial crossroads in our history, this book chronicles a life-giving response to a society bent on self-destruction. With uncommon honesty, Josh Trought lays bare the lessons of once wide-eyed beginners, now seasoned and savvy leaders in the permaculture movement. The journey presented here is inspired and instructive, though nonformulaic and a work in progress. When the dust settles upon our epoch, what will stand out are places like D Acres, which built a living alternative to the ubiquitous, me-centered society. As readers will discover, it is hard work turning dreams into reality, but with Josh as your trusty guide, your body will ache for this 'real' work."

—Jim Merkel, author of *Radical Simplicity* and founder of the Global Living Project

"Josh Trought shows how society can be brought together in harmonious fashion with ecological systems to produce healthy food, close-knit communities, land stewardship, and beauty in a sustainable way. What I love most about the D Acres model is that from the start, their intent was to experiment *and* share the learning. That's exactly what this book does, describing the philosophical and historical roots of collective living and permaculture, as well as the day-to-day work of growing and building, all with clear and compelling storytelling. Nothing short of miraculous!"

—Tim Traver, author of *Sippewissett*

THE
COMMUNITY-SCALE
PERMACULTURE FARM

THE
COMMUNITY-SCALE
PERMACULTURE FARM

The D Acres Model for Creating and Managing an
Ecologically Designed Educational Center

JOSH TROUGHT

Chelsea Green Publishing
White River Junction, Vermont

Project Manager: Patricia Stone
Developmental Editor: Michael Metivier
Copy Editor: Eileen M. Clawson
Proofreader: Deborah Heimann
Indexer: Lee Lawton
Designer: Melissa Jacobson

Printed in the United States of America.
First printing March, 2015
10 9 8 7 6 5 4 3 2 1   15 16 17 18 19

Our Commitment to Green Publishing
Chelsea Green sees publishing as a tool for cultural change and ecological stewardship. We strive to align our book manufacturing practices with our editorial mission and to reduce the impact of our business enterprise on the environment. We print our books and catalogs on chlorine-free recycled paper, using vegetable-based inks whenever possible. This book may cost slightly more because it was printed on paper that contains recycled fiber, and we hope you'll agree that it's worth it. Chelsea Green is a member of the Green Press Initiative (www.greenpressinitiative.org), a nonprofit coalition of publishers, manufacturers, and authors working to protect the world's endangered forests and conserve natural resources. *The Community-Scale Permaculture Farm* was printed on paper supplied by Quad Graphics that contains at least 10% postconsumer recycled fiber.

Library of Congress Cataloging-in-Publication Data
Trought, Josh, author.
  The community-scale permaculture farm : the D Acres model for creating and managing an ecologically designed educational center / Josh Trought.
      pages cm
  Other title: D Acres model for creating and managing an ecologically designed educational center
  Includes index.
  ISBN 978-1-60358-475-3 (pbk.)
1. Permaculture—New Hampshire. 2. Organic farming—New Hampshire. 3. Sustainable living—New Hampshire.
4. Agritourism—New Hampshire. I. Title. II. Title: D Acres model for creating and managing an ecologically designed educational center.

S494.5.P47T76 2015
631.5'809742—dc23

2014043713

Chelsea Green Publishing
85 North Main Street, Suite 120
White River Junction, VT 05001
(802) 295-6300
www.chelseagreen.com

To Betty and Bill for everything, and to Regina for

seeing it through, and to the youth for their enthusiasm,

energy, and limitless creativity; you are the inspiration.

CONTENTS

INTRODUCTION

Nestled on the north slope of Streeter Mountain is a farm known as D Acres. The farm sits on the cusp of New Hampshire's White Mountains overlooking the Baker River Valley. The forest and granite of the mountains dominate the landscape. The air is fresh and crisp; sparkling water flows down from the mountainside above. The four seasons are dramatic and intense, the white of winter briefly conceding to the lush green of summer and brilliant autumn color.

Upon arrival it is clear this is not your typical conventional farm. There are neither feedlots nor endless rows of monocrops. While there are farm animals and food crops, these systems are strategically woven into the landscape in a form that resembles nature. Instead of a farmhouse designed for an extended family, the central building is more like the hub of a village. This community multiplex both offers space for public events and serves the needs of the residential participants. In addition to farm gate sales, visitors venture to D Acres for community food events, workshops, and overnight accommodations, as well as to tour the forest and gardens. The multiplex is indicative of a design that is responsive to change and integrated into the landscape, climate, and resources of this rocky hillside. The farm design recognizes adaptation and evolution while maintaining the goals of nurturing the people and the soil. This unconventional approach to farming has evolved as a response to the dilemmas humanity faces.

Once when I was distributing flyers for D Acres' workshops and events in the affluent town of Hanover, I approached the manager of the downtown bookstore to seek a proper location for posting. As I was explaining the hands-on gardening and land stewardship offerings we were advertising, she interrupted me to say that everyone employed at the store knew about sustainability—they had backyard gardens, and some even had chickens.

While I did not stay that day to question her definition of sustainability, I think it's important to define what a sustainable future would truly look like. Consumerist culture continues to offer greener alternatives to the global market, but I am seeking contentment closer to home. To devise a sustainable future, we must construct a localized economy that provides more than seasonal produce and a handful of eggs, microfarms, and green appliances. Fossil fuels have provided a façade of permanence to the wealth and prosperity evidenced by affluent society. Evolution will require a more profound shift in how we meet our needs.

We must localize our energy, transportation, and resource consumption while creating a community that provides education and medicine for our future. To find this path we must serve as explorers seeking alternatives. To transition from the conventional paradigms, we need experiments with subsequent successes and failures. The community of people, local ecology, and mission-driven agenda of D Acres weave together to represent my attempt to undertake this challenge.

Everyone has his or her own definition of community. Is it something as glib as the people you would split your last beer or smoke with? Is community the bar where everyone knows your name? Is it your immediate friends and family, or is there a geographical basis to community, such as a city block, neighborhood, or township? Is it all of the above? There are multiple layers to community, which require depth of vision to perceive.

During college I spent leisurely time with friends drinking beer and discussing the future. Often the conversation centered on the dream of living in community. We imagined starting a farm where we would enjoy the fresh air, vegetables, and all the freedoms of the good life together. After over twenty years in pursuit of this vision, I remain undeterred.

My initial narrow perspective of both the challenges and benefits of "community," however, has been tempered by experience. The word "community" has changed for me over time. I see D Acres' community as an evolving process and as a natural system affected by its locality as well as its personnel. Communities transition as members age. There are new arrivals and births as well as departures and deaths to consider.

Community implies trust and mutual support. To my mind, we derive community by sharing a common identity. When we share at that level, it becomes instinctual to feel empathy and compassion for others and collaborate for the common good. In this sense, community is limited only by the confines and restraints of our imaginations. To me it is important to extrapolate our feeling of community in global solidarity. By recognizing the shared, finite resources on this planet, we may begin the work of perpetuating on this planet.

After twenty years I am still grappling with how best to explain a vision of community—both the rewards and imperfections of an evolving and adaptable system. My intention in writing this book is not to supply any definitive answer but rather to present a methodology of designed adaptation. In these evolving times we must respond to the environmental factors that not only shape our culture but also serve to dictate our success as a species.

My hope is that this book serves as kindling to you for the abundance of opportunity in the land-based service movement. While the book may be valuable to most anyone, my purpose in writing was to offer a compilation of information that I wish was available when I began farming. By providing a basis of understanding of the farm system, I hope that readers can use this model as a platform for their own innovation and creative living.

The Community-Scale Permaculture Farm is not a volume of statistics and scientific research heavy in academic citations. This text is designed to lay out an overall framework of a system and dissect its various constituent parts. Through this assessment I hope to convey an understanding of how the elements are integrated into a living organism.

While this book provides a wide spectrum of practical information on the physical systems designed into a community-scale homestead, it also reviews the economics and organizational particulars that we have experimented with here at D Acres. Overall, the contemporary, socioeconomic dynamic and human reluctance to change could prove to be more challenging than any other hurdle. To overcome the inertia of convention, the catalyst must be personal motivation. This motivation cannot be fear, but rather an acceptance of our role as conscientious members of the ecosystem. To reach a sustainable social dynamic we must shift our cultural philosophy from a self-serving one to that of an altruistic people with a holistic paradigm. The justification for pursuing this path is both altruistic and compassionate, providing service to others as well as inner peace.

Whatever terminology we choose, humanity's fate will be determined by our ability to sustain ourselves within nature. By systematically addressing the circumstance and designing a conscious path forward, we can face our challenges.

As a founder of the D Acres organization, I have attempted to find personal fulfillment based on scientific, philosophical, and ethical inquiry. My decision to seek this path in response to the crises

Mission

The mission of D Acres is to serve as an educational center that researches, applies, and teaches skills of sustainable living and small-scale organic farming. The center functions as a demonstration farm, offering multiple exemplars of healthy living and striving to improve the human relationship to the environment. Sharing a communal living situation, individuals come to respect and share values of interdependence and love of nature. In addition, the organization supports educational activities directed toward improving the quality of life of residents and the larger community. Our intentions are to develop a farm system that is sustainable and suitable to this climate, and therefore acts as a demonstration and experimental model; increase consciousness about people's impact on the environment by limiting our consumption of fossil fuels and other resources; reduce, reuse, and recycle, as well as emphasize local and on-site production and consumption; develop skills as a group to solve problems, organize, and pursue an agreed-upon agenda; interact with and contribute to the community at large by providing goods, services, and educational opportunities while representing the vision of the organization; provide a training center for the development of skills related to organic farming, forestry, landscaping, eco-friendly construction, and cottage crafts; and develop personal and group skills to improve economic viability through "cottage style" industry.

of our era has yielded, for me, mental and physical well-being through an active lifestyle, spiritual contentment, engagement with community, and connection to nature. With this book I hope to share the sacrifices and immense rewards of pursuing positive practical alternatives to the destructive culture of today, so that it might inspire you to take positive action, *now*!

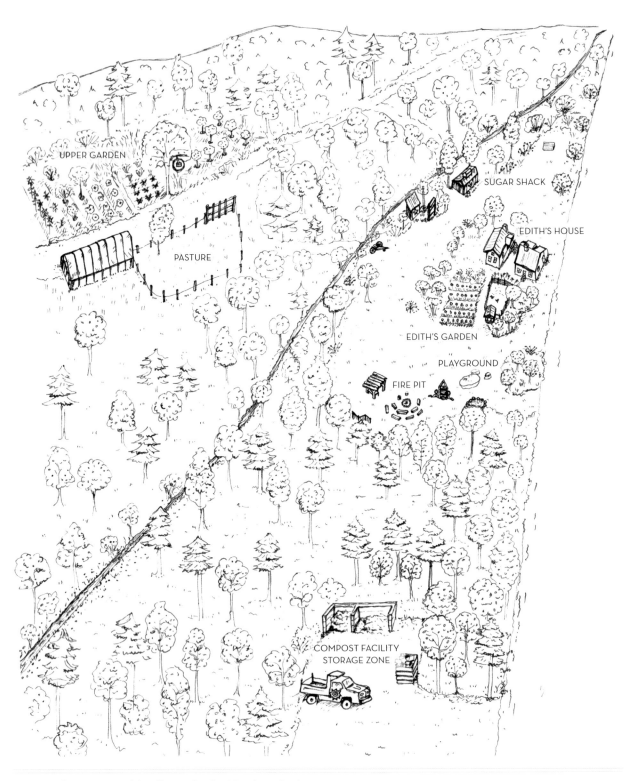

UPPER GARDEN

PASTURE

SUGAR SHACK

EDITH'S HOUSE

EDITH'S GARDEN

PLAYGROUND

FIRE PIT

COMPOST FACILITY
STORAGE ZONE

D Acres of New Hampshire. llustration by Marylena Sevigney.

LOWER GARDEN

AQUATIC ZONE

RED BARN

BLACKSMITH

OUTDOOR KITCHEN

AQUATIC ZONE

COMMUNITY
BUILDING

G ANIMAL

RESOURCE PILE

OPEN-SIDED BLDG
POTTING SHED/
BIG COLD FRAME

OX HOVEL

PIG TRANSITION
ZONE

PIGS

PIGS

CHAPTER ONE

CURRENT PARADIGM

In 1979, when I was seven years old, my family and I moved to Winterville, North Carolina. The warm, mild climate of the state provided relief to a winter-weary influx from the mid-Atlantic and Northeast. The state's inexpensive university system attracted many to the region, while retirees also found havens along the coast. The North Carolina economy was shifting from its agricultural roots.

During the ten years I spent in Winterville until my high school graduation, North Carolina grew approximately 7 percent per year. Seven percent growth has a ten-year doubling time. After ten years there were twice as many people, twice as many cars, twice as many gas stations and box stores. With this growth came a subsequent increase in traffic, noise, and other pollutions of consumer culture. The region's agricultural heritage eroded, while the commuting service class built a culture of take-out food, video games, and virtual reality.

In Winterville we were part of suburban sprawl powered by a medical center and a state university. We moved into a newly built contemporary house at the base of a horseshoe-shaped subdivision that included six other houses and backed into a dense forest. Extensive farmland stretched in all directions; drainage ditches and windrows of pine forest were the only breaks in a landscape cultivated for annual crops of corn, soy, and, predominantly, tobacco.

I spent a lot of time outdoors. Forts, obstacle courses, and explorations were part of the daily routine, and my sister and I regularly observed the neighboring fields' annual cycles of planting and harvest. The ubiquitous tobacco barns were our playhouses on rainy days. But during the 1980s the landscape of this region transitioned from farms to an asphalt-intensive lattice of houses, condominium complexes, strip malls, and convenience stores.

To an extent I was an active part of this farmland destruction. My first job was helping a childhood friend's family business, traveling around the eastern part of the state installing stale, pervasive, ornamental azaleas, hostas, and Japanese maples in suburban neighborhoods, though the experience helped instill a positive work ethic. My second job was in a shopping mall, serving cheesesteaks, cheeseburgers, and other fried foods to the indoor walkers and shoppers, though, again, this exposed me to both the realities of the food and restaurant industry and the rewards of feeding people.

On a recent trip to Winterville I assessed the current situation. The forest behind my childhood home has been cleared, and the fields beyond are filled with more than thirty houses. The irrigation ditches have become a flooding threat, heightened by the heedlessly laid concrete and asphalt. Cars now race on four lanes between stoplights as they maneuver to be the first in line at the drive-through. The windrows of pine have been mostly cut so the sounds and sights of uncoordinated growth of lights, noise, and traffic are intensified. The feeble

Handmade block print by Josh Trought

landscape trees planted in the adjacent subdivision are dwarfed by the typical houses of this era. I find it likely that these houses have identical counterparts all over North America.

Growing up, I am not sure I understood the consequences of this deruralization I was participating in, nor was I aware of other options. The status quo of comfortable consumerism was (and is) the doctrine of unrestrained growth. As I approached college I withered without direction, I was complacent without a passion, and I lacked motivation to participate in the rat race. However, years of study and travel followed, from Spain to Alaska, Washington to Morocco, Colorado to Costa Rica,

increasing my awareness of the disparities that exist between the global north and south, and spurring me into action to build my skills for the purpose of finding a sustainable future and resolving societal inequities. All of this led me to form, collaboratively and with the hard work, talents, and sweat equity of many, D Acres of New Hampshire.

Addressing the Times

D Acres of New Hampshire expands the common definition of a farm. Although we grow vegetables, raise animals, and practice forestry with sustainable intentions, our ongoing development is based on

Handmade block print by Josh Trought

spiritual and philosophical evolution. Our themes of collaboration, cooperation, and mutualism transcend the typical organization of a farm. With the goal of perennial viability for humanity within our ecosystem, we attempt a multidimensional approach to sustainability that encompasses practical, spiritual, and ethical components. We are trying to create a rural community ecology that evolves in perpetuity.

The mission of the organization addresses the cultural crisis of the times. The problems of the world are complex; no single panacea will heal our malaise. The corporate, globalized, military-industrial complex is destroying the planet. Mass media and academia offer few alternatives to the status quo. Population growth accelerates environmental degradation while further accentuating societal inequities. Climate change looms as a repercussion of our age of excess. An economy of addictive consumption has ascended to be the accepted culture of today by capitalizing on our primal competitiveness and our dependence on the corporately controlled valuation of currency and wealth.

D Acres provides alternatives to this contemporary nightmare and confronts the conventional paradigms of our culture on many levels. We place our values on such essentials as clean food and water, housing, botanical medicine, community networks, and education. We demonstrate practical approaches to reducing our fossil fuel footprint while improving our quality of life. There is a clear division separating the fundamentals of conventional culture and the alternatives that we are evolving.

The Value of Labor

The mainstream perspective demeans manual labor, minimizing the importance of people's efforts to provide their own essential needs. Success in our culture means freedom from the responsibilities of the natural world. Cars speed us from place to place; smartphone screens have replaced in-person communication; fossil fuel–derived heat and electricity provide shelter and entertainment that screen us from nature; fresh air is absent from our sterilized environments. We have distanced ourselves from nature with technology.

During the economic shifts of the 1990s the idea that a service economy is superior to manual labor was advanced and that carpal tunnel cubicle work and retail food service jobs are superior to fresh air and creating things with two hands. Our culture promotes a lifestyle in which money provides for all our needs without the satisfaction of physical exertion, as if the epitome of life is sitting on a beach somewhere tropical and sipping margaritas. From slavery to fossil fuel exploitation we continue to utilize unaccounted-for and immoral economic externalities to create the extravagant lifestyle that our culture equates to wealth. This illusion of wealth provides a materialistic, hollow, unsustainable existence removed from reality and altruism.

At some point soon our education and culture must revalue physical work. We need a higher percentage of the population to engage as farmers and shepherds in mutual relationships with plants and animals. The cultivation of food is an act of care for the people and land, rooted in history and cognizant of the future, that requires the presence of human labors. It is through these empowering

and enabling physical exertions that we can maintain our connection toward each other and nature.

D Acres reconnects our community with the physical work of farming. We invite the public to participate in workshops, tours, and hands-on activities. In addition to explaining the benefits of the permaculture processes that guide our farm system, we also share the literal fruits of the labor, demonstrating both the value and the rewards provided by physical work.

The Value of Food

Our culture's sense of entitlement is illustrated three times a day by our eating habits. The Green Revolution's cheap food produced via fossil fuel exploitation has reduced per capita spending on food products to less than 10 percent of the average United States citizen's budget today. At the same time we spend over 50 cents of every dollar on housing and transportation.[1] This proliferation of nutritionally poor food commodities has shifted our economic prioritization. Less than 1 percent of the American population claim farming as their occupation.[2] The overall number of farms in America decreased from 6.8 million in 1935 to around 2 million today.[3]

Profit in the conventional food system is derived from exploitation. The conventional food system robs

---

1. USDA Economic Research Service, "Food Expenditures, Table 7: Food expenditures by families and individuals as a share of disposable personal income"; www.ers.usda.gov/data-products/food-expenditures.aspx#26636.

2. United States Environmental Protection Agency, www.epa.gov/oecaagct/ag101/demographics.html.

3. USDA, Economic Research Service using data from USDA, National Agricultural Statistics Service, Census of Agriculture and *"Farms, Land in Farms, and Livestock Operations: 2012 Summary."* For 2012 data.

Handmade block print by Josh Trought

non-locally produced commodities such as chocolate, coffee, bananas, and sugar. Subsidization and corporatization of commodities such as corn, potatoes, and wheat have artificially lowered food prices. We do not see the care, work, and skill embodied in every piece of nutritious food supplied through sustainable practices. Although they are essential for daily life, the farmer and the land are poorly compensated for their contribution to contemporary lifestyles.

A sustainable food system reinvests profits into the community and natural resources. In a sustainable food economy both the labor and the land are valued as an essential component of our civilization. By respecting the individuals and profession of farming, we demonstrate our understanding of the skills, effort, and energy necessary for food production. A true valuation of the land provides the incentive to conserve, preserve, and enhance this resource. This true interpretation of the value of food in our economy would adjust the appropriate value of food to ensure sustainable production.

Food is the ultimate medicine for our society. When sustainable food is produced, the process constantly reinvests in the fertility of the land as well as the people. As opposed to the synthesized and processed food promoted for profit by corporate greed, sustainably produced food connects the public with fresh, unprocessed nutrition that is the basis of life.

D Acres grows much of its own food or else sources it from farms in the region. By providing an example of the agricultural process as well as offering the food prepared, we demonstrate the capacity of the local food system to the public, who gain recognition of not only the value of food but the work it takes to produce it. While the flavor of local is superior, the value of the people and the land base should also be savored.

Competition

Modern society values competition and consumption in place of cooperation, community, and

both the land and the laborers. Corporations have horizontally and vertically integrated themselves into all aspects of the food production system. The scale of corporate commodity producers has overwhelmed the competitive advantage of small farms, while government commodity subsidies exacerbate the inequitable situation. Resources and people are being plundered to supply bananas year-round in New Hampshire for $1 a pound. The standard price for commodity potatoes is $.05 per pound, requiring a farmer to utilize resource-intensive production techniques to compete. Within the economic system cheap food provides an economy of illusion.

While it may seem ridiculous for a farmer to be paid an hourly rate proportional to all the white-collar professionals, we have lost the real value of food in our society. In our quest to satiate our consumptive addiction we indulge regularly in the trade of exploitive conservation. Competition exudes from virtually all aspects of our lives. The primal urge to be faster, smarter, bigger, stronger, sexier is a response to our evolutionary beginnings. Primitive humans utilized these genetic attributes to survive. Biologically we are predisposed toward dominating and seducing others in an attempt to pass on our genes. The attributes of physical strength and intellect have been utilized throughout history to fulfill our egotistical desire for victory and compensation.

Competition is expressed in everything from what we eat for breakfast to what we do for a career and where we choose to live. Instead of focusing on sharing and cooperation, our culture is trapped by a primal instinct from which we urgently must evolve. Our culture alters the perceived relations between people to create the illusion that all of life's interactions are competitive. It is my belief that this primal urge to compete and exploit among each other is a principle component of our cultural malaise.

The framework of competition involves winners and losers. This "game" between competitors creates a structure where the victors reap the rewards while the losers are resigned to the scraps. On an individual level this game of life is defined as an accumulation of wealth and assets. Although we ridicule the naïveté of the ancient Egyptians, we are competing in a similarly unrealistic and irrational game. We are now enacting the pharaohs' legacy by amassing material possessions as if we could retain them for eternity. On a large scale this competition plays out in the resource-extraction methodology of global economics. To conceive of the earth as humanity's playground for an economically motivated game of resource extraction and consumption is not a soothing thought. This equates humans with greedy exploiters, the hungry manipulators of a benign planet.

D Acres demonstrates alternatives to the traditional competitive model. As a not-for-profit, our model is not based on growth or accumulation of currency and resources. Instead, our farm system provides service through collaboration.

Money is only an illusion of wealth, but it has come to represent value in our society. People ask, "What would you do for a million dollars?" as though this is an important question to consider. What act would you commit in exchange for the dollar value that it would translate to? When this question is considered normal to ask, money essentially defines our morality.

With money as our cultural rudder, we are caught in a vicious cycle. Money provides fluidity to the valuation and transfer of resources. Through this monetization of resources, exploitation and consumption are accelerated and enabled. Economists value the accelerated consumption as growth, and the cycle is encouraged until the resources are expended. But is there a dollar limit we can set on the value of clean air and water? For what dollar amount are we willing to exchange those intrinsic needs? Clean food and water are priceless without a value placed in currency.

The debate over a nuclear energy–powered future highlights our contemporary dollar-driven delusion. By accepting power from nuclear plants, for example, we are willing to exchange a dangerous fuel source to provide energy for our immediate needs. We are placing nuclear as a socioeconomic choice over other options such as photovoltaics. We are declaring, in essence, that dollars define the decision, with irreparable consequences for generations, because it makes current economic sense. We could choose instead to make a moral decision based on what is right and what is wrong. Can we implement a system in which the populace is allowed to make informed decisions on solar or nuclear power for our future without the constraints of the current monetary illusion? The true long-term wealth of our planetary ecology is threatened by our short-term vision and profit-driven economy. The air, land, and water are our habitat. If we agree to their privatization, we are conceding that a corporate entity can own our shared resources.

Of course, no economy can exist without offering the basic necessities of food, shelter, water, medicine,

Handmade block print by Josh Trought

and community. But contemporary Americans have very little hands-on contact with their essentials. We rarely harvest our own food, build our own shelters, sew our own clothing, or provide medicine and care to our infirm. These basic needs are provided by profit-driven corporate institutions that depend on growth to satisfy shareholders. It is a modern dilemma to be both civilized and so far removed from our essential needs. Handheld gadgets exhaust daily hours and add a vapid haze to our interpersonal relations. We drape ourselves in technological bling to hide the paucity of heart and soul in our current culture. Instead of avoidance we must renew our focus on becoming self-sufficient. If we accept this

responsibility to ensure a sustainable future, we will perpetuate a *truly* rich culture.

There must be a realization that resources cannot be transferable, or degraded, or valued in terms of currency. The protection of these resources is a legacy of our human presence on this planet, and we must prevent their monetization.

D Acres' goal is to create bounty and abundance ecologically. Instead of acquiring legal tender, we seek community capital and quality of life. Ensuring planetary health is our primary motivation rather than accumulating personal possessions. To accomplish this we share our wealth, be it information, food, resources, or helping hands. We willingly and ably redistribute our time and energies for the common good.

Competitive instinct can be redirected toward community engagement. People can compete in efforts of civic volunteerism. Farms can compete to grow more nutritionally dense food with fewer fossil fuel inputs. Communities can compete to reduce single-passenger vehicle miles. Why is our economic competition reduced to forms of consumption and resource extraction? We must translate and transcend into an economy that does not compete with each other and the earth. We must become competitors for global good. The arena is the planet, and we share the water, air, and land. Instead of shifting blame we must accept this responsibility and unify as a global team. Instead of constantly seeking an opponent, the victim, the loser, we must seek collaborative, proactive solutions to demonstrate our competency.

My late aunt Edith was pivotal to inspiring the founding of D Acres and freely offered unsolicited advice and stories of her life experience as a modern-era homesteader. She once recounted to me a conversation she had in the 1940s with an art teacher of a traditional Early American decorative technique. At the time the art practice was rare and highly valued; there was considerable interest in marketing products for sale. When my aunt asked the teacher if she felt threatened by teaching others this technique, she relayed that she was not but rather was excited and delighted by the interest. She became more well practiced and added skills by teaching. She felt inspired by the art possibilities and innovations that further interest would create. To progress as a society we must similarly shed negative competition and use our competencies and instincts to empower each other. Surprisingly perhaps, creating mutualistic relationships is more gratifying to the ego than defeating others.

It is important to fulfill our destiny as creatures of this planet to preserve a future for those that follow. Perhaps the moment has to come to realize a vision for the future that diverges from the status quo. With our brief time here, we must transcend the primal urge to compete and consume that has led us to the inequities of a hierarchical society. We must transcend our acceptance of monetized resources and share the value of the true wealth of this planet. It is time for our culture to evolve.

At D Acres we are aware of the limitations innate in the conventional system. We are an alternative to the traditions of the family farm and corporate control of the food system and are developing models that can provide continuity to farms and livelihood for their inhabitants. As our population shifts from resource-intensive urban and suburban population centers, there will be a need for alternative community ownership of farm enterprises. There is land that will need management by groups of people who are not organized in traditional corporate or family hierarchies and management constructs.

We see D Acres as part of a cultural revolution. Through research and experimentation we are creating an alternative to the consumptive, competitive society of our era. We are spreading information and practical solutions that yield hope and inspiration. Actualization provides benefits beyond educational services. By embodying this living process we choose a conscious evolution.

CHAPTER TWO

HISTORICAL PRECEDENTS

In guiding its efforts moving forward, D Acres seeks the wisdom of historical precedents, and the many efforts that have contributed significantly in our development. We do not take a dogmatic approach to any specific methodology. Instead, we combine perspectives from a wide spectrum that are aligned with our intentions. In response to the eternal and modern questions that face humanity, there is a common thread: our collective ability to respond to our environment. By pooling the experiences,

A handmade flag mounted on the Community Building deck with Mount Carr in the background

Diversity is possible in a natural, chemical-free garden system.

technologies, and traditions of humanity we can continue to implement responsive and adaptable systems.

In essence D Acres is a manifestation of distinct yet overlapping trends in society. We draw inspiration from the recent era of environmental awareness, as well as the back-to-the-land, communities, permaculture, and localization movements. In addition, there have been specific projects and organizations that have been especially influential in our journey. These past and present movements and projects all share the theme of seeking alternatives to a destructive status quo. By combining positive aspects from this diversity of antecedents, D Acres is designed for resiliency. And by drawing from the successes and failures of history, we can shape a better future.

Environmental Awakening

I was born in 1971. My parents' generation was the first to conceptualize the global impact of humanity. By campaigning to educate the public, sharing information, and developing technologies, they responded to the repercussions of industrial-era man. This chapter identifies particular catalysts that have influenced the development of this awareness. This collection of events, people, and books form a library of human knowledge that serves as convincing testimonial and a foundation for the creation of D Acres.

In 1962 Rachel Carson published *Silent Spring*, which detailed the catastrophic effects of DDT in our environment. This book is often credited with signaling the beginning of widespread awareness of the damage synthetic chemicals were causing the environment. The public began to develop an environmental consciousness that would peak in 1970, when the first Earth Day was celebrated, announcing a new era in cultural awareness of environmental issues. Two years later *The Limits to Growth* was published, which outlined the lack of viability in our continued population growth and resource consumption patterns.

Mother Earth News and *The Whole Earth Catalog* provided inspiration for those eager to immerse themselves in self-sufficiency information. *Mother Earth News* is a magazine dedicated to the modern homesteader. Although glossy and commercialized in the contemporary era, the roots of the magazine were sown in practical tips and advice. *The Whole Earth Catalog* gathered tools and appropriate technologies geared toward intrepid adventurers of the world. The tools ranged from tepee and guitar designs to advice on circumnavigating the globe. This esoteric collection of catalog items continued as a journal until the publishers ceased printing in 1999.

Back to the Land

As long as there have been cities, there have always been those who seek a return to nature. The draw toward a self-sufficient life is innate. Natural settings provide a connection and freedom that are lost in the urban environment. But there is a price to pay to escape the pollution, crime, and poor health of cities. People may have chosen the land to be free from the regulations imposed in high-population centers, only to instead be beholden to the vagaries of nature. The difficulties of this transition are immense, and there are many strong reasons that the general migration has historically been gravitating toward urban centers, which possess advantages in the availability of skill diversity, public services, and other social and infrastructure support systems.

In the latter half of the twentieth century Helen and Scott Nearing became the most renowned advocates of rural living, and icons of what is known as the back-to-the-land movement. They spent over sixty years practicing, writing about, and presenting on their lifestyle. Scott had been a vocal proponent of peace during World War I and was later professionally blacklisted for his public positions regarding child labor and corporatization. Helen was a trained musician who had spent time traveling with Krishnamurti. In 1932 the Nearings left New York City for a farm in the Green Mountains of Vermont. By fleeing the city they hoped to escape the constraints imposed by the economic system. They felt dislocated and distanced from the conventional attitudes, especially with regard to the slaughter of humans and animals. They moved to Vermont to improve their health and livelihood, which they felt were constricted by the shallowness of the city, and to develop a home economy that would supply their subsistence.

In 1950 the Nearings published *The Maple Sugar Book*, which documented their business plan for a value-added, natural resource microenterprise. *Living the Good Life*, a philosophical and practical manifesto, followed in 1954. The book defined why the Nearings had chosen to leave the city, as well as how they had coped with the ramifications. The book's insights vary from gardening pointers on blueberry cultivation to dietary preferences and the prioritization of life's tasks. *The Good Life* particularly resonated with me when I first read it while working construction near the Mendenhall Glacier in Alaska. Scott and Helen eventually moved to the coast of Maine, where they continued to publish and present their cultural critiques and alternatives. Scott published *The Making of a Radical* and lived at their homestead, Forest Farm, until the age of one hundred. Before Helen's death she founded the Good Life Center, which is still open to the public, alongside Eliot Coleman's Four Season Farm.

During the Great Depression in 1935 another writer, M. G. Kains, published a book that would inspire a generation of back-to-the-landers. *Five Acres and Independence* was Kains's practical guide

Delbert Gray was an avid farmer, artist, and outdoorsman.

to the selection and management of a small farm. The guide provided advice regarding farm selection and rural living, as well as particulars on bees, green manures, irrigation, and orcharding. In its preface Kains quoted Abraham Lincoln, who said "The greatest fine art of the future will be the making of a comfortable living from a small piece of land." *Grow Your Own Fruit* was published next, in 1940, and presented methodology for fruit cultivation in an alphabetically arranged encyclopedia format. The following year Kains published his step-by-step account of returning to the land entitled *We Wanted a Farm*. My Uncle Delbert Gray, who moved his family to Dorchester, New Hampshire, in the 1940s, was particularly inspired by Kains's writing.

Intentional Communities

While back-to-the-landers tended to seek isolation, throughout history there have been movements to form more populous enclaves. While the public perception of communities in the modern era may evoke images of hippies, intentional communities are as old as humanity itself. Most intentional communities form based on shared ideals or beliefs, typically religious or political convictions. As early as the fourth century Christians established monasteries to pursue fulfillment of their life's devotion to God. In the sixth century followers of Buddha had rejected wealth and collaborated in ashrams dedicated to a productive spiritual existence. By 1527 the Hutterian

A hen finding respite in the Mandala Garden, the kitchen garden for the main Community Building.

simplicity. In 1698 Jacob Ammon began creation of the Amish communities, built upon strict Mennonite religious practices.

In 1825 a new type of nonreligious community originated at New Harmony, Indiana. Founded by industrialist and social reformer Robert Owen, New Harmony was a village-style community whose inhabitants strove "for unity and cooperation." In the same year Frances Wright established Nashoba in a Memphis, Tennessee, suburb as a community that would provide job training to black slaves and empower them to earn their own sale price. This project was intended as a strategy for slaves to buy their own freedom without economic loss to their owners, then be transported to emancipated settlements in Haiti and Liberia.

Other examples of both religious and nonreligious communities include Boston's Brook Farm, which was established as a refuge for intellectual expression and was frequented by Ralph Waldo Emerson. The Oneida Community was formed in 1848 in Oneida, New York, based on the belief that through practicing the ways of Jesus, one could arrive at a state of perfect grace. Adherents of the single-tax theory proposed by economist Henry George founded Arden in Delaware in 1900.

Communities also formed to address social needs and inequities. In 1889 Jane Addams founded Hull House. This Chicago organization offered protection and services to the urban immigrant masses displaced by the onslaught of industrialization. Gould Farm, in Monterey, Massachusetts, was established in 1913 to provide a community environment for psychiatric treatment, rehabilitation, and recovery. While these experimental communities often focused on art, spirituality, or social justice and were therefore generally excluded by the mainstream culture, they provided models for later generations interested in initiating communal living projects.

In the 1900s models of collective living continued to evolve and integrate diversity of goals and design. Degania was founded in Palestine in 1910, the first of hundreds of Jewish communal societies known

Brethren was founded and began practicing a form of common ownership. By 1649 peasants in Britain, known as the Diggers, had revolted to communally occupy the land of the nobility.

European settlement of the New World was often accompanied by this communitarian spirit. The Puritans, who founded Plymouth Colony in 1620, were communitarians bonded by their beliefs in as kibbutzim. These religion-based organizations formed communitarian economies based on agricultural and cottage industries. In 1937 the first co-op house was established in Ann Arbor, Michigan. This innovation in student housing offered a communitarian option for college students.

In India in 1956 a rural, nonpolitical, nonreligious educational community was formed; Mitraniketan was founded to empower people "to think globally and act locally." They also focused efforts on deeper spiritual and practical development of a holistic human individual.

All of these diverse communities helped to create the framework for an explosion of idealistic adventures that constituted the communal movement of the 1960s, which is what most people think of today when they imagine communitarian living. In the late 1960s thousands of hippie communes sprung up across the United States. This commune era, however, was short-lived. Disorganization, a lack of practical skills and work ethic combined with overconsumption of "free" love and drugs all detracted from the success of the individual islands of community. As personalities interacted without boundaries and the realities of weather, taxes, and raising families came due, many communities' existences were brief and brutal. Several, such as the Renaissance Community in Massachusetts, prospered for a decade or so, growing with multiple properties, businesses, and hundreds of members only to self-destruct because of personality conflicts and unreasonable experiments in communal living. But with notable exceptions such as The Farm in Tennessee and Virginia's Twin Oaks, which continue today, most of the communities established during this era did not survive.

The concept of cohousing came to fruition in Denmark during the 1970s. Cohousing encouraged families to participate in community living by offering collective parenting, while allowing them to retain their individualities. Cohousing's intentionally designed built landscapes combine access to amenities such as central heating, while providing private individual family accommodations.

Ecovillages are communities that utilize sustainable technology and aim to conserve energy based on shared environmental consciousness. This term is also associated with projects in which substantial income is derived from cottage industries or collective business endeavors. Founded in 1930, the Icelandic children's home and organic farm known as Solheimar is considered the first example of this concept. Sirius, founded in 1978, and the EcoVillage at Ithaca are two prominent examples of enduring projects in the Northeast.

While there was a high percentage of short-lived communities during this era, these projects have yielded a perpetuation of the movement and continued experimentation. The Fellowship for Intentional Community was established to provide connections and support among existing communities. This organization thrives to this day, publishing *Communities* magazine and an online directory of thousands of intentional communities.

Localization

As our economy and culture have continued their trajectory of resource consumption and pollution, outspoken advocates have offered alternatives. The most prominent proponent of the power of the puny in the globalized era has been E. F. Schumacher. His book *Small Is Beautiful* has empowered a localization movement that focuses on local currency and land-use alternatives. The concept of localization has been particularly vibrant as applied to food. The term "locavore" gained prominence as activists challenged their own ability to eat a local diet. The resurgence of farmers' markets in the last twenty years has provided a venue for sales as well as a community connection for the agricultural movement. The Internet has spawned organizations such as Freecycle and craigslist to encourage the trading and reuse of secondhand goods.

Rob Hopkins's transition town concept has become a popular approach to addressing the challenges of contemporary urban development. The concept, which is based on a pilot project in the UK, provides a systematic framework for sustainable design within a neighborhood or town. The transition town movement has started a dialogue on larger issues such as zoning, legal tender, taxation, and property ownership that have institutionally repressed positive progress at this level.

Specific Examples

To imagine a project such as D Acres requires inspiration. While the various schools of thought, gurus, and authors provide intellectual fodder for envisioning alternatives, there are also concrete examples of projects that have influenced D Acres' development. These projects have faced similar real-life challenges and serve as mentors on this shared journey of discovery. Of particular interest to me are projects that serve the community, educate, and offer experimental innovation and research.

Founded in Colombia in 1971, Gaviotas is an ecovillage, an experiment in sustainability that inspired Alan Weisman's book of the same name, which outlined the mercurial rise of this rural community welfare project. The inhabitants of Gaviotas built their own hospital and developed renewable energy technology, such as deep-water well pumps powered by children's seesaws. Their reforestation program created a renewable cottage industry engine that powered local employment.

In 1969 the New Alchemy Institute was formed on Cape Cod to experiment with ecological design principles. The institute provided tours and documentation of sustainable living experiments in action, related to renewable energy, passive solar, and season-extending growing techniques. This pioneering institute documented research for over 20 years and incubated personalities, such as Stonyfield Yogurt founder Gary Hirshberg and wastewater guru John Todd. Also on the Cape in the seventies, Anna Edey began her Solviva Solar Dynamic, Bio-Benign Design project. Her book *Solviva: How to Grow $500,000 on One Acre & Peace on Earth* has become a cult classic among aspiring gardeners and homesteaders. Although the claims of the title are grandiose, the book is packed with energy conservation and lifestyle tips for a practical, rural existence. The book not only outlined strategies for season extension and wastewater management; it also outlined entrepreneurial strategies for selling salad greens and other value-added agricultural products.

The 1980s saw the manifestation of two Northeast institutions that developed around the personalities of sustainability pioneers. In East Corinth, Vermont, Laura and Guy Waterman embarked on a Nearing-inspired homestead from which they advocated for wildness and environmental stewardship and modeled the simple good life. Their nonelectric lifestyle focused on having minimal intercourse with traditional, currency-based economic systems. Their homestead, Barra, would eventually attract such notables as Jim Merkel, founder of the Global Living Project and author of *Radical Simplicity*. In Maine Bill Coperthwaite espoused the possibilities of a simple foraging existence using the easily transported yurt as the preferred housing option. These land-based models created by individuals inspired D Acres toward our roles as sustainability educators.

In the 1990s Hartland, Vermont, saw the formation of a community that integrated cohousing and small-scale farming. Cobb Hill combines the basic elements of cohousing, including a shared central heating system and shared community buildings, but also incorporates a working farm with dairy and vegetable production. The community coalesced through the leadership and vision of Dartmouth professor Donella Meadows, coauthor of the definitive Earth Day–era book *The Limits to Growth*. Another project that influenced our activities is Earthaven, founded in 1995 in the mountains of western North Carolina. Earthaven provided the first domestic example of a permaculturally designed ecovillage with shared ownership and governance.

Internationally, there are many well-known examples of pioneering communities as well. Damanhur in Italy has a spiritual- and artistic-based community where over a thousand members reside. They operate cottage industries and have developed an internal monetary system. Freetown Christiania is an example of the self-organizing independent spirit in the midst of Copenhagen, Denmark. Auroville in India is a spirituality-based community of over seven thousand people. This endeavor has spawned additional projects, such as Sadhana Forest, with reforestation activities in Haiti and Kenya.

Permaculture

In the early 1970s the word permaculture was coined by two Australians named David Holmgren and Bill Mollison. The two men collaborated on theorizing a permanent agricultural system while Holmgren was a graduate student of Mollison's. The brilliance of their work was the recognition that agricultural systems could be synthesized to mirror natural systems. In principle the idea was to observe the characteristics of biologic sustainability presented through its 1.8 million-year evolution existence versus basing a system on the constructs created during the 5,000-year existence of human civilization.

Permaculture was originally conceptualized as a system of permanent agriculture. The original concept observed the diversity and abundance of nature's ecological systems and sought to recreate through design an edible perennial Eden. By combining human, plant, and animal systems the original permaculture model focused on subsistence-style small land holdings. It soon became evident, however, that the principles undertaken on a small scale could be expanded to encompass not only larger farms but also urban and suburban dwellings. In addition, concepts of permaculture could be employed to address issues of economic, social, and political sustainability. From its basis as an agricultural system, the broadening of the permaculture definition now includes a connotation of permanent or perpetuating culture.

There are fundamental ethics and principles that define and regulate the philosophies incorporated into permaculture design. The three fundamental ethics are defined as care for the earth, care for people, and fairness and equity in resource distribution. Care of the earth includes all life and natural resources such as air and water. Care for people provides access to essential resources and encourages self-reliance. Fair share connotes a distribution and sharing of time and resources for the common good. The holistic charter provides an ethic for design choices, while the principles are the elements of the design.

Principles are provided by Holmgren and Mollison to encourage effective design and successful implementation. While they may be relayed in simple vernacular, such as "make hay while the sun shines," these principles reflect deeper systems theory, such as conservation of energy and storage of the natural yields of nature, valuing diversity, and the importance of slow, incremental solutions. Another principle, designing for relative proximity, is a common engineering practice; extending this principle into the realm of culture and daily life makes sense as we shift from a fossil fueled society. On their own each of these principles can be seen as practical, common sense, but the genius of permaculture systems is in combining these principles so that designs are strengthened and improved. While this process requires planning and investment, systems are designed to yield exponentially in the longer term.

One of the profound insights that Mollison offers as a principle states that the "problem is the solution." While this may appear nonsensical, its design jujitsu helps to identify the conflict at hand and fundamentally addresses the root of the perceived dilemma. For instance, if you have a problem with ground hornets stinging people in the garden, then you might want to encourage—and not remove—local nocturnal animals such as skunks, which eat ground hornets, and are themselves preyed upon by great horned owls, which cannot smell. This would

The opening circle at the first regional permaculture gathering held at D Acres in 2006

represent a permaculture solution, which incorporates natural feedback and diversity into design. It is important to realize that nature typically responds to disruption and disparity of ecological systems with alternatives.

Following the conception of permaculture, Mollison and Holmgren took divergent paths toward pursuing their passion for sustainability. Bill became an international teacher and director of the Permaculture Institute. He developed a seventy-two-hour design certification course to resonate the ideals and practices of permaculture. While Mollison became the international face of permaculture, David Holmgren found a rooted, localized expression of his permaculture vision, retreating from academia to a small homestead he dubbed Melliodora in the urban-rural fringe of Hepburn, Victoria, Australia. From this base Holmgren chose to build a career as a design consultant and conduct homestead-scale permaculture implementation with his nuclear family. From this small one-hectare plot he designed and implemented a passive solar house, gardens, and water-retaining landscape, which he chronicles in the book *10 Years of Sustainable Living*. In 2003 his book *Permaculture: Principles and Pathways beyond Sustainability* was published to more fully elaborate his expanded vision and perception of permaculture application.

After the publication of his own books, *Permaculture One* and *Permaculture Two*, Mollison left the confines of academia and began teaching courses internationally, including in North America as early as 1979. The curriculum for the design course essentially followed the contents of the

576-page design manual, which specializes in lower to midlatitude climate zones. Mollison chose the design courses as the basis for a grassroots horizontal influx of permaculture enlightenment. Designers certified by his certification courses would then be eligible to teach permaculture classes to others, building a cadre of teachers to spread the concept.

Still, the concept of permaculture defies traditional educational pedagogy by insisting that nature is our true instructor and observation is the tool to discovery. Instead of being dependent on human instructors, the permaculture school of design promotes the realization that we ourselves are just a part of the design. It is a cooperative school where natural elements such as landforms, plants, humans, and animals interact in an ever-evolving flow of life. By building on the complexity and diversity of its various elements, this continuous process yields exponential strength and abundance. This reality defies our traditional education system, in which scientific authority yields predictable and definitive results. Permaculture accepts the reality that the future is uncertain and that while complexity can be interpreted, it is not entirely predictable. Part of the difficulty of translating this concept to the general population has been the difficulties inherent in an experimental design school.

As Mollison traveled globally, this horizontal structure spread beyond his capacity to control it. Ideals spread organically without a strict process of accreditation. Expensive courses and unqualified instruction conducted in remote, exotic locales began to undermine the true purpose of this empowering, practical design system. At times eager permaculture graduates rushed into teaching, turning it into a pyramid scheme that resulted in poorly qualified permaculture instructors. Course content was unregulated, and participants' experiences varied widely based on the broad range of courses offered. During these growing pains Mollison reembraced the regulatory system and attempted to impose international copyright restrictions on the usage of permaculture concepts. This trademarking of permaculture was a severe response to the externalities of unplanned, haphazard growth and expansion.

Over time permaculture course curriculum has continued to be refined and now represents a wide breadth of opportunities. There are many different styles and objectives of permaculture courses. The content of the curriculum varies, as do the expectations of the participants and course instructors. In fact, many question whether the goal of courses is for people to earn a certificate in permaculture design or simply to certify more permaculture designers. Currently, while there may be some notion of adherence to the design manual, there is a wide divergence in classes offered as the design certification course. To me, there should be a buyer-beware mentality regarding the plethora of permaculture classes available. Tours, guest presentations, and audiovisuals are dependent on availability and the subjective interpretation by instructors of the course curriculum. The ability to incorporate hands-on activities such as gardening and earthen construction vary widely, depending on the location and schedule of the course. Courses are offered in institutional settings at universities as well as at small farms and ecovillages. Some courses have emphasis on rural, urban, or suburban settings, while others may be more focused on community building. I would also recommend a course in your bioregion so that you can build connections with the people and plants as well as have realistic insight into climatic and natural resources in that region.

Permaculture in Practice

For some the term "permaculture" implies the promise of a permanent utopian existence where sustainability is scientifically quantifiable. To me, loading the term with this kind of presumption ignores one of permaculture's basic tenets: respect, compassion, and empathy for all forms of life. The idea that we can address how we interact on this planet through a designed process is reasonable, but permanence in any form is elusive. The reality is that the problems we face as a culture will continue to evolve, and likewise our designs must correspond to circumstance. If human culture and our relationship to the earth are to evolve, so must our understanding of life as nonstatic and entropic. The point of permaculture is to be conscientious and systematic in the ethics and principles with which we approach this relationship and respond to our environment.

This process of evolution is also transpiring within the world of permaculture itself. The world of "permaculturalists" is expanding into the mainstream with its vernacular of chicken tractors, rocket stoves, swales, hugelkultur, and herb spirals. As the worldwide awareness of permaculture grows beyond the lifetimes of its initiators, how will the movement be perpetuated? If mainstream America accepts the permaculture brand, will it become a diluted version like the USDA organic certification program or just another marketing term like "green" and "local"?

Because of the complexity of the topic and the imprecision of a short explanation of permaculture, dissemination to the general population has been difficult. Permaculture implementation requires recognition of the complexities of the chaos,

entropy, and evolution that are universal. This design system, while based on our interpretation of reality, is as complex and multidimensional as nature. Ultimately, permaculture is an attempt to proactively plan for the uncertainties of the future: an infinitely difficult challenge.

Permaculture is a set of tools whose principles and processes are transferable from the garden and homestead to government. It offers a systematic approach to designing alternatives that address the crises of our era, transcending the schools of anthropology, engineering, physics, biology, chemistry, philosophy, spiritualism, agriculture, sociology, political science and economy by improving humanity's counterproductive constructions through functional, sustainable, synthesis with natural systems.

Permaculture is also sometimes referred to as ecological design. The particular words we use provide us with a common language, but it is the understanding of the concepts that is important, as this encourages greater awareness of the possibilities for dismantling the destructive culture of exploitation and empowering its replacement with the embodiment of the permaculture concept.

CHAPTER THREE

SENSE OF PLACE

Permaculture design is responsive to circumstance. The design of a farm system may vary radically depending on its location. While farming options available in the climates of Florida and New Hampshire are obviously distinct, the difference in flat land versus hillside or proximity to urban markets within the same climate can also impact farm design and implementation success.

Heading toward the farm on Streeter Road during the midst of a winter storm

Fall view to the north, with Mount Carr more prominent on the right and Mount Moosilaukee, already snow-covered, being the tallest and the headwaters of our region's watershed

Natural resources dictate the possibilities for designing and implementing a farm system. By recognizing the parameters of place you can be responsive to the abundance and scarcity of a particular location. The resources of place are not defined solely by natural conditions such as biodiversity, geography, or climate but also by culture and socioeconomic history.

In addition to place, the design reflects the people involved and their corresponding goals. For perpetuating a successful system you must seek to answer the question of how people are to be fed by the land, both physically and spiritually through design, and be conscious of the systems that sustain us.

This chapter outlines the parameters of D Acres' place. It is an accounting of observations that lead us to an assessment of design possibilities. While our organizational goals motivate design, successful implementation must be responsive to the practicalities of our specific location. Understanding of local human history, combined with the elements of biology, weather, and geography, produces a sense of place that ties us to the landscape. Our lifestyle's success is dependent on intimacy with these basic aspects of reality, which solidify our sense of place with an emotional, familial connection.

Geography

D Acres is located on a bumpy road a mile off the state-maintained road in Dorchester, New

The south branch of the Baker River flows near the property.

Hampshire, on the north slope of Streeter Mountain. There are no traffic lights, billboards, or police in Dorchester, no fire department or ambulance service. While traffic is occasionally audible from the highway, cars passing the farm are rare, with the exception of a half dozen local commuters. The sparsely populated landscape is over 95 percent forested. The natural landscape is intruded upon only by the scars of roadways and single-family houses that sporadically dot the countryside. Cellar holes, cemeteries, and stone walls littered through the woods are the remnants of previous attempts at human settlement. The view to the north encompasses the southwest boundary of the White Mountain National Forest with Mount Carr looming large to the right and Mount Moosilauke (4,802 feet) rising up from the distant horizon.

The last ice age had enduring effects on the landscape. Glaciers scraped the mountains down to their granite bedrock. Buoyant, organic material was carried to the valleys as the melting ice deserted rocks and sand on the hillside. As the one-mile thick of glacial ice melted on the landscape, the flooding was beyond conception. After this catastrophic geologic event, the landscape was repopulated by a forest adaptable to the new soil and climatic conditions. Our Northern Forest is a combination of deciduous species that are more common in the south and coniferous species that are predominant in the boreal forests farther north.

Dorchester is uniquely situated along a ridge that divides two major watersheds: the Connecticut River Valley and the Pemigewasset and Baker River watersheds. D Acres' terrain is sloped to the north, draining into the Baker River (Asquamchumaukee) watershed. Our soil is sandy, with a cap of richer topsoil that is from one to three inches thick. While the ledge protrudes in some locations, the depth of the rock below the soil varies throughout the town.

The latitude of Dorchester is 43 degrees. At this northerly location the summer days offer over sixteen hours of daylight, with the sun setting at nearly ten o'clock at the summer solstice. As the axis of the earth shifts through the seasons, this solar influx is offset when the last days of December end at 4:00 p.m. In comparison to the three hundred days of sunshine typical in Colorado, New Hampshire averages about 140 days per year. We also average forty inches of precipitation per year, although a considerable portion is stored seasonally as snow and ice.

Because of the constraints of climate and geography, Dorchester has never been a mecca of food production. This land is not a zone of immense fertility, and the area's culture has been eroded by depopulation and modern technology. While evidence of farming still exists in valley towns alongside Dorchester, the majority of upland fields in Dorchester have been abandoned to the resurgent forest.

Weather

While the weather of the White Mountains is predictably erratic, there are some seasonal generalities. For instance, certain outdoor activities are more common at specific points of the year. Ice fishers exchange places with boaters on the lakes as the seasons shift. Loggers prefer sugaring to pulling wood through the mud as the snow melts. Traditional activities such as cider making are dependent on cyclic seasonal progressions, equipment, and knowledge that define this region. Wind is common in all four seasons, typically arriving from the west, although nor'easter storm systems and hurricanes arrive from different directions. Precipitation is also common throughout the year, with the heaviest snows typically from February to March and the highest volumes of rainfall during hurricane season in the fall.

During spring, from March until June, the days grow longer and the cold dissipates from the region. The frozen ground reawakens as the trees bloom to life and the forest grows green. Temperatures range from the teens to 80 degrees during the day. The steady, cold, freezing temperatures of winter yield to frost-free conditions. The snow melts around the base of trees as the sap begins to flow. Maple sap is collected and boiled for syrup, while seedlings are started in the greenhouses. As the snow melts, the streams run high with clear, cold water. The season changes in spurts, with sporadic frost possible until June. Drenching rains as well as blizzards and periods of drought add to the reliably inconsistent weather of the season. With bare ground exposed, the period known as "mud season" begins. In some areas the meltwater saturates the soil to souplike consistency for several weeks.

In mid-April, as the ground dries and thaws, it becomes workable. On one magically warm day during this period the frogs emerge to sing a chorus for the season. By May the gardening season is fully underway, and the disappearance of snow from the peak of Mount Moosilauke finally indicates that the danger of frost has passed.

The summer months of June, July, and August are typically referred to as the growing season. The weather is less erratic, with longer periods of continual rain or sunshine. Temperatures vary, although they generally range from 50-degree nights to 70- to 90-degree days. July invariably brings on a heat wave, as well as occasional periods of little to no rain. The sun traverses the horizon directly overhead. Violent afternoon and evening thunderstorms are produced in the energized atmosphere.

By September temperatures begin to drop noticeably. By late September or early October the first

A seasonal blanket covers the Lower Garden.

frost will occur, ending the season for tender annuals such as tomatoes, beans, squash, and corn. One highlight of the fall is the beautiful transformation of the foliage in the Northern Forest. During the first weeks of October the remarkably vibrant display of red, yellow, and green color peaks. A hard, killing frost occurs by November, which freezes the soil and destroys the cell tissue of all but the hardiest vegetable crops.

The winter months of December, January, and February are cold and dark. The sun rises and sets low on the horizon, barely elevating above the tree line. Precipitation is mostly snow, although rain and other frozen variations are likely. The snow typically blankets the earth for this period, insulating the soil from the cold temperature extremes. Occasional ice storms in which the precipitation freezes upon contact can be catastrophic for the forest, as well as for human safety and infrastructure in this region. On occasion temperatures drop below 0 degrees for weeks at a time.

Natural History

A diversity of species weaves together the ecological web here in the White Mountains region. Each species has adapted to fill a role or niche based on the environment and is interdependent and mutually supportive of the others. Plant pollinators become bird food, which is digested and excreted to produce nutrition for the soil, therefore fertilizing for future cycles. By reflecting on our role in the diversity that has emerged in this environment, we can be responsive in our designs to perpetuate alongside other forms of life.

The Northern Forest tree species on our property are categorized as conifers (evergreens) and deciduous (hardwood). The conifers include the eastern white pine, hemlock, red spruce, and balsam fir. The hardwoods consist of sugar and red maple, white ash, black and pin cherry, and white and yellow birch, as well as red oak. There are also clumps of striped maple and elm. Brambles and blueberries are common in the clearings and edges of the forest. Late spring also produces a niche of plants known as ephemerals, such as ramps, wild oats, trillium, and lady slippers, which briefly appear. These plants choose this period of emergence to capture the abundant sunlight on the forest floor before the deciduous leaves cloak the terrain with shadows.

The warmth and sunshine of late spring also provide a window for other biologic species searching for their niche in the Northern Forest. Commonly known as "bug season," this intense period is often delineated as Memorial Day to July Fourth, though it appears to be shifting to earlier in the year. During bug season three species of biting and bloodsucking insects appear and inflict misery on mammals. The blackflies emerge first, with several days of nonbiting males swarming in annoying clouds that herald the appearance of their viciously hungry female counterparts. The no-see-um is the next variety of bloodsucker to arrive. Small as a pen tip, its bite is painful and results in swelling. Then mosquitoes buzz onto the scene and stay throughout the summer.

A progression of diverse bird species through the seasons also utilizes this terrain. As the bare ground reveals itself after the winter, robins move in to assert their territory. While jays and chickadees are year-round residents, most of the species adding to the symphony of the forest are migratory, such as various thrushes, flycatchers, and warblers. The songs of the birds are strongest in the spring and summer.

A bullfrog finds adequate habitat on the grounds at the farm.

Insects and amphibians are most active during the longer days of summer. At all hours during the summer months, moths, butterflies, winged pollinators, and arachnoids celebrate their life cycles and contribute to the ecological web. The ponds, streams, and rivers also host an abundance of fish, turtles, and other aquatic species, while newts and salamanders creep through the fallen leaves of the forest.

While the forest has adapted to seasonal and climatic limitations, there are also limitations to the animal life that can be sustained on this land. New Hampshire supports a lean population of just five thousand moose, whose diet consists primarily of aquatic plant species. Because hunting is carefully monitored, this indicates that limited availability of resources controls their population. Unlike the deer-laden woods of the Connecticut Valley and points south, our zone does not support a caloric gleaning from the landscape big enough for large mammals. While the oak and beech do provide some forage in the fall, the biology of New England does not produce this digestible surplus in conditions that would induce greater densities of large animals. Consequently, the woods

Moose wander the property and are seen mostly in the early spring and fall. This moose was spotted wandering the edge of the Upper Field near the Skinny Shack tree house.

support niches for solitary moose, hibernators such as bear, and predators such as coyotes and bobcats. There is also an assortment of smaller mammals that reside year-round, including fox, raccoons, fisher cats, and the resilient red squirrel.

Though the forest is lacking in calories, it is rich in carbon. As trees reach the end of their life cycle, their biomass is constantly being recycling by members of the fungal kingdom. The most visible element of this powerful regenerator is the flourish of mushrooms, the visible reproductive organs that spread the spawn of the fungus.

Human History

Prior to the European migration to America the Abenaki inhabited the river valleys of our region in familial settlements. Small groups would come together seasonally for hunting, planting, and fishing. The arrival of Europeans as early as the mid-1500's brought epidemics of illness that decimated the native population, and military conflicts ensued until they were largely displaced. Yet the rugged conditions of interior New England remained inhospitable to the colonists, and it required 150 years to pass until the early settlers considered occupying the Baker River Valley. Governor Benning Wentworth granted land by petition that ultimately established the charter for the Town of Dorchester in 1772. The land was settled by Anglo farmers migrating for the opportunity to set up agricultural systems based on traditional European models of agriculture. Timber was cut to provide charcoal and building materials for the expanding urban industrial economy. As the forests disappeared, the fields were cleared of stones, which were utilized to build fences. Sheep were imported, and the wool commodity became the monocrop of the region.

Dorchester grew rapidly during this period of settlement. By the mid-1800s there were over twelve hundred residents and eleven water-powered mills, as well as several schoolhouses and post offices. During this era the initial deed was made to the Eliot family for the property on Streeter Road. The Eliots and their neighbors made valiant efforts to survive and thrive on these mountainous slopes, but unfortunately several factors led to the community's decline.

The geography and climate were inhospitable to the style of farming these settlers had imported from the more temperate climate of Europe. The length and severity of the winters tested the capacity of the settlers to survive. As the forests were cleared and the sheep ravaged the remaining vegetation, the thin soils were further depleted of fertility, resulting in a downward spiral of health on the farms. Global economic shifts also negatively affected the value of wool, and the era of industrialization invited workers to the factories of the urban core. In addition, western expansion enticed people with the offer of free, fertile land in the Ohio Valley and beyond. Ultimately, the Civil War struck a nearly fatal blow to the community of Dorchester. As young men traveled to fight, the town was bankrupted by this loss of residents. The Civil War years led to an exodus of funds and people from which Dorchester has yet to fully recover.

Because of these political, socioeconomic, and environmental factors, the thriving growth of the colonial period was short-lived. When my aunt and uncle moved to Dorchester in the 1940s there were only ninety residents. But with the growth of the Post–World War II era, specifically the spread of rural electrification, there was a second wave of settlement. Current development is empowered by the advent of the Internet and telecommuting economy. With each period of growth the nature of land ownership has changed as well. What began as wilderness punctuated by subsistence homesteads became by the early twentieth century sizable tracts of land owned by timber companies. Now in the twenty-first century timber companies continue to divest from these large parcels, as wealthy individuals purchase these tracts. Thousands of acres have been consolidated in this manner within Dorchester and neighboring towns.

For the first half of the twentieth century the primary employment available was work in the woods. The forestry industry remains a substantial natural resource economy. However, in the second half of the twentieth century families began to migrate back to the land. The population of Dorchester is a typical mixture of rural Americana: retirees who have migrated from the population centers, families who have chosen to escape to the region for the recreation and lifestyle opportunities, and also local people, some of whose families have lived in Dorchester for generations. The cultural collision of new arrivals and locals brings together an assortment of interests, values, and belief systems to the town. There are multi-million-dollar homes on the same street as trailers. There are immaculate, neat, and tidy weekend cabins and seasonal summer residences as well as ramshackle homes with yards of trash and debris.

The general personality of the town's residents is reflected in the large percentage that refuse political party affiliation, instead preferring to register as independents. The town government is a representative democracy in which a selectboard

Entering the D Acres homestead

comprising three individuals steer day-to-day deci-sions, while local ordinances, laws, and regulatory changes are subject to approval by the citizenry at annual town meetings. Dorchester is not a bastion of "progressive" politics or a thriving hub of liberal affluence and wealth such as Boulder, Colorado, or Burlington, Vermont. The populace is a juxtaposition of socioeconomic classes that share strong-willed, stubborn, and fiercely independent personality traits. While resiliency is necessary to endure the four seasons of New Hampshire, people who move to the mountains share a love for the landscape and the privacy and solitude that they

provide. Successive waves of people have chosen to flee the amenities of the population centers for the freedom and opportunities presented in the rural community life.

Onto this backdrop the D Acres project has emerged. The realization of the project is constantly shaped by the environmental factors discussed above. Conceived as a response to human-induced crisis, D Acres seeks to forge a conscious path to refine our role within the ecology. Instead of divined dominion and ownership, humanity must design methods of coexistence and perpetuation as partici-pants in the ecological design of the future.

CHAPTER FOUR

HISTORY OF THE PROJECT AND PEOPLE

In 1948 Edith and Delbert Gray, along with their daughter Patricia, moved to Dorchester, New Hampshire. Delbert was born and bred in the

Edith Gray at age 91

Northeast Kingdom of Vermont and had worked in the woods to pay for his schooling. Edith emigrated from England with her family and settled in New Bedford, Massachusetts, before World War I. She later left Massachusetts for Hartford, Connecticut, where she soon met Delbert Gray.

The Gray family bought the Streeter Road property as a traveling respite for the visits to Delbert's family in the Mount Mansfield area. Advertised in a hunting magazine, the property of approximately two hundred acres was bought for nine hundred dollars. When rural electrification arrived in Dorchester in 1948, the dilapidated barn and cape house became the full-time accommodations for the family. Edith would later recall that she spent the first year crying from loneliness, a city mouse suddenly deep in the country.

Delbert Gray had been a successful insurance accountant in cosmopolitan New England before he returned to the Appalachian hills of his youth. He tried various schemes to subsidize the farm enterprise, where he worked in the gardens, and with poultry, oxen, pigs, and horses. He built furniture and upgraded and maintained the farm structures. Blacksmithing and sugaring were two of his favorite hobbies. He found success in such civic organizations as the Plymouth State Fair

Committee and was also a town selectman and tax collector.

Edith began her art career with a class in decorating chairs in 1938. She became accomplished in numerous mediums within the American decorative tradition, including gold leaf, mother of pearl, reverse on glass and velvet painting, and sewn and braided fabric, and was a member of the League of New Hampshire Craftsmen. She taught adult classes in her backroom studio until just months before she passed away. Her time in Dorchester improved as she adapted to rural life, developing expertise in gardening, fiber, culinary arts, and food preservation.

Patricia Gray followed in her mother's footsteps and was an accomplished artist. She graduated from the University of New Hampshire and moved to Connecticut, where she predeceased her parents. After her death Delbert and Edith decided to pass the property to the son of Edith's brother, Bill Trought. Bill and his wife Betty, my parents, built careers as medical professionals in eastern North Carolina but had spent substantial amounts of time on the farm in Dorchester.

After driving through a flashing red light with a license examiner in the front seat, Aunt Edith lost her driver's license at age eighty-nine. She had lost her husband and daughter ten years before. Now her rural New Hampshire independence was threatened because she was dependent on the single passenger transportation system. As the son of Bill and Betty, I had traveled to the property throughout my lifetime. Through the years our family had fantasized about living the good life on the farm in New Hampshire. When I received the news that Edith needed assistance, I felt compelled to migrate for both of those reasons.

D Acres Begins

When I arrived in the White Mountains to embark on what would become the D Acres project, I was twenty-six years old. The idea my friends and I shared was to start a market garden and develop a community-supported agricultural model to fund a summer-season farm. My level of experience was humble even though our plans were grandiose. I had no essential skills in gardening, marketing, management, or animal husbandry, and I was ignorant of the idiosyncrasies of the inclement New England weather.

The crew that arrived with me in October 1997 was a motley mix of characters. Brenna was a San Juan Isle gardener and homesteader looking to design and experience a permaculture homestead. Charles was a chef and gardener who had taken the permaculture design course with the Bullock brothers at their renowned farm on Orcas Island, Washington. Crazy Jimmy was an experienced home builder who had joined the team in the Virgin Islands, where we had partnered for rock climbing expeditions, then traveled together to Alaska. My sister Dara had basic gardening skills she had gleaned from backyard gardening in North Carolina and her time spent farming on Orcas. I brought an attitude of confidence and enthusiasm along with rudimentary construction skills.

This cadre of people was indicative of the mix of individuals who have brought collective energies to the farm throughout D Acres' history. People arrive to learn, educate, and actualize, offering their skills as well as positive enthusiasm to the project, and we have always depended on ability to adapt and fill different roles in the organization. While the initial participants all knew each other prior to arriving, the idea to bring together a collection of people with a diverse range of farm-related skills and interests to collectively accomplish land-based service has remained consistent. While the goal is noble, we also recognized early on that the obstacles presented by this process are complex and ongoing. There is always potential conflict in decision making among different personalities and philosophies. Limitations posed by funding and inexperience have been stressful, though they've also yielded innovation.

When we arrived that fall we envisioned transferring the lifestyle and socioeconomy of our

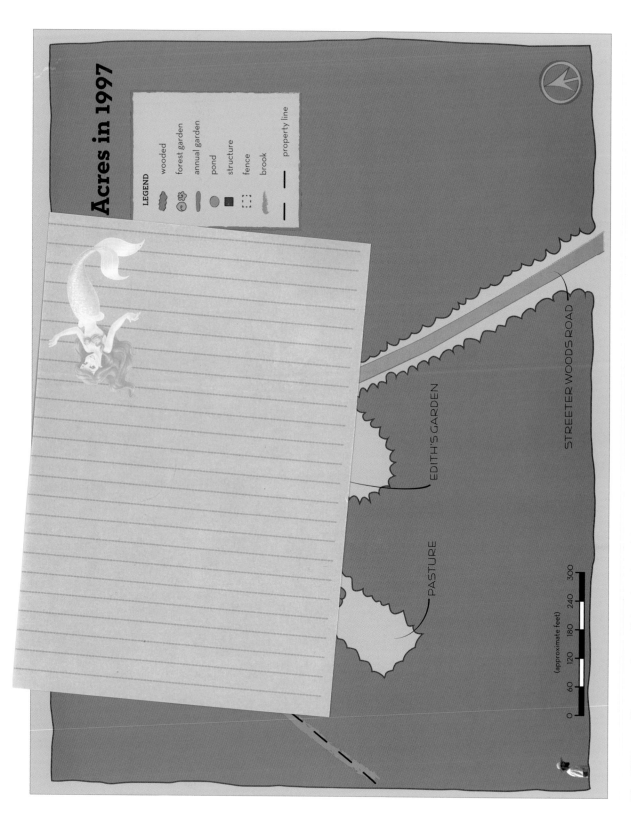

The 1997 map illustrates the traditional scale and layout of Edith and Delbert's homestead. Illustration by Marylena Sevigney.

experiences on the West Coast back to the East. The progressive, affluent islands of Boulder, San Francisco, and Seattle had been fertile grounds for local food aficionados and upwardly mobile, recreationally focused youth. The weather of the West was mild: On Orcas the ground had not frozen for two winters, and Boulder was renowned for its three hundred days of sunshine per year. The typical remaining forest in the Northwest is predominately rapidly growing softwoods whose species dwarf those of East Coast forests ravaged by two centuries of human occupation.

The shock of transplanting to the East was intense. We were aliens to both the landscape and the society of Appalachia. The conservative culture of northern New Hampshire feared the threat of a potential commune with the assumption of a hippie lifestyle. We lacked the social support network of other idealistic youth who were abundant as collaborators in the West. There was intense skepticism of organic food production, and potential clients in this rural location were slim. A more practical problem was the weather. Frost-free growing in Dorchester is limited to ninety days per year, with frozen ground from November until April. Icy weather and cool reception were the first inklings of the many obstacles ahead.

Our initial efforts were put into preparing soil to plant garlic. As we dug into the sod we discovered the soil was typical of the fertility abyss that exists from Maine to Vermont. As we moved in, we shoveled out the outbuildings that had collected the dust

Charles and Brenna shape the first garden beds using leaves and compost as mulch and a swale of rocks as a heat sink. The first crop is garlic.

from twenty-odd years of disuse. We assembled a rudimentary kitchen and office meeting space in the former stable area of the red barn. We slept in the barn; moldy, cold, wet tents; vehicles; and Edith's guest room. Even as Edith reminded us, we were cognizant of our lack of knowledge and carefully sifted through the diverse array of broken parts, half-finished projects, and lawn art that had accumulated at the farm. We quickly learned that when we threw objects away we would likely need them the next day for an unanticipated project. Edith was supportive of family and our agricultural ambitions, though our idealism clearly lacked experience. She was quick to offer advice when she felt necessary; the name of the farm resulted from an incident in which she reprimanded Brenna, who reacted by singing "D Acres is the place to be" to the tune of the theme song from television's "Green Acres." At least now we had a name. While we would later utilize the acronym **D**evelopment **A**imed at **C**reating a **R**ural **E**cological **S**ociety, the name stuck because it also recognized Delbert and Dorchester. Plus it sounded better than other alternatives such as 3D and D-Cubed.

We fiddled with the Farmall tractor we had inherited and added a woodstove to the Red Barn. As winter set in, Brenna and Charles returned west and Jimmy headed to Philly for winter work. My sister and I began further renovations to the Red Barn, making improvements to the living area and the kitchen. As the season moved in with ferocity, Dara, Edith, and I settled in. Winter in New Hampshire has never been considered easy. Two adult siblings freezing in a barn, without running water, with an elderly aunt, and isolated from peers is not everyone's ideal. In February Dara chose to head west to intern at Jerome Osentowski's Central Rocky Mountain Permaculture Institute (though she would come back), and I almost moved on from this project.

And our errors were historic. In the spring, when Jimmy and Charles returned, we built outdoor structures using poplar for beams, which is particularly susceptible to rot. We turned the soil with Delbert's old tractor using the plow, an invasive technique that removed the rich, fertile top sod layer. Fortunately we were able to reuse this sod in water-conserving swales. Our farm stand on rural Route 118 attracted approximately five customers every Saturday that summer—if you counted the regulars. We acquired four old, scraggly hens from the feed store in Plymouth, though our ineptitude in the transfer resulted in the fifth's being freed to the streets of a college town.

But despite the mistakes, frustrations, and reservations, the situation at D Acres provided an opportunity that could not be realized elsewhere. I had reasons to remain at the farm. The familial relations I enjoyed with Edith were a strong incentive to continue. The history and sense of place encouraged me to invest further, while the long-term security and stability of living on the land also attracted me. D Acres presented the freedom to implement hands-on experimentation in sustainability. Family responsibility and the opportunity to pursue my philosophical passions were compelling reasons to forge ahead.

Progress and Adaptation

Over time the gardening strategies implemented at D Acres have adapted to the realities of our locality. Our perennial focus turned to traditional crops such as blueberries and apples. To provide for season extension of annuals we began experimentation with hoop houses and cold frames. It was during this period when we made the acquaintance of a local logger named Jay Legg. He was a wealth of information for our forestry, construction, and gardening trials who had also weathered the tribulations of being the newcomer in Dorchester. We gratefully accepted his expertise and assistance as we worked to remediate the erosion impact of the previous logging on-site and learned the basics of chain saw operation.

Through observation we assessed the realities of our situation. Our land base was marginal soil, three times cut and run ragged by sheep in the 1800s. In addition, the damages issued by the last glacial period had left piles of conglomerate stones and sand with minimal organics and subsequent nutrients. Our clientele for farm products was limited. Frugal locals planted their own, and the human-scale permaculture to which we gravitated was not scalable for restaurant and wholesale levels.

While we initially idealized a farm enterprise based on revenue generated by market vegetable sales, the reality of our circumstance encouraged other avenues of focus. Indebted by the gift of the land, we felt the need to make the farm accessible as a service to the community. There were few opportunities nationally for on-site experimentation and education related to sustainable farming and lifestyle. We decided to make tangible plans for an organization that would steward the land while providing service and education to the community.

To support these operations I used funds that I had inherited from my grandparents. We also solicited Bill and Betty to help with the initial purchases that enabled the agricultural and community programs. In general, we attempted to control costs with frugality and common sense, though we often wasted money because of our inexperience.

Our quarters in the Red Barn felt inadequate to these future plans. Its water was supplied by a twelve-foot-deep, mice-infested dug well by the roadside, and its floors were uneven and difficult for elderly to access. The Red Barn would not suit our goals to be publicly accessible and sanitary or reflective of a positive and plausible alternative to conventional consumerism economics.

A 1999 midsummer view of the Community Building

Beth Weick continuing the New England agricultural tradition of boiling down maple sap into sweet syrup

We began formulating designs for a community building around a round table amid the inadequacies of the Red Barn. In the barn's stalls we discussed an ideal building that would serve our on-site operations as well as the community functions we hoped to create. A family friend, Shelley Pripstein, was invited to draw plans for the Community Building. In the fall of 1998 we began the foundation under the supervision of Bob Guyotte, an experienced heavy equipment operator and concrete installer.

It became increasingly clear that our capital investment in dollars and sweat equity would require years of hard work to repay. With our land base, goals, farming philosophy, and location it was not feasible to start up an economically viable operation without significant initial investment. The fertility of the land would take years to bear fruit, while the thousands of dollars invested would not be recouped from vegetable sales alone. We began our business operations as an LLC devoted to construction of the Community Building. This business represented the first steps in the process of morphing our idealist vision of an educational farm into a legitimate organization. Throughout the Community Building construction we spent approximately $250,000 in materials and subcontractors' fees for the construction of the Community Building. We garnered fees from the Trought family for D Acres' construction personnel time, which were later siphoned into our service-based, educational enterprises. The start-up capital provided for operational expenses and created the infrastructure necessary to further our mission.

Building a Community

During this period we began recruiting strangers, in earnest, to live and work at the farm. Instead of just to friends and family, we began to do promotional efforts locally and nationally. While we would continue to invite personal acquaintances, this pool of people with the capacity to relocate and commit year after year in Dorchester was limited. We made a flyer to canvas local health food stores and colleges, and placed an ad in the World Wide Opportunities on Organic Farms (WWOOF) farm listing. We hoped to attract people based in the area as well as folks from farther away wanting to experience rural New England farming.

People are a crucial component of our farm system. Their motivations for participating vary, as do their levels of commitment, experiences, skills, and aptitudes. While the roles and positions within the organization transition overtime based on both individual interests and the organization's needs, the people provide the purpose as well as the results.

Bella and Clover were part of the farm system between 1998 and 2003.

By harnessing the energy of responsible, motivated idealists we transcended the traditional family farm dynamic and became a part of the long lineage of collective communitarian endeavors.

There is always a large pool of candidates interested in enjoying the summer months on an organic farm. However, matching candidates' expectations with the realities of living at D Acres has created a substantial amount of contention through the years, especially early on, when our ineptitude as farmers and managers in this unconventional system became fodder for conflict. While the people supply the purpose and produce the results that meet the goals of the project, the challenges of coexistence and mutual support remain the greatest D Acres has faced.

Still, over time the continuity of long-term, skilled staff has reaped rewards for the organization, and we've been able to look beyond the farmstead and offer service to the broader community. The arrival of Abby in the spring of 2002, fresh from a post-graduate program in community organizing at the School for International Training, added experience in nonprofit administration, retail, and grant writing. Through Abby's persistence and diligence. and our accountant Ronda's expertise, we achieved 501(c) status and hosted a yearlong calendar of diverse hands-on sustainability workshops on topics from food preservation, gardening, and earthen construction to forestry and herbal medicine. This allowed us to begin seeking grants, partnerships, and donations in earnest. We also applied for a commercial kitchen license and began offering value-added food products. These two accreditations helped build economic stability and diversify our income stream.

Zoning

As our operations grew, however, a contingent of local government officials decided that our farm activities were expanding beyond the agricultural and residential purposes the land was zoned for. Rather than listen to the rationale that dictated the complexity of small farming in this era, the town government declared we had violated all possible zoning exceptions, excluding the construction of a hospital. We were classified as a school because we offered classes; we were classified as retail because we sold farm products; we were classified as a warehouse because we kept an inventory; we were classified as a restaurant because we served farm food; and we were classified as a manufacturing facility because we produced wood products from the forest. Because of the diverse extent of our activities as a farm the town told us that we required variances in zoning. We were reluctant to seek these variances because of the increased tax burden and code upgrades these variances would have required. To prevent us from proceeding with our activities, the town issued a cease and desist order to curtail operations or face a fine of one hundred dollars a day. But after a prolonged legal battle and the expansion of the state's definition of agriculture, the issue has been resolved.

To develop a harmonious relationship within the broader community it is important to consider the culture and regulations that affect community-scale developments. In addition to the dictates of local regulators, community-scale operations often face prejudices, stigmas, and skepticism by the general public. Since the activities of the communes in the 1970s, the perception of collective efforts as enclaves of drug affliction, poor sanitation, and laziness has been pervasive. Local property owners were also cognizant and wary of the potential for children to be born at D Acres, which would add subsequent costs to taxpayers. In addition these tax-conscious individuals were aware of the capacity for a nonprofit to remove itself from the property tax role by declaring exemption. We have attempted to respond to these concerns from the community as we've become aware of their perceptions. We invited neighboring property owners to tours and open houses so that we could explain our intentions and answer their questions. We also became involved in local committees and government positions to provide a service presence in public affairs. In addition we have clarified our intention to pay our share of services for schools, roads, and other public services.

In one sense the story of the D Acres farm is part of the family farm tradition, with my aunt passing on the land to her kin. But how we've occupied the farm has been quite novel and unique. Within this crucible of experimentation, we have continually adapted to changing circumstances while adhering to the goals and philosophies of the organization. The story of D Acres' efforts has been forged by observation and adaptation as well as the collective communitarian efforts of many nonrelated individuals.

In some circumstances the creation of community-scale organizations may be a step-by-step process planned from start to finish. My experience with D Acres, however, has been the result of a process of evolution dictated by place, people, time, and variables as unpredictable as the future itself. Our organizational philosophy has been to recognize not only the existence of evolution but its ecological necessity in response to continual changes in the environment.

The history and personnel of an organization form a foundational element, but the evolution functions on a continuum that changes with time. While an organization is predicated on its history and its personnel, these relations are not stagnant. Through cycles people arrive and move

LEGEND

- wooded
- forest garden
- annual garden
- pond
- solar hot water
- dehydrator
- solar collector
- compost
- structure
- P pig house
- outhouse
- T tree house
- S solar shower
- hoop house
- fence
- stage
- bridge
- stairs
- P parking
- sand play
- tire rocket
- seesaw
- fire pit
- brook
- — — property line

(approximate feet)

0   60   120   180   240   300

BLACKSMITH

SUGAR SHAC

EDITH'S HOUSE

EDITH'S GARDEN

PASTURE

UPPER GARDEN

This present-day map shows how we've progressed from the original homestead to community-scale operations.
Illustration by Marylena Sevigney.

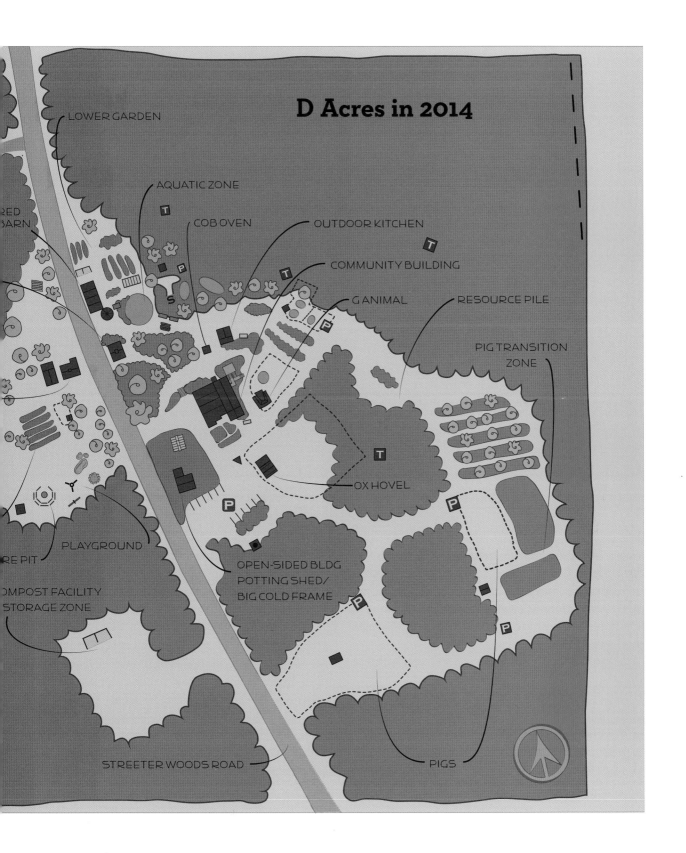

D Acres in 2014

LOWER GARDEN

AQUATIC ZONE

RED BARN

COB OVEN

OUTDOOR KITCHEN

COMMUNITY BUILDING

G ANIMAL

RESOURCE PILE

PIG TRANSITION ZONE

OX HOVEL

PLAYGROUND

RE PIT

OMPOST FACILITY
STORAGE ZONE

OPEN-SIDED BLDG
POTTING SHED/
BIG COLD FRAME

STREETER WOODS ROAD

PIGS

Micki Visten sharing her vast plant knowledge

away, children become adults, new people are born and eventually die. Regulations, as well as the officials who enforce them, transition alongside the communities they serve. This evolution is inevitable, and our designs must be responsive to that reality.

End of the Beginning

In the fall of 2003 Edith passed away in her home of fifty-five years. Although she had been bedridden for the month before, our mission to allow an ultimate, peaceful departure at her home was accomplished. Her presence is still felt here through her artwork, her tradition of home cooking, and her practical life philosophy.

Edith's passing, along with my parents' recent arrival as full-time residents, marked the end of the incubation of the D Acres project. This incubation allowed for the organization to identify a role responsive to the needs of the larger community. From its initial seeds the project spread roots in the land and began to branch its service into the regional community with greater assertiveness.

CHAPTER FIVE

DAILY LIFE

When we first arrived at the farm, my interpretation of an ideal farm life had been flavored by reading the Nearings' book *Living the Good Life*, in which they described twelve-hour workdays divided into three separate segments. They advocated four hours of bread labor, such as gardening or firewood collection, four hours of professional activities, and four hours dedicated to fulfilling civic responsibilities. We have followed this ethos, though our days are not as neatly segmented.

Instead of following the standard paradigm in which life is divided into work versus vacation, we understand that the farm is a twenty-four-hour-a-day ecology. Instead of working for a living the adage we follow is to work *as* a living. Instead of a wage, the provision is subsistence, and our existence is viewed as part of a vibrant, living system. Over the years we have adopted a shortened workweek of organized group activities, as well as a seasonal approach to labor, to manage the exertion of this perpetual rhythm.

As a farm our schedule of activities follows the flow of the seasons and changes to reflect the personnel and priorities of the moment. But to truly conceptualize the patterns of activity at D Acres would require time spent as a part of the system. Even immersion for several months offers but a snapshot, while a year of participation represents simply a single cycle. Farming operations involve a multitude of daily, seasonal, and long-term decisions that dictate the activities. Combining the seasonal uncertainties of farming operations with the challenges of a service organization in a communal living situation can be daunting. The prioritization and implementation of tasks has a tremendous impact on the project's overall viability. Time spent on fruitless labor drains the organizational and individuals' resolve, while success and accomplishments feed the positive energies. For planning purposes it takes years of observation to predict the patterns based on the complex interactions of people and place.

The D Acres' lifestyle is rich in seasonal experiences and activities. Unfortunately, immersion in this unconventional paradigm can also lead to feelings of chaos, ambiguity, uncertainty, and frustration. Without agreed-upon standards of work ethic, sanitation, safety, and personal welfare, there can be no method of establishing reasonable expectations. Through continual dialogue for enduring success, it is crucial to match the realities of the farm operations with the expectations of the participants.

A typical workweek involves a plethora of task possibilities, ranging from leading a tour for a school group, to slaughtering a pig, to weeding the garden, to office tasks and cleaning the Community Building. Smaller tasks take minutes to perform, while bigger projects such as the potato harvest can extend for weeks. But everything we do varies tremendously with the seasons. Garden tasks change from ordering seeds at a desk in December

Daily tasks are seasonal; summer duties center on gardens, with harvesting scheduled at least twice a week.

to the hands-and-knees work of planting and weeding throughout the summer. The nature of the work also varies as radically as the New Hampshire climate. While grants, marketing, and educational services are intellectually challenging, logging and gardening require high levels of both mental and physical exertion. By extension, tasks such as turning compost or hauling water are accomplished primarily through positive attitude and physical efforts.

We attempt to live with the flow of the seasons, not only in our work but also in our pace. The spring thaw brings with it energetic enthusiasm as the days grow longer. We work diligently through the summer so the fruits of our labor can be cherished through the restful winter months, maximizing

the daylight hours in the gardens, and keeping up with educational programming and overnight accommodations. Prudent seasonal prioritization of work activities defines success in agricultural activities. While the climate dictates our seasonal rhythm, organizational priorities are developed and implemented by the participants. Frozen terrain is a significant limiting factor to winter activities, and we utilize that opportunity to mimic our mammal relatives by practicing techniques of hibernation. During the shorter days of the winter, our focus shifts to indoor tasks such as woodworking, fund-raising, planning, and administrative tasks. In general the work expectations of the wintertime allow for personal endeavors such as travel, art, and study. Winter provides solace and a respite from the grind

Josh works with volunteers to harvest potatoes.

We sequence and stage much of our work in a progression, evaluating progress throughout the day. For example, in the garden we need to weed, then seed, then mulch. So in the beginning of the day much of our energy would be directed to weeding, then later some seeding, and finally the mulching. This process requires constant evaluation so our goals can be met and we can arrive at a logical stop for the day.

In addition to seasonal work, there are also consistent, daily responsibilities throughout the year. Cooking duties, cleaning, and animal chores are divided at weekly meetings, which vary in duration and complexity based on the seasonal activities and number of participants. We generally meet on Sunday or Monday for one to three hours to define a plan for that week. Our task list is based on prioritizations made during annual and seasonal planning sessions, as well as on precedence. The actual duties are divided through a dialogue that primarily considers the organizational priority but also is responsive to individual interests. Some projects are self-directed, while others are coordinated efforts by the entire group.

The workweek can extend to all seven days. Activities typically begin between 6:00 and 7:00 a.m. and end between 5:00 and 10:00 in the evening. The daily schedule encompasses self-directed as well as supervisory roles. Group workdays are generally Tuesday, Wednesday, and Thursday. Fridays and Mondays are for housekeeping, meetings, and personal projects, while Saturday and Sunday are generally for community events, individual pursuits, or collective outings.

On group workdays we finish animal chores at an agreed-upon time, usually 8:00 a.m., in order to then begin the collective effort and work until noon. At noon we break for lunch, and in the afternoon we generally work another four-hour block. By meeting at specific times and sharing our efficiency in organizing, activities improve.

Our farm system processes are the creation of the people who steward the project. This power

of the days when the ground has thawed, rest that is crucial for this system to function.

Changes in the weather dictate much of the work. For instance, outdoor garden plans must be altered in the advent of a thunderstorm, and a forecast for five straight days of rain changes our plans for that week. The weather can also vary tremendously year to year. Following a cold winter more time will need to be spent cutting firewood, while a dry summer will require the high prioritization of irrigation projects. The quantity and capacity of personnel also plays a role in what we plan to accomplish. If we have large numbers of personnel it may be more stressful to manage numerous projects instead of focusing on a singular group effort.

Chris O'Leary, Emily Rumberger, and Jeff Reinhardt enjoy one another's company while spending the afternoon harvesting raspberries.

is derived not only from physical attributes but also from knowledge and enthusiasm. For this to manifest, group members must be fluid in switching between the roles of instructor and supervisee, which is necessary to manage our activities.

As an organization we recognize adaptability as a crucial component of survival. By being conscious of the perpetual evolution of the organization we confront ongoing change conscientiously and address areas where adaptation is necessary. This process involves an investment in the present; a recognition of the distinction between what is practical reality and what is hypothetical and theoretical. By instituting a system that recognizes the utility of feedback cycles, evaluations, modification, and refinements as well as radical shifts in philosophy, we continually evolve.

Rest and Relaxation, Health and Well-Being

Safety is a priority for everyone working on the farm. We are one half hour from the nearest hospital, and ambulance and fire service both arrive from the volunteer organizations in neighboring towns. Ensuring one's own safety requires awareness and observation of impending circumstance as well as precautionary measures, so that accidents and injuries on the job can be prevented. It is important to acknowledge one's limitations regarding proper tool usage or physical exhaustion so that informed decisions can be made. But despite our intentions, injuries and accidents do occur. And when they do, it is important not only to remediate the circumstance but also to take the time and measures necessary to regain health. By following a regime to properly heal, the participant regains his or her capacity to rejoin the collective effort.

Proper rest is important throughout the year. The pace and proximity of this collective effort is demanding. While an extended trip may serve to rejuvenate, it is also necessary to monitor our personal energy reserves on both a day-to-day and a week-to-week basis. To be successful on the farm through all four seasons, it is necessary to maintain personal health and a positive attitude. It is important to develop an annual rhythm that is responsive to the farm system and our climate. With a vision that recognizes the ebb and flow of our physical bodies, and our spirits and emotions, we can pace our efforts so we meet the needs of the present without depleting the reserves necessary for the future. Recognizing the rewards and accomplishments of our efforts as well as taking breaks to replenish our health reinvigorates our motivation. Sharing our successes and mishaps, as well as communicating our emotional, philosophical, and spiritual journeys, reaches deep into the core of what defines our species.

In our day-to-day life we attempt to both accomplish necessary tasks and enjoy the process. We incorporate art, recreation, and communitarian philosophies into our activities so we can fill our time here with richness. Through a process that defines expectations and evaluates the results, we are continuing a farm system in which day-to-day harmony creates an annual melody, a farm system that will endure and evolve for generations of participants to come.

Work Standards

Beyond what crops are grown and animals are raised, every farming operation has a culture. Based on the economics and the farm philosophy, as well as the ages, personalities, and experience of the participants, each project develops its own rhythm. Standards for participation in terms of household cleanliness, punctuality, and time spent on collective tasks define how they are to be accomplished. Are some tasks done hastily and with shoddy planning while others are meticulously tended to? What are the expectations in terms of sanitation when the definition of "clean" varies between individuals?

The specifics of how we accomplish the goals of our operations is not only the who, what, and when but also how that decision is derived. What mechanisms do we have for collaboration within this complex and evolving system?

Roles and Responsibilities

Over the years we have utilized different terms to denote the level of commitment, responsibility, benefits, and decision-making capacity of the participants at D Acres, such as WWOOFer, intern, apprentice, staff, resident, and participant. These descriptive categories are not intended to limit or constrain the roles that people can play within the organization but rather to provide a framework for determining expectations.

The responsibilities vary for each individual. While we are collectively responsible for achieving the directives of the mission, how we accomplish these goals depends on the personnel, the season, and the specific organizational direction at any given time. One of the primary responsibilities of all staff and residents is to serve the educational component of D Acres. From promoting the opportunities and inviting people onto the land to the actual delivery of information, the staff perpetuates the farm's educational mission. More experienced staff generally act as organizers and instructors for the daily activities at the farm and supervise the less experienced short-term residents. On the weekend everyone is responsible for managing workshops and public events. In addition to their work on the

Josh has given countless tours to various groups throughout the years. This is a group of Plymouth State University students enrolled in a spring environmental science class.

farm, both long- and short-term staff serve as liaisons to the public and represent the organization in the larger community. In carrying out our mission-based activities, each staff member and resident is a steward of D Acres' ideals and an ambassador for our ethics.

Staff must be able to fill the variety of roles required by this type of entity. In addition to the arduous tasks performed in the forest and the field, staff is also responsible for bookkeeping, marketing, fund-raising, and grant writing; maintenance and upkeep of the facilities; daily cooking and cleaning; and welcoming the public. As educators the staff answer to the public via presentations and workshops, as well as through tours and our on-site

learners program. Staff must also manage the efforts of coworkers and volunteers. By coordinating the farming operations with our community outreach education programs, they are the primary liaisons for the implementation of the D Acres mission.

Paid employees generally have been recruited for a specific purpose or, as with residents, have matriculated within the program into a role that is mutually beneficial to both the individual and the organization. To offset their salaries, employees are expected to contribute efforts directed at revenue generation for the organization. As an example, personnel employed as youth educators seek grants or donations or solicit fees to provide for their salary and program costs.

Monetary compensation is minimal by North American standards. What makes this arrangement acceptable is the cumulative benefits of participation as a group. By collectivizing these resources staff and residents combine overhead costs for amenities to which they would otherwise be deprived or be forced to incur the cost of individually. Supplying room and board is the most fundamental obligation of the organization. Incomparable food and lifestyle in a beautiful setting is provided for participation. Benefits also include usage of the library, vehicles, tools, Internet, phone, and facilities at the farm. In effect, D Acres provides every day-to-day necessity, with the exception of health insurance. Living at D Acres also provides a social support network. In addition to the friendships established among on-site participants, the staff can fluidly become a part of the extensive social network in the area.

Long-Term Residents

To build upon prior organizational experiences requires some continuity in personnel. Long-term personnel have stated goals and expressed interest in remaining a part of on-site activities through several seasons. They provide the operational backbone committed to insure multiyear plans are implemented in pursuit of the organizational mission. Since short-term participants are not expected to bear the burden of lengthy project commitments, they are allocated less power in organizational decision making. With higher levels of commitment, experience, and investment, there is a corresponding increase in decision-making capacity as well as responsibilities and benefits.

To ably represent the organization, long-term staff must be committed for the duration of seasons and years, versus mere days and weeks. Staff must be prepared to be responsible for the planning and implementation of organizational activities. This requires not only knowledge to implement but also commitment to fulfill plans. In addition, that time invested through practical experience and observation is the only way to achieve a profound understanding of the strength and diversity of D Acres' particular natural and agricultural systems. For this model to succeed there must be recognition by the participants that continuity is crucial. It is important to maintain an organizational memory that tracks experiences as participants cycle through roles of responsibility.

While there is an extensive community of supporters, the staff ultimately must hold themselves responsible for organizational operations. By supplying input and direction throughout the planning and implementation process, staff members are the primary designers as well as installers of the organization model. Their performance directly corresponds to the overall success and vitality of the organization. With responsibility there is also accountability.

According to the legalities of our nonprofit status, the board of directors is responsible for enforcing the fiduciary and philosophical standards of the project. The director position serves as the representative to the board and addresses budget and personnel concerns. The director is responsible for making time-sensitive decisions as a liaison of the board. By channeling the responsibility of leadership and decision making to an individual, the board has organizational accountability.

Specialists

To fully develop the possibilities of our long-term residency program, D Acres integrates several skilled positions into our programs. We have the potential to increase our capacity if efforts are undertaken with skill and focus in specific areas. The nature of specialized work is dependent on and reflective of the interests of long-term personnel residing at D Acres at any given time. For example, a blacksmith has, at various times, been assimilated into our current program seamlessly because we have amassed equipment and an organizational history of educational workshops and demonstrations. Further examples of fluid, specialized positions include herbalist, wellness coordinator, woodworker, metal

Joe Vachon working on a blacksmithed project in his tepee

fabricator, education coordinator, aquaculture specialist, alternative builder, seed saver, perennial gardener, administrator, fund-raiser, videographer, and many more.

Tradespeople

D Acres pursues ongoing opportunities to collaborate with experienced tradespeople. The program has been devised so that people with professional experience can seek a hiatus from their traditional jobs and join the D Acres team for a designated time period. This program allows us to provide education to a population from someone with specific expertise that is useful to the organization. For example, our initial website and logo were devised by Monika Chas, who at the time was in the process of developing a freelance advertising and marketing business. As a resident of D Acres during the winter of 2002, Monika found solace at the farm, changing pace from the office frenzy, and was able to offer her valuable skill set to the organization. Tradespeople such as plumbers, electricians, or metal workers are particularly sought after for specific on-farm projects.

Artists in Residence

The artist-in-residence program is designed to provide an opportunity to the artist, the organization, and the broader community. The intention is to share the resources and organizational capacity so that a productive, self-directed artist can make a living on-site. The organization benefits from the artist conducting workshops and demonstrations to educate the public. The organization also is rewarded by productions of crafts or art specifically intended for farm purposes. Goods ranging from aesthetic sculptures and murals to practical axes and mugs have been produced by artists in residence.

It is important to negotiate a specific agreement with an artist with regard to expectations. Often a discussion involves the number of workshops and classes offered or actual products produced that can then be used to offset the costs imposed by the program. The amount of time required for planning and implementation is also important to consider for project success. Generally, a three-month commitment is necessary to develop a program that will benefit the organization, the artist, and the general public.

Our most successful venture of this nature thus far involved Joe the Blacksmith. Joe arrived at D Acres to participate as the resident blacksmith in 2006. While he had only taken a few college classes, he was passionate about working with the forge. He set up a tepee as his shop and began making products from recycled steel. Over several years Joe taught many workshops and demonstrated his craft with a portable forge throughout the area. He eventually earned a position as a demonstrator at Loon Mountain, where tourists flocked to his educational mountaintop shop. The time spent at D Acres as an artist in residence allowed Joe to incubate his passions until he was able to take flight as an independent artist.

While Regina Rinaldo's primary staff focus is on the culinary arts, she came to D Acres as an accomplished fiber artist. She has received instruction at both the prestigious Penland School of Crafts and John C. Campbell Folk School. Upon her arrival Regina set up a weaving studio in my parents' basement, creating woven and crocheted items. Equipment and supplies have been procured through Edith's collection and donations from the community, as well as from Regina's personal supplies. Regina has sold crocheted hats, woven placemats, and table runners on-site and through her participation at Cardigan Mountain Art Association and Artistic Roots.

The intention of having the positions of specialists, tradespeople, and visiting artists is to provide a self-directed space for individuals to practice their expertise for the benefit of the organizational mission. Artists generally set their own hours, though they are responsible for household chores and attendance at meetings.

Short-Term Participants

The notion of the short-term worker program is dualistic and based on a willing, mutual exchange. One of the initial rationales for inviting shorter-term participants was pure need. My sister and I had been schooled in expensive institutions of higher learning, yet we lacked basic skills such as food production, culinary arts, and shelter construction and maintenance. While we were intellectually depressed by the global socioeconomic system, we were completely incapable of providing for ourselves and creating alternatives to the destructive cycle. Although we had both chosen to continue schooling until graduation, we were disenchanted by the time and money that had been expended.

When we were freed from the academic world we began seeking hands-on opportunities and autodidactic methods of learning. In essence the college years provided the confidence and fundamental learning skills that enabled us to begin a journey in pursuit of personal education. Toward this end we began exploring opportunities in service and learning. Inexperienced, though enthusiastic to learn, we volunteered, bartered, and traded work throughout the country. This process afforded us the opportunity to travel and seek experience by participation in actual projects.

Similarly, D Acres' program serves multiple purposes. For many participants it has served as an introduction to community and agrarian life. For others who are unable to commit to this lifestyle full time, the program allows them an immersion experience in agriculture. In addition this program serves the organization as our recruitment and screening tool for longer-term residents. The

A group of Boy Scout volunteers at the farm helping with a sheet-mulching project

short-term commitment serves as a trial period for both the organization and the individual. While only a small percentage of short-timers resolve to stay and become residents, this potential provides a constant, revolving pool of candidates.

However, the program is not a scheme to receive free labor. In offering our home to visitors, we require their commitment to education and to the D Acres community. Ours is not the conventional education of a typical classroom experience; the seasonal agriculture program is a full-time immersion. A successful program addresses the underlying goal of making the experience enjoyable and rewarding for all.

When we first moved to Dorchester we began the process of advertising opportunities to participate at the farm. We used a desktop computer to design a flyer soliciting folks interested in organic gardening and community living to contact us. In addition to posting the flyer throughout the region, we created a listing for WWOOF of New England. After being overwhelmed by the response to this broad advertising, we decided to focus our future recruitment efforts on a specific segment of the population.

Interns and Apprentices

For many years D Acres operated a program for interns and apprentices. During that period we accepted over two hundred short-term participants from across the United States. We felt college-age participants would offer the greatest possible benefits to the organization. Schools offered youthful and idealistic recruits who were available during the summer months. Institutions of higher

D Acres of New Hampshire
Permaculture Farm & Educational Homestead

D Acres of New Hampshire is a nonprofit Organic Farm & Educational Homestead that offers hands-on learning experience designed to increase our capacity to live sustainably and act as stewards of the Earth. We offer two residential programs that give the opportunity to become immersed in a model of community-based, ecologically-oriented rural living.

INTERNSHIPS

minimum of 26 hrs of work per week, $20 fee per week, 6 week minimum

Interns are involved in managing a modern day organic farm, working in the gardens, with the animals, and in the forest. Interns also work on ecological building and wood-crafting projects, as well as in the kitchen preserving the harvest and baking. Interns must have some experience related to sustainable living or farming.

APPRENTICESHIPS

minimum of 26 hrs of work per week, $100 per week

Looking to get college credit? Interested in learning rural & sustainable living skills? Let us guide your learning experience. Individuals are responsible for getting approval from their school or university if credit is desired. Choose from focus areas listed below for an in-depth learning experience. No prior experience is necessary to apprentice.

Projects include:

No-till organic gardening	*Graphic Design & Videography*
Sustainable forestry with oxen	*Youth Education*
Animal Husbandry	*Woodworking*
Ecological construction	*Renewable Energy*
Food Preservation & Culinary Arts	*Cottage Industry*

Program fees include room and board (tent platforms or treehouses), and organic meals. In addition to 26 hours of work, interns & apprentices participate in the communal contract —approximately 5-8 hours per week of cooking, cleaning and community meetings. The remaining time is at your disposal for reading, hiking, biking, swimming, or rock climbing at the nearby Rumney Crag.

PO Box 98 Dorchester, NH 03266   603.786.2366   info@dacres.org

www.dacres.org

2010 flyer

learning already spoke the language of internships and apprentices, so we utilized those terms to integrate ourselves into this world of academia.

Every year, primarily in the summer months, from one to ten college-age participants would arrive at the farm. These participants would come to gain knowledge regarding sustainable living. Our intention was to provide a classroom in which necessary farm skills could be practiced by the students. Apprentices and interns participated in the larger group tasks of the workweek, as well as in household chores and public events. Because of their lack of experience and motivation, these individuals were included as participants in the farm activities, though they were not depended upon to produce results.

In recent years the intern and apprentice program has shifted its focus from college-age transient experimenters toward more experienced agriculturalists and communitarians who are passionately mission driven. It is our hope that by redirecting our outreach this way, we can more directly harness positive energy in pursuit of our project goals. While the work remains the same, a more seasoned group will be capable of self-direction, requiring less of the long-term staff's time spent micromanaging and training. Instead of idealists without experience, we hope to draw from a pool of people who recognize the opportunity to pursue this collective endeavor with enthusiasm.

International Travelers

We enthusiastically receive international travelers as participants at D Acres. We have hosted individuals from New Zealand, Scotland, Ireland, Mexico, Brazil, Israel, Denmark, Germany, Italy, France, Australia, and China. When these individuals are unable to commit to a six-week stay at the farm, we have made exceptions to share the possibilities of cultural exchange.

Language is generally not a substantial problem as most travelers speak functional English. These travelers usually have a specific interest in agriculture and community, although many lack experience.

Group photos of D Acres staff and residents in different seasons and years

In general these travelers are affluent because of the intense financial restrictions of accessing this country. While there is no guarantee that travelers will provide any serious accomplishments of tasks, they do add immensely to the experience for the residents of the D Acres and Dorchester area communities. International visitors help bring ideas and perspectives from around the globe to our rural community. By sharing their diversity of global experience, these visitors bring a wealth of knowledge difficult to obtain through the mainstream media portals. This process of welcoming these guests also serves the global community when our international guests share our ideas and ideals upon returning to their homeland.

The tasks we give these visitors depends on the seasonal activities and organizational needs, as well as their own aptitudes and interests. Our intention is not to maintain our farm operations through this labor but rather to value the intercultural exchange and experience we are able to provide. While we are flexible, we do expect the visitors to contribute an agreed-upon commitment of time to be spent on farm tasks during their stay.

Applying to D Acres

Experience, personality, and commitment shape the possible contributions residents bequeath to the community. Defining roles at D Acres requires ongoing dialogue among the participants. To accept the responsibility requires an enthusiasm philosophically to engage in efforts directed toward the mission of the organization. Individuals are often attracted to the organization because of its capacity to empower their efforts without the restrictions imposed by starting their own operations.

Still, some individuals are interested in utilizing D Acres as a stepping-stone on their individual journey toward a project in which they would secure private ownership, while others see D Acres as a destination for a lifelong commitment to communitarianism. D Acres represents an escape from the typical constructs that impose absolute value to individual ownership and nuclear family dynamics. By offering an alternative format D Acres can actualize the concept of collective responsibility. This requires individuals who are looking beyond the conventional forms of government, trade, and financial markets. Instead these individuals have chosen to put their faith in collectively organized, rural occupation and emancipation of the landscape.

An effective program utilizes targeted advertising to fulfill specific needs of the organization. The needs of the organization and roles of the individual must be clearly communicated. Our process begins with filling out an online application, which is then distributed to the staff and residents for review. If we have space available, the prospective applicant is invited for a visit and an interview. If geography prevents an on-site visit, we will opt for video calling or a telephone interview. During the interview process it is ideal for multiple people from D Acres to be present and speak with the potential candidates, so we can benefit from different insights and perspectives as we ascertain the viability of the potential relationship. With multiple community residents serving as ambassadors in recruitment, it also serves to strengthen the applicant interview process with organizational continuity and knowledge developed through historical experiences.

Evaluating candidates is a difficult process. Judging how an individual will succeed at D Acres is not easy. We choose to accept new people by evaluating both the applicant and the status of the organization at the time. At times we reject candidates based on their lack of experience or understanding of community living. We also evaluate individual motivations for participation and how those goals overlap with the capacity of the organization. While we would like to offer an opportunity to all that apply, we have chosen to deny access when we feel a new applicant might detract from the well-being of the existing staff and the overall goals of the organization.

By defining expectations and setting standards we construct a framework for personal evaluation

While not everyone comes with the knowledge to prune kiwi vines, blueberries, apples, and other fruit trees, it is an essential spring task that is shared and builds skills.

and reflection. When people express interest in participation with the organization, they must evaluate their own commitment, experience, and capacity to excel in the various organizational roles. During the application process the current residents must perform their own evaluation to determine the viability of the applicant. This documentation, combined with the interview process, serves as a basis for establishing mutually determined expectations so that miscommunication and conflict regarding responsibilities are minimized. This process is not intended to inhibit flexibility in roles or responsibilities. Instead these discussions serve as a baseline starting point to begin the dialogue in this evolving discussion.

Upon arrival, housing is generally allocated on a first come, first served basis. Tree houses are primary dwellings, though outbuildings and tents are utilized when necessary. The transition to living on the farm requires an adjustment, and we have instituted an orientation procedure to acquaint new arrivals with the system.

Challenges

The difficulties that come with long-term commitment from staff and residents are varied. For some the social isolation of the community detracts from the experience. For others the intensity of communicating within a small group can become a distasteful crucible. Another potential detraction is the limit of readily available cash-earning potential on-site and within the region. While the position of staff or resident may imply some prestige in certain agricultural and community cliques, there is a general distrust of communities within contemporary society. Friends and relatives visiting may relay concerns that the organization is a cult. The demands, sacrifices, and constraints imposed by this lifestyle draw the concern of loved ones swimming in the mainstream.

Maintaining positive relations as couples and work partners has also proven difficult for residents. The intensity of the work schedule and complexities of community living can degrade quality time for couples, and the long hours and pressure of the responsibilities have strained personal relations.

To maintain physical and mental health, staff and residents must proactively seek preventive measures. The workload is demanding. Burnout is a common dilemma for staff. The grind of the farm work and the drain of endless visitors are exhausting. The transitional nature of personnel at D Acres can also be emotionally draining to the longer-term residents. While it is interesting to meet and engage with folks who are flowing through the project, sometimes the energy invested in orientation and supervision does not feel reciprocated.

Community obligations are a balance of positive and negatives. While we collectively benefit from our efforts, the communication and levels of sincerity required to live in such close proximity is demanding. While the specialization of community responsibilities such as paying bills provides freedom to particular individuals from some specific domestic duties, the quantity of household work is unfortunately not reduced by a collective undertaking. In fact, because of the variability of effort and skill invested in shared tasks, the burden of maintaining standards often falls on the more competent, responsible participants. Unequal chore distribution is an ongoing point of discussion. It is difficult to pass judgment between peers regarding performance within this model.

By investing time together as a community we reap the benefits of profound communication and understanding acquired through sharing experiences, reflection, and dialogue. With a goal of staff retention we attempt open discussion and communication designed to address the staff's needs. Issues such as ownership and equity, regarding both financial and decision-making capacity, are a crucial component for increasing retention. Security provided by tenure, salary, and health insurance would help solidify the duration of staff commitment. Consequently, we are focused on refining the employment agreements to advance the vital personnel component of our operations.

CHAPTER SIX

MEETINGS

When we began D Acres, I was naive regarding the complexities and difficulties of maintaining positive human interactions while undertaking a collective project. I assumed that all conflict would result in acceptable compromises and that clashing personalities would eventually work to harmonize. While we have enjoyed tremendous success encouraging that positive progression, there have also been historic failures. Poor communication, perceived slights, and jealousy are common and can hinder organizational efforts if not addressed.

Background

My friends and I first began considering the formation of D Acres in dorm rooms or over a beer at the pub. The content of our discussions was mostly theoretical; conflicts of opinion did not require immediate or agreed-upon resolutions. When we moved onto the land in Dorchester our meetings shifted from hypothetical content to time-sensitive financial, infrastructural, and personnel concerns with real-life ramifications. Because of our lack of organization and experience, these meetings were difficult to endure. Meetings had minimal structure; we did not have a proper mechanism to address issues that required efficient or clear resolutions. Agenda items were not grouped by category, and issues were haphazardly introduced for discussion. Often these poorly defined discussions never led to tangible proposals of action and ended without concise responsibilities or timetables being accepted by individuals. Without a clear format new participants were deterred from engagement in the process; without a voice they lost interest and enthusiasm. The tone of the sessions swayed as discussions of proactive solutions could drift toward antagonism and condemnation based on people's moods and personalities. Without a rudder to steer the direction of discussion, meetings were lengthy, and the time spent was stymied by the ineffectiveness of the process.

To improve the dynamics of our meetings we researched other groups' processes, starting with *Robert's Rules of Order*, the traditional, widely accepted format for legislative and committee work in the United States. Its archaic parliamentary style was developed by Henry Martin Robert, who served as a military engineer during the Civil War and after; the book uses the postbellum terminology of that era to define meeting structure. Robert's Rules are precise, standardized, and codified for resolving conflict through democratic process. However, they are also inflexible and too rigid for evolving circumstances; the complexity of the book's language and procedures is foreign to most. The resulting structure enables those familiar with the rules of discourse a methodology to expedite decisions in their favor while preventing the formation of true compromises.

Unsatisfied with the format of the antiquated Robert's Rules, we sourced several short guides on the consensus process. *Wise Fool Basics* by K. Ruby has a brief though helpful section dedicated to consensus. We also obtained the more comprehensive *Introduction to Consensus* by Beatrice Briggs. In pursuit of experiential practice, I traveled to participate in the consensus process advertised by La Caravana Arcoiris, an ecovillage in Ecuador.

During my stint with La Caravana my ideology about consensus confronted the realities of actually making decisions by consensus. Ideally, consensus strives for a completely flat decision-making process that recognizes and equally represents each participating member. Consensus formats in practice are grounded in reality. While inclusion and transparency are important components, value is also placed on experience, commitment, investment, and privacy as principal elements of successful group processes. Consensus process offers limitless possibilities for solutions for the common good, but these solutions require determined engagement and commitment. Without committed participants willing to sacrifice personal prerogatives for the common good, the horizontal structure of power does not truly exist.

When I traveled to La Caravana I naively assumed that they had achieved an egalitarian, utopian practice of ideals. And in many ways it was. It was an honest, brave, and determined attempt to practice the values and ideals of consensus. The general meetings were opportunities for open discussion and feedback with rotating facilitation. It was empowering and gratifying to participate in forums where diverse opinions were expressed at length, with persuasive detail and eloquence. During my tenure with La Caravana, I truly felt invested and appreciated in the process of day-to-day group collaboration. La Caravana was a remarkable example of consensus potential. However, it also highlighted limitations and the reality that the consensus process evolves perpetually with each communicative interaction.

La Caravana's nonhierarchical decision-making process was limited by the transience of its membership. The group's participants had vast differences in investment and experience. Its membership fluctuated through the years as the organization traveled through South America. Founders and long-term investors were granted organizational status as "*tripulantes*," or stewards. These pillars of the project collaborated to make decisions regarding personnel, budget, and organizational activities. Founders who had traveled with the project for many years could not be expected to relinquish decision-making authority to me as a new arrival. Likewise, because of my limited investment, I could not be expected to provide the day-to-day leadership or contribute to longer-term organizational planning.

I also witnessed how the success of the consensus process in fulfilling the mission of an organization ultimately hinges on the commitment of the participants. At times I felt that the short-term, transient members of La Caravana, such as I was, put unreasonable pressure on the stewards to accommodate our individual preferences. I felt I exerted improper influence on the direction of the organization even though I was not as firmly committed to the future as others were. I felt that the transients often resisted the existing governing structures and rebelled against perceived organizational authority. We demanded immediate gratifying attention and accommodation for our short-term sacrifices or else we would abandon the organization. Short-term participants did not acknowledge and respect the investment, commitment, and sacrifices of the longer-term members. The lack of understanding and knowledge of process by transient members, such as I, undermined La Caravana's focus and drained energy from their mission-driven work.

While the consensus format offered by La Caravana was not the egalitarian vision that I had imagined, it was conceived as an attempt to equitably distribute the decision-making capacity of the participants with fairness and structure. The group had essentially formed two tiers of decision making in which the tripulantes retained budgetary and personnel jurisdiction. By dreaming up new creative approaches and refining organizational processes, the group continually sought systematized decision making that was reasonable for the circumstances presented. My experiences with La Caravana have shaped the group processes here at D Acres. I am not fixated on achieving a utopian, nonhierarchical system but rather a system that evolves with reason, transparency, and fairness. Invested, experienced, and committed participants should be allocated higher positions of authority to match their responsibilities while still continuing to be inclusive of the broader community. My experience with La Caravana reminded me to remain vigilant to disconnection between tiers of authority, abuses of power, and potential clique behaviors of subgroups.

Communication Philosophy

Group communication is essential to maintaining healthy relationships and ultimately crucial for the perpetuation of humanity. While we are enamored of the possibilities of a true consensus format, it can only exist where there is equal commitment to the cause, and no hierarchy. D Acres has a hierarchy that respects experience, commitment, and investment and the reality that there will always be different levels of commitment to the organization. Participants must be aware of their own role, and that of everyone else within the group. Long-term success requires participants to perpetually define and retain awareness of roles and expectations. Our goal is to provide a format that encourages positive decisions within this context.

We primarily utilize a regularly scheduled, face-to-face, agenda-oriented meeting. These meetings are designed as open forums where the discussion of ideas fosters shared group philosophies that bind participants in a mutually supportive network of understanding. Meetings are also designed to help efficiently plan and direct the implementation of our programs, projects, events, and endeavors. They are

a time for us to both resolve the miniscule details of the operations as well as address broader organizational questions. With a clear process and fair, equitable structure, meetings enfranchise participants and allow them to be proactive in a mutually beneficial cooperative effort.

Dissent is a necessary component in all decision making. Meetings are opportunities for participants to speak openly from their hearts. Our collaborative format incorporates dissenting opinions and perspectives through a spirit of compromise that transcends the problematic, winner-take-all, democratic seesaw characteristic of contemporary debate. To prevent the silencing of minority opinions through peer intimidation, we must remember how our roles in a holistic process promote the shared vision. One fundamental precept is to shape problems into solutions by utilizing the opinions of the minority. True compromise considers the perspectives of all parties.

To voluntarily commit to this process requires recognizing an agreed-upon mission and shared goals. A goal can be as broad as survival and happiness, or as finite as organizing an annual event. Shared purpose unites the group to act holistically toward the common goal. Part of the intention of the organization is to encourage the development of a work ethic and empower our populace to lose their sense of personal entitlement by becoming immersed in the community farm system. Developing a system that fosters inclusive decision making challenges our sense of individuality. But overcoming this resistance can be empowering and rewarding.

Time management is an essential part of this transition. Summer only has so many days. With seasonal farming experience comes awareness of the limits of time. As the internal time clock of seasonal and yearly progression is reconnected, people change how they value their relationship to work. When we develop a shared valuation of time and efforts, our capacity to communicate as a group simultaneously increases. Effective meetings save

Beth, Regina, and Scott convening a meeting in the kitchen, to discuss weekly plans

time, conserve energy, and improve the overall efficiency and diligence of the organization in pursuit of its mission.

Structure

Structure is necessary for successful meetings. By normalizing and codifying meeting structure based on the needs of the group, we provide a comfortable, safe environment to allow for a fair decision-making process.

The regularity of meeting times, decorum, conduct, and structure creates a predictable pattern with shared expectations that builds capacity for the group to make difficult decisions and resolve conflict. Consistency in structure is vital to ensure a

productive session. Continuity is also vital to project completion. Typically we review the prior week's notes before proceeding with new agenda items. By revisiting the minutes from the prior session we are able to evaluate project progress and develop plans for the future. This process is perpetual and requires prompt, comprehensive attention to detail. A running agenda that reflects updates and progress provides the platform for systematic task completion. Without regular check-ins that force discussion, conflicts can go unresolved. By planning and communicating the breadth of organizational activities on a weekly basis, we maintain a format for new arrivals that helps them connect to the activities of the organization.

For the group decision-making process to be effectual each individual must invest in the process.

Active participation, displayed by body and spoken language that signals engagement and understanding, is essential for ensuring a successful meeting.

Agenda

It is important to use an agreed-upon agenda for the meetings, and to consider how new items are introduced. For many years we chose to use a flip chart format to introduce agenda items prior to meetings. During the week participants would add topics to the chart, which was posted in a visible location in the Community Building. When adding to the list members picked whether they felt the item was an announcement, a proposal, or a discussion and chose a time slot during the meeting that they felt would be appropriate to address it. By establishing the classification and time allocation preference prior to the meeting, other participants could come prepared.

Classification of agenda items can organize efficient meeting structure. Announcements do not require the group to make decisions, while proposals are made in the hope of reaching definitive judgments. Discussions are open ended and informational, addressing issues relevant to the group without preplanned courses of action. By classifying each agenda item, clarity is lent to the immediate goals revolving around each issue. Defining each topic as an announcement, discussion, or proposal allows the group to understand their role in the dialogue.

Prudent group decision making also requires an investment in time. Important decisions are best made with ample research and discussion rather than shooting from the hip. Several meetings may be utilized to introduce an idea, research options, and revise plans of action for a particular issue. Project completion also requires continuity of information sharing through updates and feedback. Time between sessions allows group members the space to reflect and develop creative, innovative, and proactive solutions for issues and projects that are being

considered. While flexibility is necessary to make improvisational choices in certain circumstances, an agreed-upon process of collaborative, deliberative decision making should result in the strength of a shared and thoroughly reviewed resolution.

Fluidity is encouraged; there are times when decisions are expedited to conserve time and energy. In general it is important to retain consistency in discussion and agenda format to maintain organizational efficacy. And since potentially useful ideas are often introduced during dialogue that are not relevant to the subject at hand, it is helpful to record these thoughts, so they may be referenced at other points in the meeting.

At times the group may choose to table a discussion of a particular topic. Certain projects may justify separate meetings designated for a particular focus where discussions can be attended in smaller working groups. There are also times when the specifics of a particular project are beyond anyone's expertise, in which case concessions should be made to educate the group as much as possible, or arrangements should be made to meet separately to develop a plan of action. There is balance between efficiently conducting the decisions of the moment and endless discussion aimed at informing decision makers. We must constantly assess the relevancy of further research, debate, and discussion in relation to prompt, concise decision making.

Facilitation

It is important for all the members of the group to practice the role of meeting facilitator, which helps them garner empathy and gain experience that will help them understand and relate to the process in the future. Rotating facilitation also engenders a depersonalized, goal-oriented meeting environment. The facilitator serves to direct the group toward making a decision rather than influencing the group toward any particular decision. Therefore, the facilitator should articulate when he or she is speaking for personal interests rather than in the facilitation role.

Rainy Day Plan
pighouse → woodchips
sap, water ghouses
build soil

	Tuesday	Wednesday	Thursday	Friday	Saturday	Sunday	Monday
Daily Work	BOIL (DL) Proctor 9:30-3 tour, lunch work: sap ghouses WOODSHOP — GARDEN / GHOUSE WORK	BOIL (MP) seeding pruning	FORESTRY A.M. (DL) pruning (fiber)	BOIL Chores food prep 9 am (MP)	Sewing Wkshp 2pm	Vol. Day 10-4 (BW) Snowshoe (DL) CMAA workshop	Staff gen. mtgs.
Lunch Prep	RR						
Fires	DL			MP	DL		
Hostel / Hostel Buddy							
Meetings	8am - garden						
Office	RR	JT	AM Betty (RR)				
Dinner	6pm RR	MP	BW	DL/MP			
Animal Chores — Pigs	DL			RR	BW		
Ducks	MP						
Chickens	JT			MP			
Oxen	RR / BW			BW	DL		
Garden — Harvest							
Water Buddy		MP		RR	DL		
Pig Food / Town Run			RR		DL		BW
Evening Activities	7pm Howard workshop	JT to PB Cain	JT @ Putney	JT @ Prescott Potluck / Open Mic	Member Dinner 5pm		

A copy of our weekly planner, during sugaring season in 2011

Facilitation requires practice and diligence. It is important to maintain a level of impartiality and neutrality that encourages participation of all opinions and to introduce agenda items in such a way that does not predispose or prejudice the opinion of the group or inhibit full discussion of possible outcomes and resolutions.

The facilitator must also use objective interpretation and judgment to orchestrate the meeting, interceding if the discussion drifts and adjudicating when dialogue becomes antagonistic or insulting. He or she must also encourage shy or new members to express themselves, while reining in personalities that tend to dominate discussion. A responsible facilitator officiates the proceedings by being consistent and impartial, while also informing and educating. But while facilitators encourage a diversity of opinions to reach group goals, they must also be cognizant of the group's prioritization and actualization of purpose.

Meetings can be lengthy, and the mental focus necessary is arduous. Therefore it is crucial to plan meetings at an appropriate time of day with a reasonable duration. Facilitation should be concise;

a timekeeper may be enlisted to limit the length of discussions. By streamlining meetings according to the prioritization of the agenda, groups can accelerate the meeting or choose to table issues for further future discussions. Food can help stabilize blood sugar, which without it can falter, resulting in erratic moods and lack of mental acuity.

Accountability is a crucial aspect of a successful meeting format and organizational success. By successfully taking account of activities, we can measure progress and determine the next course of action. Accountability is ensured through facilitation that clearly articulates the responsibilities that are delineated by discussion. When an agenda item identifies an issue that needs to be resolved, the facilitator orchestrates discussion of the prior experiences and mediates the possibilities at hand. As dialogue presents the scenario, the next step is to identify action that should be taken. By seeking direction and ideas from the assembly, the facilitator seeks the best option that is acceptable and productive for the group.

The facilitator serves as a historian for each of the agenda items by reading the notes or providing background information. His or her role is to ensure all information has been presented regarding a given issue. As resolutions are articulated, the facilitator encourages debate regarding the merits and weaknesses of the various options. To move forward, the facilitator sets the stage so that the group can arrive at a compromise through open dialogue.

Frame of Mind

By accepting the consensus approach, one is accepting a particular collaborative frame of mind and should be willing to transcend the typical, egotistical approach to success. Instead of personal victory, winning is defined by the success of the mission-driven agenda. At times the weight of the decision is borne by those with the experience, commitment, rationale, and investment related to that particular decision, as evidenced by our deferring most decisions in the kitchen to the kitchen manager.

Meetings can be intense environments of expression. The pressure of important discussions and decisions affects the emotions of everyone in the group. While passion is important to the success of the organization, it is in the best interest of the whole when members of the group maintain respectful etiquette and refrain from personal attacks and counterproductive contributions to the meeting. High levels of investment in group decisions can lead to emotional explosions. The lightning and thunder of group dynamics—screaming, profanities, and sobbing—can be indicative of a significant trend in poor communication that needs to be addressed. On the other side of the spectrum, silence is understood as acceptance and agreement to proposals that are voiced. Without verbal articulation of opposition or reluctance, the individual agrees to abide by the resolution.

Ideas presented to the group are investments made by the individual. Refining a detailed proposal to be brought to the group can require significant planning. This expenditure of time and energy is not without risk to the presenter's humility. By defining a proposal to the group, the individual concedes that his or her idea could be rejected or accepted and refined based on its merits in relation to the common good.

While an individual may not agree with every solution, each should be tolerant of the decision of the group and abide by it. The biggest and most profound leap of faith in this acceptance is that when a group united by a mission-driven agenda follows consensual process, it will benefit both the individuals and the group more than any unilateral attempt would have done. We are greater than the sum of our parts.

Upon arrival the primary role of participants is as observers, asking questions and providing insight. As their experience grows in the organization, their responsibility to report on projects will grow. The meeting experience will also provide the knowledge and skills necessary for individuals to attain a more substantial role as a meeting organizer. Through this

Dave Jacke facilitates a regional meeting to plan the first NE Permaculture Gathering at D Acres.

maturation provided by knowledge and experience, participants can become more able stewards of the process.

When the structures fail, however, this process falters. The basic framework is a well-reasoned, consistent structure of process applicable to any group of participants, which depends on effective facilitation and accurate note taking. Ultimately, meetings depend on the participants' knowledge of consensual process as well as their capacity to support the mission. Participants must be actively engaged and should follow through with the tasks that come out of the meetings, so that the group can move forward.

Communication Evolution

Over the years the structures and procedures of D Acres' group meetings have evolved. New methods have been introduced, while others have become obsolete. As an example, early in the group's evolution it seemed necessary to publicly post agenda items prior to the meetings to prepare the group and avoid major surprises. At this point we have chosen not to post agendas prior to meetings. For the most part, we've become so well versed in our activities that our agenda topics offer few surprises week to week. We do, however, continue to avoid adding hastily improvised items if possible. As our system of planning matures and our experience grows, we have been able to anticipate future deadlines so that discussion may begin weeks prior to actual decisions being made. We recognize that surprises can be used as a weapon in the consensus process; the overall philosophy is to improve the process rather than use information as a destructive tactic. Consequently, sensitive topics are often discussed at length by the vested participants prior to formal meetings.

The ongoing evolution of our meeting structure results from continued evaluation of the process by both experienced and novice participants. The fluidity of the process is necessary as the personnel and project goals shift with time. One summer, for example, a group of short-term residents chose to exclude themselves from discussion of longer-term organizational issues and instead chose to focus on the week-to-week planning.

Certain groups have chosen to restrict and divide discussion based on their personal interests and duration of commitment. Using this feedback, long-term staff members also meet separately once a week. While the agenda is similar to the larger meeting's, the intention is to limit the time spent on material irrelevant to all participants when we meet as a group. All personnel are still invited to sit in on the long-term staff's discussion, with the exception of confidential or inappropriate topics.

Our meeting structure will continue to be refined. Each interaction of communication is an opportunity to improve relations and our cooperation. By structuring the process we create a stable, hospitable environment for the systematic progression of a mission-driven agenda. This codified structure allows for flexibility and fluidity in process based on common sense and reason. By orienting the group to decide and implement the action necessary to achieve group goals, collaborative process fosters results for the common good. Ultimately humanity must realize that our capacity is greater than the sum of individual parts. We must provide a forum to realize the possible potential of human cooperation.

CHAPTER SEVEN

BUDGET

Every organization requires a method of accounting for its income and expenses. While there is considerable subjective interpretation regarding prioritization of time, the objective analysis of a farm's sustainability is often defined by dollars. The primary goal is to remain solvent and to be able to continue the work. In meeting our budgetary needs, our emphasis is not to accumulate more funding but rather to utilize existing funds most wisely. By continuing to evaluate the costs and benefits that prioritize the work and play of our lives, we strive to develop a lifestyle and an economy that is sustainable for the participants and the organization.

This chapter reflects how the income and expenses of D Acres are directly tied to the mission and activities that occur in any given year. In addition to the weather, there are many variables that affect revenue and expenses. Through the various stages of our development, we have radically shifted where we derive funds from and for what purpose. While we now have some expenses that are predictable every year, there are always surprise bills and inflation. Likewise, while we depend on several consistent revenue streams, as our focus, personnel, and infrastructure change, so does our income. Ultimately we are attempting to develop an organization that endures by both minimizing expenses and creating diverse revenue streams aligned with our service goals.

The budget at D Acres has evolved based on our organizational goals. The impetus and immediate priority for my migration to New Hampshire was to provide a safe, comfortable environment for an elderly woman in her rural homestead. While Edith was quite competent at eighty-nine years of age when we arrived, she was becoming increasingly frail. Her capacity to drive a vehicle was suspect, and her memory was failing. Though she had hired help to assist her with cleaning and yard work tasks, the family was concerned for Edith's well-being.

Rather than considering a future for Edith in an assisted living facility, we chose an option designed to yield multiple benefits. By providing economic assistance to start the farm operations, our family was able to provide a high quality of life for Edith for the last six years of her life. It was my aunt's wishes to remain at home until her death rather than be institutionalized. To honor her wishes we diverted funds that would have supported her in a nursing home so that assistance to provide for her well-being was instead invested in her farm estate legacy.

Considering that the cost of a nursing home can be up to seventy-five thousand dollars a year, we saved a tremendous amount of money and allowed Edith to enjoy a superior lifestyle in her home. The security of having trusted family on-site was also extremely valuable to my parents, who were residing in North Carolina at the time. For the several years before D Acres' inception, for example, my

The Community Building was initiated as an investment for the family as well as for the farm's operations.

mother had been called to visit Edith with increased frequency because of her declining health and to reconcile her finances. Once I arrived at D Acres, we were on call to meet Edith's needs for the remainder of her life.

When we arrived we had minimal initial cash expenses. We cleaned up the Red Barn and moved right in. With our housing costs controlled, our initial expenses were mostly for food and agricultural supplies. Our appetite for seeds, rootstock, and gardening equipment was voracious, but we needed to acquire the necessary tools to invest in the farm enterprise.

During this period Bill and Betty covered such overhead costs as food and electricity, while I utilized savings that I had inherited from my grandparents' demise to finance our agricultural expenses. As these funds were expended I began to evaluate how to proceed and meet our overall objectives. After our first growing season we began formal discussions regarding construction of the Community Building.

At the time Bill and Betty were in the first stages of moving to New Hampshire and enjoyed hosting friends and family. Bill also had visions of working in a wood shop during his later years. The concept of a building that could facilitate Bill and Betty's retirement migration and serve as a recreational space for our many relatives and friends was proposed to my parents. This building would provide a space that was in proximity to Edith, while ensuring the privacy she needed from our family visitors.

What Is a 501(c)(3)?

The 501(c)(3) is a designation assigned by the Internal Revenue Service for entities incorporated for charitable purposes. Achieving the designation requires approval by the IRS following submission of a substantial fee and an application. The application details the intended operation of the organization and identifies a board of directors. Most grant agencies will not consider appropriating funds to organizations without this designation. Likewise, donations of products, services, and money cannot be deducted from personal taxes unless the organization is designated by the IRS. Recently a new classification of corporations with charitable intentions has emerged. The L3C classification allows tax-deductible donations and private ownership. When we applied to the IRS the L3C classification was not an option, though it may be suitable for similar operations at this time.

As we undertook the construction of the Community Building, the nature of our income and expenses increased beyond elder care and market garden start-up into the field of construction management. We provided labor toward the construction process to subsidize our educational farm endeavors. Instead of paying for a general contractor, the family invested in our entity, which provided those services. The multifunctionality of this arrangement is a concrete example of stacking functionality through a single process. In constructing the building, multiple parties, with various overlapping intentions, benefited. The Trought family benefited from property improvements and available accommodations, the fledgling organization received start-up funding and a four-season base of operations, while participants gained construction experience. This process involved many compromises, but the investment has surpassed expectations for providing returns to both the investors and the general public.

After the construction phase was completed in the year 2000, we were able to shift our focus more fully into the land management and educational components of D Acres. At this juncture we began to pursue 501(c)(3) status from the IRS so that we could solicit tax-deductible funds to amplify our service and educational work.

Over time the organization has shifted from a growth model to one in which perpetuation is the focus. The basis for this change is a result of the organizational and individuals' trajectory. The median age of the staff has increased through the years. As we have aged there have been shifts in priorities for the maturing people and project. This philosophical shift is reflected in our budgetary expenditures over the first fifteen years. As we have become established, there has been less emphasis on advertisements for well-attended events and services. We have also purchased durable equipment and built the fundamental structures on-site. After this initial investment period our focus is shifting toward maintenance and personnel. Revenue must be channeled to insure the buildings and equipment are maintained. Investing in our personnel is also crucial to our perpetuation. We must insure the funds are allocated to provide for the needs and security of the personnel who have committed life energy toward organizational operations. By investing in the people and the on-site infrastructure the model can continue to perpetuate.

Over the course of years we have chosen distinct areas of emphasis to invest in and expend funds on. While we have a prioritized agenda and a "wish list" of organizational expenditures,

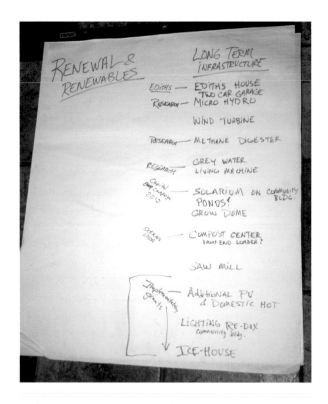

Each year special projects and plans are proposed. These projects guide our fund-raising letters and grant proposals.

we also are fluid in responding to opportunities as they are presented. Every year at D Acres we produce a budget that anticipates our revenue and expenses. Right now our minimal expenses are fairly predictable. We know by what we have spent in prior years the approximate annual costs of our operations. Bills such as insurance, food, utilities, and rent can be anticipated.

To provide income for the organization, we rely on a broad, diverse, unpredictable revenue stream. A baseline of expenses can be anticipated, but it is difficult to predict minimum amounts of income or from whom they will be derived. With the exception of the Troughts and some reliable alumni and community members, our income streams have fluctuated. To compensate for this uncertainty, we plan for each year by setting a low range that will cover basic operational costs. Once that figure has

been established, any additional funding is directed to a prioritized list of possible expenditures. In the event that the minimum budget is not obtained we would resort to savings, emergency cost control, and fund-raising.

There are several broad categories of revenue sources that we identify and seek every year, including contributions and grants, educational programming, fund-raising events, product sales, and services.

Contributions and Grants

It is important for the 501(c) entity to demonstrate public support. Therefore, in addition to providing service to the community, D Acres must solicit contributions and grants from the public sector, which account for up to a third of our revenue. Within contributions and grants are several subcategories, including those made by individuals, families, and businesses. These donations can be cash or in-kind gifts such as vehicles.

Membership Programs

For many years we operated a membership system as part of this contribution program. In exchange for an annual fee or barter, we offered several different levels of membership. The student and low-income rate was fifteen dollars a year; individuals paid twenty-five dollars, and family memberships cost thirty-five dollars. Contributors who donated more than one hundred dollars were recognized as supporting members. Membership benefits included a 25 percent reduction on workshop fees and a 10 percent discount on lodging. We also invited members to an annual dinner in which they received an organizational status report and were solicited for feedback.

The intention of the D Acres' membership program was to encourage a method of establishing support from the community, for members to forge a link to the broader community and serve as representatives from the constituency. Unfortunately, membership also implied to some people an exclusive club for select individuals.

Recently we have established the "many hands" program, which recognizes not only financial contributors but also volunteers and participants and fosters shared community ownership. We are optimistic that we will continue to evolve this community ownership concept. While we have yearly turnover in participation, with new acquaintances arriving and others moving on, at this juncture this program depends on roughly 75 to 125 regular contributors to the organization. Of this number 10 to 20 provide consistent volunteer participation, and we are also able to collect several thousand dollars in donations per year.

Foundations and Private Grants

Foundations provide another sizable portion of our income in the contributions and grants category. There are thousands of entities that offer grant funding to nonprofit organizations. However, grant funding is competitive. Many foundations only represent the narrow interests of wealthy families, while the remaining ones have scant funds available for the tremendous needs of the public. The omnipresent concerns presented to funding agencies include children's welfare, health research, elder care, homelessness, and the environment as well as cultural activities such as the arts. These are all grave concerns for which funding is crucial.

Consequently, acquisition of grant funding can be a difficult proposition. Because of the competition, a 10 percent success rate for grant-writing applications is considered successful. Fulfillment of grant-application guidelines can be a rigorous bureaucratic process, thick with paperwork. The money generated is neither "free"—there are usually strings attached on how you can use it—nor predictable income. Funding for operational costs and overhead is rare. Funders' priorities are typically for new, exciting projects rather than the common and necessary costs of farming operations. While that stimulates innovation, we must be cautious not to overextend our organizational capacity in pursuit of funding. We submit approximately two hundred grant applications per year, which provide for 10 to 25 percent of our revenue.

The uncertainty involved with grant writing makes organizational budgeting difficult. The time spent on submitting paperwork could well be spent on the educational purposes of the organization. The grant-writing process in itself tests our capacity to implement specific projects or programs. Gambling with organizational time in this manner has led us to an annual schedule: Winter is utilized for writing grants, while the summer is the season for project implementation. To address the uncertainty of receiving grants, we estimate a minimum and maximum that we anticipate, then make expenditures between those extremes as grants are approved.

Our success in this category is indebted to our nonprofit status. Granting institutions and individuals are more inclined to donate to an organization that has obtained 501(c) status because of the tax implications and benefits provided, as well as the transparent accountability ensured through publicly available financial information.

Federal and Public Grants

In addition to private foundations there are also federal and state grant programs. These funds are generally available for conservation improvements, tourism, and research purposes. At the federal level we have dealt primarily with the U.S. Department of Agriculture regarding improvements to the farm and farm environment. Through the USDA, the Natural Resources Conservation Service (NRCS) has several programs aimed at improving agricultural operations. We have received funding through the Environmental Quality Incentives Program (EQIP) for pasture and fencing improvements, as well as construction of a hoop house. In addition we have recently completed working with NRCS to construct a compost facility.

In dealing with either agency there is a lot of bureaucracy. While the individuals employed to administer the programs can be pleasant, they are often bound to guidelines imposed by Concord and Washington. While some of the regulations are intended to be beneficial, the stipulations imposed by these contractual arrangements can be frustrating. For instance, during our NRCS-funded compost facility project, we were required to obtain concrete waste blocks from over two hours away instead of those available locally because of a required design component, a groove in the blocks that helps them stay stacked like Legos. In my opinion, because the wall we wanted to build would be backfilled, the additional V groove was unnecessary. Regardless, the engineer who designed the concrete pad with a block wall insisted that we obtain the blocks he had specified. While this added costs and sent our money out of the local economy, we had to comply with this stipulation or funds for the project would not have been reimbursed.

Grants are often considered the elixir of the nonprofit world: free money. For me it is important to share my view on their actual price, and the ethical quandary presented by that perception. In gambling our dependence on these funding streams, we are submitting to partnership with the bureaucracies of government and of Wall Street financiers. Most private foundation assets are stock market shares held in perpetuity as capitalization for the global economic system. The cost of deriving benefits from these sources should be considered carefully when

When inviting experts in their field, such as Michael Phillips, we literally "stack functions." Collaborating with the local university, D Acres' cosponsored a free presentation. On the following day Michael shared his knowledge with the Permaculture Design Certification class, and we opened that session to the public for a workshop fee.

prioritizing organizational efforts. It is also important to note that many governmental agencies do not supply funding to nonprofit entities.

Educational Programming

Education programming is another income category. There are various subcategories to this section. Programs range from fees for workshops to our overnight accommodation options. Also included are public outreach and education programs such as the Local Goods Guide project. This category amounts to approximately a third of our yearly revenue. Educational programs are budgeted individually so that we can track each program's viability.

Fund-Raising Events

We have regular community events in which we solicit donations from the general public. Our monthly community events include farm feast breakfasts and pizza and movie nights, though we have also hosted a soup night in prior years. These events are intended to highlight the local seasonal food possibilities and bring the community together for an educational event. The Sunday breakfasts include tours, public speakers, and activities such as yoga and Reiki. The pizza nights are opportunities to screen thought-provoking documentaries of environmental or human rights subject material.

We have also hosted a larger annual fund-raising event known as Farm Day. For this event typically over two hundred people would attend for a pig roast, farm tours, and bands. While we generated revenue, goodwill, education, and recognition, the cost of putting on the show was significant. We leaned heavily on volunteers and staff for the event planning and implementation. Attendance was affected year to year by the weather or other competing regional events. Hence, when considering gala fund-raising events, be aware of the overhead costs and energies diverted to stage such events.

Sales and Services

The sale of farm goods and services provides an educational model as well as revenue to support our outreach efforts in educational programming. By demonstrating effective ways to generate revenue utilizing the resources provided by a farm-based model, we offer solutions to the quandaries of rural land management. These cottage industries, further detailed in chapter eighteen, provide an example for revenue generation models that are applicable throughout the rural landscape.

While we sell fundamental farm commodities such as cordwood, potatoes, and vegetables in season, our emphasis is not on traditional wholesale

Farmers' markets have been a method for not only exposing the public to the organizational mission but also garnering revenue from farm sales.

On-farm sales, such as selling potted seedlings or transplants, further generate income for the organization.

production models. We focus instead on value-added products. These are produced from infrastructure, skills, and resources provided by our landscape and our organizational capacity. For example, instead of selling raw lumber, our forestry program adds value by using all parts of the tree and preserving the highest quality wood for enhanced-value products such as furniture. Instead of our energy going quickly up in smoke, our efforts are embodied in higher-value goods. To accurately reflect the success of any cottage industry it is important to document the array of costs of production, from raw materials to personnel time and packaging to the point of sale.

In essence adding value to our existing resources and stacking the functionality of our services is our fundamental mechanism for fiscal success. Instead of competing with other farmers for market share, we expand the markets by identifying unexplored niches and potential within the rural economy. The diverse streams of income integrate and act in conjunction. As an example, the work of the garden provides an attraction to tours and visitors, reduces food costs for residents, and is utilized to garner revenue through meals and food events. The concept is to be diverse yet integrated so that new enterprises do not exhaust energy and resources but rather complement existing operations. By building a web of interconnected income sources, you can create a resilient and viable income generation capacity to meet the costs of operation.

Expenses

The annual expenses of the organization are easier to anticipate. After many years of operation we have a baseline of expenses ranging from utilities to seed and feed that continue to escalate with inflationary pressures. We can also anticipate there will be costs that we are unable to predict in our initial budgetary effort. By allowing for a cushion and maintaining awareness of the annual flow of income and expenses, we attempt to maintain a positive cash flow balance throughout the year. Not only do we consider the annual budget totals, we also must consider the cycle of cash flow throughout the year and periods such as the winter, when bills persist with minimal income.

Inflation and growth of the organization have escalated costs through the years. Going forward, we must carefully design expansion of services so that additional costs are not incurred unless we can raise additional revenue to balance the new expenses.

Personnel costs are a substantial element of the expenses that the organization incurs. In addition to the wages that are garnered for the staff, payroll tax is included in this component. Our payroll cash outlay amounts to between 25 and 35 percent of our total expenses. The costs of living include food, lodging, and amenities that are provided through the organizational umbrella within occupancy expenditures.

Animals also represent a portion of the costs to the organization. In addition to the time and effort spent with the animals, the yearly budget must allocate funds for their feed and care. The yearly amount of grain and hay varies with the number of animals on-site. The oxen require regular veterinary and hoof care. The budget must also consider any butchering services that will be rendered through the year. The expense of the oxen is three to four thousand dollars per year, while the chickens' and pigs' annual costs are a thousand to two thousand dollars.

Advertising is another category of expenses. Our budget varies, depending on the year and the promotional techniques with which we are experimenting. We look for innovative and effective opportunities to maximize our ability to reach the public. We decide on strategies yearly and spend to meet the costs of the plan. We generally spend two to four thousand dollars per year on advertising strategy. In addition to paid advertising we print materials for promotional and fund-raising purposes. While we often supply the printing in-house, we occasionally outsource printing projects. In either case the expenditures for printing must be budgeted.

Another component of the expenses we incur are our dues and subscriptions category. Within this category are the professional association, magazine subscriptions, and vendor fees for conferences that we attend to promote the educational outreach of the organization. Insurance is a component of our annual budgeting as well. The worker's compensation and the policy that provides for the liability of the organization are nearly five thousand dollars per year each. In addition, the policy that provides coverage to the board of directors is almost a thousand dollars per year. The other insurance that the organization currently covers is for our vehicles. In total we pay nearly fifteen thousand dollars per year for our insurance coverage.

Occupancy is another facet of the expenses paid by the organization. There are various licenses for our business operations that are due on an annual basis. We also have an assortment of semitypical household bills. For communications we expend revenue on the telephone and Internet service, as well as the fees associated with hosting our website. Food purchases are also a component of our occupancy expense. While some of the food purchases are allocated to visitors' meals, the majority of our food expense provides for the sustenance of our residents. We also are billed by the propane and electric delivery companies. While we minimize our usage we still purchase three to four thousand dollars a year in industrially produced power.

Every year we require professional services. A basic service that is necessary is assistance with accounting and compliance with tax codes. We also pay professionals, such as the coinstructor for the permaculture class, who is paid as a subcontractor for his services. In addition, we typically hire several experts in fields of sustainability to come for public presentations throughout the year. The provision of this expertise and these services must be budgeted based on yearly planning and objectives.

We also account for purchases that we make for resale. For example, when we purchase materials for specific crafts, such as wood finish, these purchases are attributed to this category. There are regular purchases made for kitchen, woodcrafts, and other ongoing cottage industries. We also expend funds on postal services. While there are a steady stream of outgoing bills and correspondence, we also mail products and track the funds we expend to mail grants.

We also expend funds for repairs and maintenance of the buildings, vehicles, and equipment. The vehicles rely on gas and a registration payment to ensure safe passage. The forestry program requires tool maintenance and upkeep. At minimum, new chains and bar oil are essential to supply the wood needs of the farm. We also purchase tools and equipment for the kitchen, office, and wood shop as existing equipment needs to be replaced or upgraded.

For our gardening operations we purchase seeds and bare rootstock as well as soil amendments and straw. We also purchase equipment such as hoses, tanks, and pumps for irrigation purposes. After some initial investments, we generally repair or utilize donated hand tools in the garden. On occasion we will also purchase hand tools for our farming activities. We are reluctant to spend money on anything new and prefer to rely on revitalized older equipment.

Planning for large-scale projects such as the pond excavation requires the organization to consider multiple budgetary angles.

In addition, we pay taxes. The Trought family pays property taxes as owners. The organization pays payroll taxes on salaries, as well as taxes on the rooms and meal services that we provide. We also pay fees for vehicle registrations, and while New Hampshire does not have a sales tax they do garner fees for licenses, road signs, and unemployment services. So while the word "nonprofit" implies tax-exempt status, we contribute more taxes than some Wall Street corporations.

CHAPTER EIGHT

GOVERNANCE

Beyond what crops to grow and what animals to raise, the authority to manage and direct the activities of people and the land can be a major point of contention. While benevolent dictators may exist, the goal of an organizer is to create a structure that is fair, open, and transparent. This sets a framework that can be predictable and inspire confidence in participants. Ensuring the organization has the flexibility to change in response to needs, while consistently pursuing the mission, breeds a sense of security and organizational integrity. The strength and trust in this foundational platform helps create organizational longevity and sustainability.

Governance implies overall control of decision-making processes. At D Acres our goal is to provide a mechanism for long-term governance of publicly owned agricultural lands and land-based community organizations. The two areas of focus are ownership of the land where we reside and ownership of the corporate entity D Acres for which we work. To function within the existing legal framework, we must identify who is responsible for both facets of the organization. That said, property rights are entitled to the ownership, our liability must be accountable, and our decision-making process must be defined.

While we continue to move toward progressive alternatives in organizational management as a legal 501(c), the board of directors is legally responsible for organizational activities. Within this structure we promote a leadership format that recognizes the commitment, experience, and investment of staff and other community members. While we support ideals and theories of egalitarian democracy, consensus decision making, and utopic anarchy, our leadership tends to be a hybridized version that incorporates aspects of many governance models while retaining the legal status of a nonprofit 501(c). The difficulty is that D Acres has consistently attracted individuals who are disaffected by conventional organizations and forms of government. These independent, antiauthoritarians provide challenges to even the aroma of hierarchical precepts. It is hard work to demonstrate and convince participants that responsibility and authority are not the same thing. Governance is not a privilege but rather a responsibility to each other, to the land, and to our mission of service to the community.

Ownership of the Land

Our legal and cultural norms prevent us from exploring all the avenues of organizing people on the land. We are currently limited by the provisions of ownership. While we all share the air that we breathe, water and land can be owned based on legal provisions. The construct of ownership has inhibited communitarian values since Columbus claimed the New World for the king and queen of Spain.

While this scenario of ownership ensures that someone will be identified for the responsibilities

The main cluster of buildings that make up D Acres sits on approximately five acres.

of taxes, there are also several identifiable liabilities associated with ownership. The risks of ownership include liabilities for injury or harm, as well as any potential illegal activities. In exchange for these responsibilities, private ownership implies the right to utilize natural resources or develop the land within the parameters of local ordinances and government regulation. This ownership construct also provides the mechanism for transfer and sale of the property.

In our current legal system property owners govern decisions related to the usage of the land, including choosing tenants and activities on the land. Landowners are permitted to do whatever they choose within the constraints of zoning and government regulation.

The property at D Acres is currently owned by the Trought family and leased by D Acres for organizational use. As a family we have looked into the possibility of conservation easements and other deed restrictions for the farm. These types of legal agreements have become common practice as restraints against future development. In some cases, such as with easements enacted by the Nature Conservancy, the entity takes ownership of the land, while in others the land stays in private ownership but the entity enforces the deed restrictions.

There are several land trusts and conservation entities that enforce deed restrictions that would be willing to consider the D Acres property. I, however, am skeptical of these land management entities. Their overall emphasis has been based on preservationists seeking to maintain vistas rather than actual working landscapes. Easements have been adopted as triage for the extreme problem of land loss from development. While habitat preservation is commendable and development of farmland for suburban sprawl and box stores must be curtailed, deed restrictions are a short-term fix. This Band-Aid will have long-term ramifications and unforeseen complications that are difficult to resolve.

Land trusts are designed to be property managers. They hold title and decide how the land will be managed. The primary focus of these organizations has been the prevention of development for habitat preservation. These entities seldom pay taxes on the property, thus removing its value from the government's revenue. They can also choose to sell or transfer title of the property. Over time, however, the type of land considered by land trusts has grown beyond that with scenic and habitat value to include working farmland. As these entities have become farm owners they have discovered that the role of the farmer on the land is different from that of a research biologist or logger. While biologists and loggers have a more transitory relationship to the landscape, the farmer is generally more permanent and stationary. This relationship to the land adds complication to the role of a land trust as an absentee owner of farmland.

Furthermore, deed restrictions are permanent. By writing in a restriction that limits the development rights of the land, the value of the land is reduced. At times landowners have sold conservation easements to provide needed cash and to reduce the taxable value of their property.

Deed restrictions, nevertheless, depend on enforcement. Over time the landowners who chose the restrictions will be replaced by new owners who may not share their same values. The new owners will then challenge the restrictions that are enforced by the entity. The time and costs associated with negotiations and litigation will become a burden to the enforcement agency as the property ownership shifts. Another likely outcome is that the restrictions will make the land inaccessible to working-class ownership. To accomplish the goals of the initial intentions of the easement, the covenants typically have strict restrictions on development and construction. Without possibilities for a flexible business plan that allows for farm evolution, the possibilities for success are marginalized to the agricultural entrepreneurial working class. As this restricted property only has appeal to wealthy preservationists, they will become the owners of this devalued landscape, displacing those who would work the land.

Both land trusts and easement management entities are beginning to experience problems that will plague them into the future. For their business model to survive they must be subsidized by continual land acquisition, which pays for their operations. Eventually, the burden of overhead costs will force the consolidation or dissolution of these land management entities. The board of directors of land trusts who govern farmland will act as feudal lords for a rootless farmers' class. As the restrictions are no longer enforced or the titles become controlled by a smaller segment of the population, the initial relief offered by easements will dissolve and the long-term implications will manifest.

These mechanisms for land preservation also have property tax implications. With restrictions on the use of the land, the property is often devalued in the tax assessment with a corresponding tax reduction. While this is helpful to an aging farmer faced with high taxes and development pressures, the local schools, roads, and services are degraded by the reduced funding. If the ownership of the land is transferred to a nonprofit land trust or 501(c), that entity could choose not to pay property taxes.

If the land trust that owns property becomes insolvent, the property should be transferred to another similar entity. Most bylaws stipulate dissolution clauses for the assets of the organization. Language often used states that the board would donate all remaining property and assets to an organization with a similar mission. This vague language

Managing a community-scale operation into the future requires effective governance to oversee investment in both personnel and physical infrastructure.

exists because entities are unwilling to commit to such an agreement a priori without full disclosure. For instance, in the event of dissolution, D Acres is unwilling to precommit to donate its property to the Society for the Protection of New Hampshire Forests, not knowing the state of affairs at that future juncture at SPNHF. Likewise, the society would be unwilling to accept the potential liabilities of an unsolicited gift without reviewing the status of the assets. There would be specific concerns regarding ongoing or impending litigation and liabilities that could affect the solvency of the arrangement. Over time a state of limbo would ensue, in which deed-restricted or publicly owned lands would not be properly managed and overseen. As the future unfolds, the shifting responsibility of land management will

create unforeseen complications and frustrations. With the options that exist with various deed restrictions or public ownership models, it is challenging to envision what is best suited for the many contingencies that the future holds.

D Acres has met with local land trusts as well as the SPNHF to assess the possibilities. We have considered the option of creating a land trust whose mission is simply the management of the D Acres property. It is my belief, however, that the land trust and easement enforcement agencies have a flawed business model, as there is no method to cover the increasing costs of doing business. Thus, D Acres is seeking alternative forms of ownership.

We are aware of the limitations innate in the system of property ownership. This system of private control of the lands relies on intergenerational family farming or corporate management to survive. As an alternative to the traditions of the family farm we must develop models that provide continuity to the farm system and livelihood for the inhabitants. As our population shifts from the resource-intensive urban and suburban population centers, there will be a need for alternative community ownership of farm enterprises. There is land that will need management by groups of people who are not organized in traditional corporate or family hierarchies and management constructs.

The reality of private ownership necessitates transfer with death of the owner. I have considered eventualities of my wishes to perpetuate sustainable agriculture, education, and community service here on the land. As an individual I am seeking an alternative to selling the land or transferring it via inheritance to individuals not interested in agricultural pursuits. It is my intention that the efforts invested to this point serve the future on-site endeavors. Without a deed restriction it is impossible to define the activities that will occur in the future. Because of the limitations of deed restrictions thus far, we have chosen to seek alternatives for long-term property management. Our alternative idea is to set up a publicly owned entity that owns the project and collaborates with the staff and residents to pursue the mission of the organization, including the livelihood of the inhabitants. This organization is intended to survive the inevitable changes in leadership that will occur at the operation through future generations of farmers.

Decisions and Accountability

The legal system demands that a structure be set up to direct accountability in cases of an accident and civil or criminal wrongdoing. The structure of a nonprofit provides a hierarchical map to determine the accountability of organizational activity. While ultimately any negligence is a liability, the structure provides a framework that identifies who is responsible at each level.

The board of directors provides accountability to the general public for the actions of the organization. The executive director is required to meet the directions and fulfill the responsibilities imparted by the board of directors, while the staff are resolved to work toward meeting the director's expectations. The staff serve to set standards and evaluate the role of the shorter-term residents. Ultimately this entire group is responsible to the public, while the board of directors serves as the direct oversight authority.

The board of directors is the legal, official owner of the organization. The bylaws have been approved as part of the application for 501(c) status. The bylaws of the organization stipulate the governance process in documentation so the public's interests and ownership of the organization are preserved. Details of the document include conflict of interest provisions and stipulations regarding how the composition of the board is determined. The board of directors is also insured against litigations by a directors and officers insurance policy, which is typically available to nonprofits.

Decisions are made by group process at D Acres, though the ultimate responsibilities are designated. The most crucial decisions regard prioritization of activities, budget, and personnel. By intertwining these themes we work to provide a business plan and personnel management. This process happens through communication and collaborative efforts to meet shared goals. Each year the staff and board work together to develop a budget for the year. This budget is approved at the beginning of the fiscal year and is then compared to the actual numbers at meetings throughout the year.

The planning of the year is a conversation based on our fiscal status and fund-raising plans, as well as the capacity and desires of the workers. By including those who are investing sweat equity into the project we derive a course of action based on communicated expectations from those in a position to complete the tasks. While the board is capable

The Red Barn has been serving as a living space since the very beginning stages of D Acres.

of offering advice and direction, the balance of the labor is contributed by certain individuals whose opinions must be valued.

Board of Directors

The board of directors is the legal mechanism for governance of 501(c)-type organizations. The composition of the board of directors is intended to provide for transparent, public ownership of the entity and must consist of a minimum of five members. These individuals are intended to be representatives of the greater community and as such are generally not members of the same family. In their role as owners, directors are required to proclaim any potential personal conflict of interests, such as business dealings with the organization, that may affect their decision-making process.

Legally, the board holds fiduciary responsibility to the U.S. government to fulfill the mission of the organization as designated. This commitment implies jurisdiction over all elements of the organizational operations. It is the responsibility of this self-perpetuating body to ensure the sustained endeavors of the organization.

Our board of directors meets between four and twelve times per year. The frequency and intensity of board commitment fluctuates relative to the organizational need and involvement on the part of individual board members. The agenda of board

A view of the fire pit knoll from the rooftop of the Red House.

meetings consists of an update of organizational activities and financial status. Other topics include fund-raising and organizational development efforts. While the board attempts to be a consistent presence on the farm, the board intends to limit its involvement in the personnel and day-to-day management of the organization. Board members also serve as liaisons to the community; they provide face-to-face outreach as ambassadors of the organizational mission. They must be articulate and promote the organization's activities to potential supporters.

Originally, when we started the nonprofit, the board comprised long-term friends and neighbors. During the first ten years we have had a complete turnover on the board. The original board has transitioned out, while new faces have emerged as the organization's leadership. Currently, the board of directors comprises a mixture of former residents, long-term community participants, and professional advisors. They often are unfamiliar with each other when they begin and volunteer to attend many hours of meetings together.

Is long-term tenure on the board a good idea? While long-term board members provide experience and historical awareness of the organization, this must be balanced by the enthusiasm and fresh ideas of new members, who enhance the board capacity. The concern about long-term board members is that they can allow the organization to become ineffective and stale in their leadership role. But if board members maintain active engagement with the organization they can continue to provide benefits from participation for many years.

To me the diversity of a board can be its greatest strength. By attracting representatives from the broader community we gain insight into the effective prioritization and management of our operations. The board also serves as envoys for the organization in their own business and social spheres of influence. The board has immense capacity to reach into sectors of the community beyond that of the staff. By representing a broad range of people from our society the board provides a true representation to serve the public.

When the board is effective, members have the positive attitude that enables effective communication and goal-oriented activity, but this is dependent on the time and energy they are able to contribute. While the board meetings are an important component of their yearly responsibilities, the members can also contribute by doing the legwork of fund-raising and other administrative tasks. As active members of the organization who have frequent interaction with the on-site personnel, the board members become partners in the service mission of the organization.

The challenge for our board is immense. They are attempting to navigate uncharted terrain in the management of a farm-based service organization that utilizes a horizontal leadership format. They recognize that the investments, commitment, and experience by long-term residents merit their inclusion as co-owners of the organization. Unfortunately there is no legal provision for such inclusion. While the circumstance is recognized, there is no alternative to alleviate this situation. Without historical antecedents, however, it has been difficult to define clearly the role that each board member should play. Balancing the role of assertive leadership with offering empowerment of ownership and self-governance to the organizational personnel is difficult.

At times we have sought to stock the board of directors by recruiting members with specific skills, networking ability, or experiences. For example, we sought a lawyer to help provide legal advice and individuals involved with academic institutions for their educational credibility, experience, and connections.

While there is a separation between the oversight of the board and the day-to-day working at the farm, at times the board has been called upon to be more involved. On both of these occasions they were contacted over grievances brought by a former participant. To resolve this situation the chair met with the complainant to fully discuss the circumstances. After this conversation the chair met with the staff to evaluate measures to rectify the situation and develop policy to prevent reoccurrence of problematic situations.

We have struggled to maintain an active group on the board. While we have had spurts of interest and enthusiasm, maintaining an active continuity has been difficult. The inevitable evolution of the composition of the board has also affected the organization. As board members cycle through, it is difficult to maintain organizational history and progression on an agreed trajectory of prioritization and project implementation. Often, discussion reverts to the same issues without forward progress on measures designed to provide resolution. The board of directors have also been stymied by indecision regarding their roles in the organization. By acting simply as consultants, they can be ineffectual as disciplinarians, while micromanagement of the tasks leads to dissension from those who do the work.

Choosing the right board members has been a difficult process. Some people are attracted to the board as a status symbol or prestigious figurehead role. Membership on the board is also viewed as a résumé builder and a means to acquire tenure or other career advancement. While such professional ambition can be helpful, experience conducting meetings and passion for the mission of the organization are also important components of effective board members. This passion also implies an ability to relate to the residential community and an understanding of the work that is performed day to day by the organization. Often the people identified to serve on a board are already busy with many activities. It is important to ascertain their capacity to commit to the board role.

There are difficulties with applying the corporate mechanism of board ownership to a project designed to offer alternatives to conventional models. While they may share some ideals with the residents and staff, board members generally live a different lifestyle from the D Acres inhabitants. The philosophies and ethics of the board may be vastly different on some levels. As an example, a board member once told me that in his household retaining a job was more important than solidarity with a fired worker. While this opinion may be held by a majority of Americans, the sentiment flew in the face of why I had chosen my path to D Acres. Another board member advised me to lead at D Acres by mimicking the behavior of a pack of dogs. His rationale was that the toughest, meanest dog leads by example. This primitive analogy, based on questionable natural study, was not only divergent from our philosophy but is inconceivable for maintaining positive social relations within a group.

The board's role as an authority to direct action by the staff has caused friction when construed as dictates. On occasion the board has stipulated efforts be made regarding fund-raising or grant writing. While at that moment they volunteered time to complete the project, the staff were ultimately left with additional work.

In general, we have had the most success when the board and residential cadre mingle to discuss the mission outside the rigidity of board meetings. By developing a rapport that values our shared commitment and belief in the organizational mission, our relationships grow and flower. By sharing this experience the organization develops the capacity to transcend restrictions of the legal structure, so that the partnership of managing the land and the service organization flows. This fluidity ensures our capacity to overcome obstacles such as personnel turnover and budgetary limitations. The board serves as our best option at this juncture to act as community owners of this project.

Governance Methodology

It can be helpful to form committees of board and community members that are responsible for assessing different facets of the organization. Current committees include personnel, finance, marketing, mission, ownership, and land management. The committees comprise community members with specific interest or knowledge in the field. The composition of the committees allows board and staff to extend their capacity to reach higher levels of expertise and experience in particular areas of the organization. While they are called upon to draft policy and procedure, they also assist in implementation and integration of the programs. They provide specialization and expertise to the board on certain topics.

The land management committee assesses the physical infrastructure at D Acres. This committee is responsible for advising the board on maintenance and new construction projects at the farm. The committee helps prioritize the major work that needs to be done on the grounds and supplies the necessary plans and budgeting. The committee is composed of interested board and staff members, as well as building and land management experts within the community. The ownership committee is organized to focus on a plan to ensure the perpetuation of farming and community activities on the land.

The mission committee evaluates the overall work of the organization. This group attempts to investigate the efficacy of the organization, seeks feedback from the community, then helps plot the course of future activities. The group consists of interested staff and board members. The marketing committee, meanwhile, focuses on advertising and outreach to the community.

While the finance committee helps prepare the annual taxes and budget, there is an important subset of this committee's responsibilities known as fund-raising. The fund-raising committee looks for innovative, entertaining ways to generate revenue for the organizational mission. Typical examples are designed to provide supporters with education through events that highlight local food and musicians.

The personnel committee develops policies and documentation that aim to improve the residential experience. They may also serve to help mediate or resolve interpersonal conflicts that develop within the organization.

This expansion of the board framework helps address the limitations of the structure of the traditional board of directors. The expansion allows for focused research and expertise on decisive issues. These committees enhance our success by increasing the flow of information without diminishing the core leadership provided by the board of directors.

We are optimistic that we can define ways for the membership of the organization to become more involved with the ownership. By empowering individuals to recognize the capacity of an organization in which we share ownership, the membership program serves as a means toward broad-based community involvement in the governance of D Acres.

Our organization is attempting to define a structure of governance. While these efforts at times feel like flailing, idealistic, and naive first steps to offer alternatives, it is heartening to take those steps. The empowerment and inspiration that is available once we unleash the potential of alternative models provides momentum for further transformation from the limitations of convention. While I am pleased with the progress that we have made thus far, our current form of nonprofit is the same corporate structure as organizations such as the Appalachian Mountain Club or the Nature Conservancy; to truly innovate and empower this type of land-based farm and community service work, we need to further evolve constructs of ownership and management. For a new model to manifest, there must be a transformation. Instead of white-collar disconnection and misappropriation by commercial interests, we need service organizations managed by the public for the people. This transformation involves new legal structures to provide for public ownership of land, as well as a shift in our own perceptions of the possibilities. While the legal constructs can be accomplished, the societal leap necessary to accept the responsibility of shared ownership is a challenge we have yet to accept.

CHAPTER NINE

ANIMALS

The raising of animals by humans is a responsibility that must be undertaken with compassion and commitment. Our philosophical approach is based on the concept of mutualistic relations. In our coexistence with these animals, we endeavor to provide a healthy, comfortable environment throughout their lifetimes.

When embarking on a journey to create mutualistic connections with domestic animals it is important to consider and reflect on the motivation for this acquisition. What is the goal of raising the animals? The purpose may be to raise them as pets, or for breeding stock, or simply to put food on the table. Whatever the motivation, the responsibility of raising animals requires considering how best to care for those entrusted to us. Once a reasonable interpretation of best management practices has been determined, the goal is proficient implementation of that regime. Depending on the personnel, organizational opportunities, and land management strategies, the number and species composition of our animals on-site has varied dramatically. Our ability to meet the food needs of the residents and visitors at D Acres has depended on an ongoing yet evolving relationship with the animals.

With regard to the health and well-being of the animals, environmental factors affect the animals' development. Conditions of confinement produce animals that are unhappy, restless, and disgruntled. Fresh air and access to grass and forage positively improve the animals' health and well-being. Ultimately we are attempting to raise animals that are hardy, athletic, lean, and vigorous versus weak, obese, lethargic, and unhealthy.

Over the course of nearly two decades, we have raised and cared for oxen, sheep, goats, pigs, and poultry in working partnerships. We raise animals by the dozens rather than the hundreds, and while we do harvest and consume them occasionally, the protein source they provide is a bonus rather than the purpose of the relationship. Likewise, while we do sometimes sell packaged meat, the true objective of raising animals is achieved throughout their lifecycle rather than just from their demise. Motivation from our philosophical commitment to nurture the land, as well as its animals and people, has guided our animal management decisions. The role of the animals has multiple functions within the farm system. While all provide fertility, each species occupies a distinct niche and contributes to the farm system differently. For example, the oxen provide muscle power for the forestry work, the ducks reside in our aquatic environments, and the pigs serve as our natural rototiller. Our goal is to recognize the natural inclinations of the animals and match them to perform the work of the farm.

As we move forward, the quantity and species we choose will continue to evolve. This evolution is based on our chosen diets and land management goals, as well as the skills and interest of the on-site personnel. Chickens and pigs rarely remain on-site for more than three to five years, while the oxen have worked with us for over a decade. As our need to clear new agriculture fields diminishes, our work with the pigs will also be minimized. With fewer pigs we may then have time to consider milking goats, which can range on marginal hillsides that would be destructively eroded by pigs. It is important to consider these various temporal and spatial concerns along with the infrastructure requirements of different animals rotated through the farm system.

Although D Acres has existed for seventeen years, our knowledge of how to develop a mutualistic farm system is still fledgling. Through the process of observation and interaction, we continue to study ways in which to cohabit the landscape with domestic animals.

On a daily basis we interact with our animals through chores. These chores vary with each animal, although the essence of the job remains the same. We are charged with meeting the animal's feed and water requirements and providing a suitable habitat. Daily chores are a good time to connect and discover the nature of each animal. By observing normal, healthy animal behavior we can better distinguish signs of illness, such as lackluster energy. Chores also provide the opportunity to evaluate the animals' shelters to ensure their bedding is dry and ample and that they are secure from predators.

To provide for the happiness of our animals we seek to match their niches in nature with the needs of the ecological farm system. Pigs, for example, have evolved over eons to be pioneers, disturbing the topsoil with their powerful snouts while they search for roots and insects and paving the way for the succession of new plant and animal species. Thus, on level terrain we utilize the pigs to create garden space, while on steeper slopes less conducive to field crops and more susceptible to erosion, we may choose to pasture goats or other browsing animals.

Pigs

During our very first full growing season, we raised pigs. But we were only playing farmers by doing what we perceived farmers do: raise pigs in spring for sale at the fall market. From start to finish it was a fiasco. We purchased two piglets from an auction for the going rate of eighty-five dollars and set up an enclosure of fifteen by twenty feet with a squat straw-bale house in the Lower Garden adjacent to the chickens. A plywood automatic feeder was constructed to hold fifty-pound bags of grain for the animals.

The pigs quickly became acquainted with their home, reducing their pen to mud within two weeks. The house was leveled by their assault on the straw bales. They were relentless in their pursuit of spilling their water, and their automatic feeder was in constant danger of being toppled. We added electricity as the pigs began to undermine the perimeter fence and purchased a hose-end device so the pigs could satisfy their thirst.

In the fall the fattened pigs resisted our attempts to evict them humanely. After an unpleasant afternoon of wrestling, they were loaded into a trailer for transportation to the USDA facility an hour and a half south in Goffstown. Once packaged, they were sold per pound to an affluent couple.

The piglets had been full price in the spring when the market was bursting with eager buyers. Organic grain was expensive, and our self-feeder attracted raccoons and other varmints to gorge on Midwestern grain. In their small confinement zone the pigs ambitiously rooted through all of the area available. Ultimately the pigs' ability to turn over ground was underutilized while draining time and resources. The model was outdated.

The first season was an experiment in the standard model of pig farming in which piglets are raised on grain for a season to fatten them for market. The animals are given minimum exercise and maximum food and water. Conventional production in our food system relies on this type of feeder-pig

While it may have been a cozy straw-bale structure, the rambunctious nature of pig life soon made this building style obsolete.

operation in which thousands of pigs are imprisoned. This dominion over the pigs left us without an appetite for pork. The quality of the meat from the confined animals was tender and fatty. While this nutritionally vapid consistency may have been desirable to certain consumers, we realized this process of fattening the animals to near obesity was destructive to the health of the animals, people, and the landscape.

Economically, the per-pound sales barely covered the costs of the grain. Our tiny confinement area was nutrient rich without having appropriate strategies to share that fertility. Much like the large-scale factory pig farms, our enclosure's foul aroma was precipitated by the excess volume of pigs per square foot. After that season we took a break from raising pigs.

But since that first unsuccessful experiment, we have raised over a hundred piglets from inception and several generations of boars. Our eventual success with the pig has been based on a better understanding of the animal's niche within the farming system.

Partnering with Pigs

The pig is one of the most successful animals in Western civilization. Its ability to swim made this portable protein package an ideal travel mate for adventurous explorers to the New World. Its adaptability allowed it to thrive in new terrain and

A single pig in a vast cleared space that was once forested; she along with three other pigs have plenty of forage and will completely clear this space in preparation for gardens.

climates. Intelligent animals, pigs seek food by rooting into the ground for plant and insect nutrition.

These tendencies make it a valuable farming partner. The wide range of nutrients this detritivore consumes is transformed into a rich blend of urine and feces. Their rapid growth allows for a quick transition from birth to butcher. The swine's natural talent as a cultivator also can be utilized for transforming land. Its squat powerful body, strong shoulders, and plow-shaped head make the pig nature's tractor. By ripping up nutritious abundant plant species, similar to a kitchen gardener dividing daffodil bulbs, pigs encourage them to multiply. These animals need sufficient ground space and terrain to forage as they do in natural environments.

After the first experiment with pigs we waited several years before we were tempted to try it again. We were intrigued by rumors of pigs popping out tree stumps baited with grain. We also learned that neighbors were raising pigs on food scraps sourced from the local municipality. Ultimately we needed additional land areas cleared of forest for pasture and food production, and pigs were elected as an alternative to the tractor. We reinvested in a relationship with pigs in which we valued their natural tendencies and were willing to engage in a patient, mutualistic approach to animal husbandry.

In spring of 2004 we purchased six Yorkshires from Applewood Farm up the Baker River in Warren, New Hampshire. These six Yorkshires became the foundation of our pig program. We created

After birth the sow and her piglets have plenty of space to eat and nurse.

Beth assists these newborns to nurse. The sow has recently released the placenta and afterbirth.

These piglets may only be a day old, but they have found a warm getaway spot from their mama. Behind the farrowing rail is a heat lamp and a fresh-water dish shallow enough to avoid drowning, as well as fresh hay bedding.

an expansive area to the east of the Community Building and organized with Plymouth State University, a local grocery, and several restaurants to arrange pickups of their food scraps. We also decided to invest in a boar based on the price, quality, and availability of piglets. The price of piglets was approaching a hundred dollars in the spring, and the quality of conventionally raised piglets in the area was dubious. Their availability at local auctions also fluctuated year to year. By midsummer we located a Gloucestershire Old Spot boar in southern New Hampshire who had been raised in a petting zoo environment, and we welcomed him home to Dorchester.

This phase of our breeding program produced several litters until the hips of our timid Old Spot gave way and he had to be put down. During this period we learned the essentials of managing the life stages of pigs from birth to butchering. Primarily, pigs need adequate housing for birthing. It is always preferable to provide a dry environment isolated from frozen ground and winds. Our breeding structures were built to ensure separation from other

pigs, climate control, and proper nursery conditions. During birth the four walls, floor, and roof provide protection from the weather. It is also necessary to provide a farrow rail space that is inaccessible to the sow. The farrow rail is where the piglets can seek separation from a moody mom with their private water and heat lamp. This structure also allowed us to continually add bedding of wood chips and hay. The bedding, combined with the pigs' feces, urine, and uneaten food scraps created the basis for excellent compost to use elsewhere on the farm.

Food

Our food system relies on partnerships with our local community. The kitchen and produce departments of local businesses separate salable food from food that would otherwise be landfilled. This

Bill Erikson has hauled in scraps from local restaurants, the Plymouth University dining hall, and the Hannaford supermarket, allowing for a grain-free pig diet.

meat-free food waste is picked up three times a week in reusable five-gallon plastic buckets. We collect approximately thirty buckets weighing twenty-five pounds apiece from the university cafeteria when it is in session. The local grocery collects between ten and twenty banana boxes weighing approximately twenty-five pounds apiece per day. Local restaurant quantities vary depending on their usage of fresh ingredients. Food from fast-food outlets is neither high quality nor abundant. As the size of our animal population has fluctuated over the years, we have attempted to compensate by providing additional weeds or hay as necessary. The quality and quantity of food varies seasonally. The amount of protein-rich buffet food from the university cafeteria peters out during summer and winter vacations. The grocery food is heavy in oranges, potatoes, green peppers, and exotic fruits, which are not favored by pigs. On rare occasions we have resorted to purchasing grain to supplement shortages.

Fencing and Land Clearing

Fencing is necessary to ensure animal safety and prevent destruction of orchards and field crops. Pigs are unmanageable if they are constantly digging up the kitchen garden or escaping. We train pigs by setting weaned piglets in a house with electric wire across the door. The wire is backed by a woven fence or board, which prevents the pigs from attempting to cross the threshold. Bursts of electric current elicit yelps from the little ones until they learn to avoid the hot metal wire. We set at least one strand at eye level once the animals understand the hot-wire system. It is important to consider the

These pigs have been fed a large pile of weeds from the garden. While they will eventually consume everything, for now they prefer to make a warm nest.

This breeding pair is separated from the potato field and a hedgerow of fruit trees by two strands of electrified wire. They have plenty of space to roam without temptation to dig through the gardens.

animals' ability to heap soil and debris onto the wire, which can ground the electric charge. Pigs are inclined to remain in a comfortable enclosure with their needs met and can be trained easily to respect electric fencing. We have had pigs refrain from crossing disconnected electric fence for days once they've been domesticated to their locations.

The pig is nature's tractor. Every day, all day, if they are not resting, they're working, resolute in their determination to clear the land of edible plants. They will resist eating noxious species such as elderberry and prefer not to eat pioneers such as mullein. The animals will browse in a wide range, so ample room is appreciated, and subdivisions can be incorporated into fence design.

One of our intentions when we purchased the pigs was to see their supposed ability to remove stumps from the former forest. Rumor existed that if you drove a rod into the soil below a stump and then poured grain in the hole the pigs would dig the stump out of the ground. I attempted this various times with large and small pigs, and while the pigs will dig the grain out they will not remove stumps of fresh-cut wood greater than six inches in diameter. They will, however, definitely prevent the native hard- and softwoods from reclaiming the forest ecosystem through their intense browsing and soil disturbance. This prevents the forest from regenerating rapidly with sucker wood, such as red maple, which capitalizes on existing root systems.

Value-Adding

Pigs have been a fundamental, multifunctional component of our farm system, allowing us to

This forested area in our Upper Field was cut, and in 2008 seven pigs were fenced in to finish the area. The largest stumps are still present, while smaller shrubs, trees, and greenery were completely picked over by the end of the season.

These young feeder pigs have limited access to their pasture; the fencing will be extended as they require more forage, thereby clearing the sloped area surrounding several ponds.

expand our food production, and we are now in our fourth generation of breeding pigs. They replace the need for costly machinery while also adding fertility to our future pastures and fields by metabolizing food scraps. Pigs systematically clear the forest to allow for pasture, field crops, and orchards in a biome where slash and burn and bulldozer technology have been the mainstay. Importantly for D Acres' educational mission, pigs are also amiable ambassadors to visitors of the farm. The proximity to nature's creatures empowers an understanding of our food system. Children get a sense of birth and the challenges of motherhood while gaining knowledge of the food web. Youth disconnected from nature come face to face with the lives of animals that our culture recklessly exploits daily, which we view not simply as a commodity but as a working component of a sentient farm ecology.

As the era of rapid expansion of arable land here slows and our capacity to provide sustenance through a plant-based diet grows, the presence of pigs will diminish. Their roles as pioneers in the ecology will be reduced as the food production system evolves.

Oxen

Draft animals were common at D Acres during the time of Delbert and Edith. My aunt and uncle were carriage horse hobbyists and raised a team of oxen to work in the forests and fields. These animals provided the companionship and soil fertility that my aunt and uncle required for their rural existence.

When I arrived in Dorchester the smells of these work animals still emanated from the Red Barn. As we worked to clean out the barn of the remnants of hay and equipment used for their upkeep, we located tack, carriage parts, and yokes used for their operations. The olfactory and visual reminders of the farming traditions stimulated our interest in the reintroduction of working animals to the farm system.

These animals were not immediately necessary for us to proceed for several reasons. Because of the constraints of our soil-building philosophy we had no intention of utilizing animals for plowing and soil cultivation. Our nascent forestry program was inept and unskilled and thus was largely dependent on the skills and mechanized equipment of our friend and neighbor, logger Jay Legg. Consequently, we

Young Henri and Auguste, being walked by their original owner Steve, during the Dorchester Old Home Days parade in July 2001.

plodded along with four-wheel-drive pickup truck technology until presented with a unique opportunity by our neighbor Steve.

Steve had been raising a team of Jersey oxen since they were weaned from a herd in Vermont. He named them Henri and Auguste to honor the French artists Matisse and Rodin. He was training them to help pull rocks and wood as he constructed a utility barn on his property. His property had limited pasture, and the animals were confined most days in a pen of approximately 150 square feet, right beside the rural highway. Restricted by work obligations in Boston, Steve had little time available to tend to the animal chores. At the time they were less than two years old and still growing. He contacted me to inquire if D Acres would be interested in using the team.

I visited Steve, and we took the oxen out for a walk. Their confined conditions made them unruly and boisterous, but we agreed to purchase the animals and began constructing a dwelling for them in the fall of 2002. Steve sold us the animals for three hundred dollars, with the yoke included. His only request was that we not eat the animals.

I visited the oxen regularly as the construction of their dwelling progressed, so we could begin to familiarize ourselves with one another. During one visit Steve demonstrated their capacity to pull wood in his backyard. The process was confusing and chaotic as Steve attempted to explain and implement some unorchestrated logging. The power and unpredictability of the adolescent bovines became apparent as they moved over the undulating terrain. I witnessed the omnipresent potential for accidents while working with these magnificent animals, which reinforced the seriousness of attempting to relearn these traditional skills.

When the dwelling was complete we walked the oxen over a mile alongside the rural highway with a cart in tow to D Acres. Once the animals were on-site I traveled to South America to study the techniques and methods teamsters were employing for field crop cultivation in the Andes. When I

returned I was committed to working with the oxen to train them to meet our needs. But my enthusiasm was tested immediately as I attempted to impose my will upon the animals. Henri and Auguste quickly informed me that for the team to be effective I also had a lot to learn.

Learning the Ropes

My initial attempts to manage working effectively with the oxen were embarrassing. The animals required cues and cajoling that were foreign to me. My brutish attempts to dominate the animals with my presence were met with their stubborn resolve to ignore or misunderstand my commands. It became a crucial test of faith as to whether or not we were capable of effectively integrating these animals into our farm system. To avail this situation I endeavored to spend more consistent time with the animals during feeding and grooming. They became familiar with me as I learned their habits and tendencies. As we walked the property in search of fresh grass, I learned their likes and dislikes for certain species, as well how to effectively orient them through the maze of the farm.

My ability to coordinate the oxen became a semi-formalized series of habits. Staff shared teamster responsibilities while attempting to standardize how we interacted with the animals. Specific regimes were employed for yoking and hitching implements. We spent hours officiating the oxen in a training regime that emphasized a controlled, rehearsed progression of management and minimized new procedures for the animals. We tried to develop predictable behaviors that were both safe and effective.

As this system of management developed, the oxen eased into their routines. We began effectively using them to transport rocks with a sled and to pull cartloads of their manure. We acquired knowledge of woods work, hitching and pulling log-length wood and brush piles. We learned techniques to navigate the team through the woods' paths and to stack piles of firewood for eventual processing. The oxen responded positively as we adjusted systems based on our experiences.

As our relationship has developed through the years I have achieved a genuine, earnest rapport with these animals. They watch our activities through the day and have come to recognize each person with whom they work as individuals. We feed and care for the animals through regimes that connect us in daily harmony. Henri and Auguste are sensitive to the tones of our voices and express affection by gently licking our skin with their powerful, rough tongues. I respect and express gratitude to the oxen for their sincere efforts in the woods as they attempt to complete the weighty tasks of our day-to-day workload. When we discipline the animals for poor behavior my empathy for them

grows. Their persnickety behaviors remind me of the need to maintain structure in our working interactions. Our role as teamsters has shifted philosophically from that of disciplined instructors into organizing cooperators. Oxen respond well to positive reinforcement, and my banter with them has grown to include steady positive feedback. We have become a working unit with mutually understood roles and responsibilities.

Feed and Care

Oxen are large animals that consume an immense amount of vegetation. I have heard estimates that it requires approximately ten acres of pasture to provide for one team of oxen. The ideal would be to intensely graze five acres per year and grow silage

The oxen are utilized to pull the heavy pig food barrels to the feeding location.

for storage in the other five acres while rotating the system year to year. However, our property is severely limited in terms of pasture availability. The field space that my uncle reclaimed from the settlers' prior farming attempts amounted to scarcely three acres when we arrived. We filled a majority of this space with gardens and farm infrastructure by the time we purchased the oxen. Consequently we have attempted to use rotational techniques to feed the oxen fresh greens on-site. In the pasture of the Upper Field there are two separate zones cordoned off with electric fence. We utilize a neighbor's field down the road to stake the animals. We also walk and stake the oxen throughout the property. This process is time consuming, although it provides an opportunity for new teamsters to become acquainted with the oxen. This regimen acts as lawn-mowing maintenance to patches of grass and otherwise inaccessible grazing areas and requires vigilant oversight of the animals.

Compared to horses, oxen are superior in their capacity to subsist on hay. While they are more efficient in their consumption of vegetation, they are incredibly voracious. The animals will consume huge amounts of green materials until they literally must stop to chew their cud. Determining the proper food intake for a twelve-hundred-pound animal requires observation over a long period. While we would like to feed the animals to their hearts' content we need lanky, agile animals, not fat, wide animals. For our animals we estimate that one bale of hay per day or four hours of grazing on pasture is sufficient.

After grazing, Auguste and Henri are returned to their permanent pasture.

A midsummer haul of hay from nearby Rumney. Bales are stored in the loft of the Hovel, the Open-Sided Building, and the Red Barn.

We provide Henri and Auguste with supplemental food such as squash, apples, carrots, and onions from the expired grocery store vegetables that we obtain year-round. With these large working animals, it is important to evaluate feed needs over a long period of time to provide the necessary nutrition without wasting expensive feed. We look at the flesh on the topside of the hip bone as an indicator of proper weight. The flesh should not belly between the two bones so the animals have the strength to do their work.

To meet the needs of the oxen we purchase five to six hundred bales per year and provide them with a nutrient-rich salt lick. Although they enjoy the hay much less in the summer than greens, our pasture is inadequate to meet all their needs during the frost-free growing season. We attempt to buy the hay in the field as locally as possible. We have a long-term relationship with an elderly farmer down in the valley. He contracts his hay field to another neighbor, and depending on seasonal availability we purchase the hay as it is baled in the field. The quality of the hay has continued to decrease through the years, however, as the fields are being overtaken with galium and milkweed. On a typical day in June, July, and August we pack the truck and trailer from the bales spaced uniformly in the fields, loading 70 to 120 bales onto the vehicles, to be driven up the hill and stored in the various dry locations. The price and quantity of a bale of hay varies by producer, though the economics of hay is marginal. We are generally buying hay in the field for four dollars a bale, while hay available at the local feed store is over six. It is likely that our farm's source of hay will be converted for the more lucrative short-term rewards of a housing development within the next ten years.

It is important to make sure hay is stored in an adequate location. To avoid mold make sure the hay is elevated from ground contact. Hay purchased in the field should be adequately dried; otherwise the heat generated can be hot enough to cause a fire. Improperly dried hay, as well as hay exposed to high moisture, will mold. These dusty bales are semipalatable to the bovines, though large amounts will cause illness and respiratory issues. Tarps and outdoor storage can be difficult to manage, so it is important to find appropriately accessible and dry locations for this voluminous product. The reality of storing five to six hundred bales on the property from July until they are consumed is a definite space consideration.

Hay is available in bales or wrapped rounds. The hay is cut from grassy areas two to three times per season. The available protein content is highest in the vegetation before seeds appear on the stems, so the harvest is timed appropriately based on the weather and crop maturity. Hay producers follow the adage to make hay when the sun shines because lack of precipitation is essential to producing quality hay. The hay is cut and dried in the field over twenty-four to seventy-two hours. Tedding the hay can accelerate the process before the baler equipment scoops the hay to compact the material into string-tied fertility. This baling process enables large volumes of organic material to be transported and stored from one farm or field location to another. Baled wet hay will not be quality fodder and is sold for two to three dollars per bale. Wet hay and hay cut from marginal lands particularly at the end of the season when it is full of undesirable weed species is referred to as mulch hay. This mulch hay can be used in gardens, for compost, or as animal bedding though it is recommended to be conscientious of the potential to introduce weeds. Bales range in weight and quality depending on the farmer and the season variations. Hay from second and third cuts has higher protein content. Round bales wrapped in heavy plastic can total the mass of twenty-five square bales, making them awkward to transport without heavy equipment. In our system square bales are easier to manage, move, and portion to the animals.

Oxen require an abundance of clean, fresh water to meet their health needs. When working in the summer they can require well over ten gallons per day. This quantity of water must be supplied throughout the year, and keeping the water from freezing can be technically challenging in the winter. While water heaters are effective, the animals can be destructive, and the system must be well designed to prevent fire danger and for ease of utilization. We have erected a frost-free hydrant routed underground from the Community Building water supply to meet our winter water needs at the Ox Hovel.

Temperament

"Oxen" is a technical term for steers that have been trained to work. Steers are cows that have been castrated, while bulls retain their virility. In general, the oxen are predictable in their behavior. While horseflies can be an extreme nuisance and undermine the concentration of the team, loud equipment such as chippers and chain saws do not affect the animals. In such challenging, noisy situations, hand signals and footwork are more effective than voice commands. The oxen are reliable under a variety of working conditions; however, sudden noises and quick movements can frighten them, and unfamiliar dogs will agitate and distract them. Dead animals, such as slaughtered pigs, increase the oxen's anxiety, though ours have been attracted to fresh blood. While the oxen are fairly consistent in their responses to stimuli, working with these animals requires constant vigilance.

Auguste and Henri, led by Louie Holland, pull a pig in a wooden crate, moving it to a new pasture location.

In terms of general farm utility, oxen teams are comparable if not superior to horses for several reasons. They are generally available as inexpensive culls from dairies and would otherwise be raised as veal. The typical price for a male calf is a hundred dollars, while foals can cost over two thousand dollars. Trained teams can also occasionally be sourced from 4-H participants who raise the animals as part of their youth agricultural education. Oxen feed, hoof trimming, and veterinary services are also less expensive than those of horses. Oxen do not need expensive shoeing, can subsist on hay, and are not prone to illness. These calm, benevolent creatures require unsophisticated tack consisting of yoke and bows versus the complicated accoutrements of equestrian operations.

Chickens

The domestication of chickens can be traced to the roots of human civilization. Throughout history people have captured or rescued young chickens to raise for their eggs, feathers, and meat. Chickens have now grown to global ubiquity through two distinct, preeminently bred attributes: meat and unfertilized eggs. These compact, vibrant creatures need minimal subsistence and produce eggs rich in dynamically accumulated energy. The chicken is the prototypical, quintessential example of how

Young hens, also called pullets, in a fresh enclosure

an animal can be assessed through the lens of the permaculture vision. In Bill Mollison's *Permaculture: A Designers' Manual* there is a graphic representation of how the chicken interacts with its ecology. In managing the chicken in this way we seek to serve the best interests of the birds, people, and the overall system.

In the past different human cultures and climates produced a diverse array of domesticated chickens. This genetic selection created an abundance of breeds with an array of sizes, colors, and traits, including disease resistance and foraging ability. Modern chickens, on the other hand, have been bred to be pure protein producers. Modern meat birds grow so quickly and disproportionately

that they are unable to bear their own weight. Genetic breeding for increased egg production has induced calcium deficiency among these animals as they donate abnormal amounts to meet our human need.

We attempt to provide humane conditions for chickens at D Acres, including access to sufficient food and clean water, as well as housing that offers shelter from low temperatures and predators.

Our first interactions with chickens at D Acres met with our typical naïveté. We arranged to buy five barn chickens from a woman at the local feed store. One of these hens quickly eluded us in Plymouth to partake of its ancestral, free way of life. We were able to subdue the four hens that remained and brought them back to Dorchester, where we built a house and our first chicken tractors. Upon our arrival, closer inspection revealed the age and poor health of the critters, who pecked at one another mercilessly, and several developed open wounds, which we attempted to treat with a liniment that seemed to further induce cannibalistic frenzy. In the fall we chose to kill the birds with the help of our neighbor. From our initial experience with the barn chickens we discovered that chickens were not inclined to labor in small confined chicken tractors. In addition to the problems posed by the holes they dug to take dust baths, they were miserable, difficult to move in and out, vulnerable to predation, and mischievous in the garden if left unsecured. We sought to integrate free-range management and the use of chicken tractors into a workable system.

Predators

Our subsequent attempts at raising chickens have been met with mixed success. We have had consistent problems with predators. The White Mountains are home to diverse, opportunistic hunters who are eager to enjoy homegrown poultry. Because they have been bred for specific human needs, chickens are unable to avoid the ruthless and relentless predation pressure because of their relatively slow mobility and ease of capture. At times birds approached by a predator will remain motionless, pattering the ground with their feet in an indecisive dance of survival ineptitude.

I have observed chickens fleeing in panic as a crow flew over their enclosure. While they do seem aware of and responsive to the danger of flying birds, on one occasion an owl entered the chicken house through their own access at dusk and killed an inhabitant. But by our allowing fruit and seed bearers to proliferate along the fence line and enclosures, predatory birds are generally discouraged. The branches and leaves not only provide shade and nutrients but also protect the chickens as a barrier to swooping airborne predators.

Mammals really seem to enjoy the taste and accessibility of chicken. These animals generally develop two strategies in their attempts to befuddle the chicken farmer. During the summer raccoons and foxes will return repeatedly, sometimes during daylight hours, to harvest chickens one at a time for days in succession. Other animals such as mink, marten, and ermine are likely to visit during the winter when they are hungry. These critters will viciously kill all the animals in the vicinity in a frenzied quest for blood. We have not had troubles with bear yet, though I hear they would victimize an entire flock if inclined. While trained dogs are helpful in preventing predator attacks, the neighbors' and visitors' canines can also endanger the animals. It is important not to let dogs develop habits of chasing and playing with the fowl.

Our method to address predation has been prevention and relocation. We confine our animals at night in housing designed to be inaccessible to predators. Morning and nightly chores allow us to review the condition of the flock and allow them free access to the outdoors during daylight hours. We have learned to use sophisticated latch and bolt technologies to outwit raccoons, who have easily mastered hook-and-eye systems. We use fencing that confines the chickens to prevent their intrusion into our garden beds, which also prevents easy access by the four-legged carnivores.

Predators are persistent once they have identified farm chickens as a source of food. Once a predator has been successful it will continue to return. To relocate predators we use a have-a-heart trap. The intention of the trap is to capture the animal alive so it can then be transported far away to a suitable location. These traps are effective if other food sources are limited and the bait is enticing. We have also used other means of eradicating habitual predators. A rifle or shotgun is sufficient for most encounters, though this does require hunter safety experience. A rifle is more suitable for targets at a distance, while a shotgun is more effective for aerial shooting and moving targets.

Partnering with Chickens

Chickens interact with their environment on many levels. These inquisitive birds are adaptable and effective in their roles within their ecology. Foraging birds provide the landscape with fertility through their energy-rich manure deposits, which accumulate in their digestive tracts as they scavenge the landscape in search of vegetation, seeds, and insects. Chickens scratch and peck the ground in search of food, which enables biologic succession to occur as opportunistic plants and animals fill the niche they have made available. They also eradicate pathogens from the landscape by consuming the feces of larger animals such as cows.

But chickens can also be a destructive element in our gardening system. Chickens that escape or wander unrestrained enjoy a frolic in the abundance of our gardens. The access to insects, tender greens, and small fruit is heaven to chickens, who meander through the beds, pulling back the mulch and ripping up seedlings. The natural inclination of the escaped chicken leads to a poorly timed intrusion into the garden system. To efficiently utilize the chicken's compulsion to assist us with our garden tasks we have experimented with several variations of what is commonly referred to as a chicken tractor. The idea is to confine a chicken to a certain territory so that in the process of demonstrating chicken tendencies and habits such as browsing they effectively contribute to the goals of the farm. In effect the chicken performs many functions that would otherwise require mechanization, such as fertilization and cultivation.

We utilize the chickens extensively to metabolize our garden debris. Specifically, weeds such as freely seeding lamb's-quarters or readily rooting quack grass are unwelcome in our compost piles. So we design adequate means to feed the chickens this debris in their enclosures, making sure that their fencing height and entry widths are appropriate for handling large loads of biomass. It is also important to maintain easy human access to enclosures so we can check for indiscernible errant egg laying.

We have had mixed experiences utilizing chicken tractors. While the chickens have demonstrated their natural proficiencies, we have had difficulties with predators and appropriately refining the timing of rotation of the animals to reap maximum ecological benefits. We have experimented with constructing portable boxes that can be moved throughout the garden for more precise utilization. These lightweight constructions are equipped with a nesting box and an access portal sufficient for daily entrance and extraction of the birds. These devices are not overnight accommodations, however, due to the predator pressure in this region.

We have also created mobile housing units to move flocks of chickens around the farm. These mobile housing units provide overnight housing, laying boxes, and predator protection. Built to be hauled by oxen or four-wheel-drive vehicles, these sleds and wheeled accommodations have separate ramps and personnel entrances for daily access. After being moved the chickens are then confined with fencing to an enclosure of our design. Enclosures should be adequate in size for the intended use. Fresh air and dirt are important for chicken health. Dirt baths help chickens keep themselves free of lice and parasitic insects. The fencing height is geared to the age and breed of these semiflightless creatures. Six feet is more than adequate, and depending on the specific birds and terrain, five feet may be sufficient. We move these tractors to areas into which we would like to concentrate the chicken's inclinations to peck and scratch the earth.

Instead of the haphazard browsing technique that these animals are inclined to do, the chickens are confined to specific locations to accomplish a task. Generally we put the chickens into areas that we would like to cultivate in the future. The chickens denude the enclosure of the available nutritive vegetation and fertilize the area. They clean the area of insect pests and aerate the soil. The density of birds and the amount of time they are utilized on an area varies, depending on the purpose of the tractor. We have attempted to allow birds to range freely in select remote locations of the property at a distance from our gardens. Unfortunately, this system resulted in increased predator attacks. Predators, including domestic dogs, are more inclined to attack unfenced animals. Vehicular traffic along roadways is also a safety concern for free-ranging chickens.

Generally the introduction of birds into established garden beds has limited rewards. While the chickens appreciate eating salad greens directly from a bed that has gone to seed, their scratching is destructive to the layered and shaped permanent beds. Garden beds are shaped flat on their tops for drainage and direct seeding vegetables, and the chickens wreak havoc on these forms, which must then be reshaped by human hands. Longer-term chicken activity will compact areas of soil as they dig potholes for sanitation in the form of dirt baths. Two distinct ways to partner with the chickens are either to rapidly rotate or to incorporate the chickens in a longer-term spatial enlistment. By rapidly rotating tractors through areas, we are able to maintain and improve field space while supplementing the chickens' nutrition without compromising the regenerative capacity of the existing vegetation. In a longer-term spatial enlistment we are trying to change the species composition of the area and use the chickens to remove the existing vegetation and fertilize the landscape. Chickens can sufficiently remove grass species, although pigs are more suitable for removing brambles and woody plants. In the longer-term chicken tractor system, we restrict them to the area until they have annihilated the area of vegetation, which enables a transition to a designed plant species introduction.

It can take several months for chickens to effectively clear an area, depending on the population density of the birds. Once the chickens have removed the vegetation and the area is smoothed of their constructed undulations with shovels and garden rakes, we proceed with one of two options described in chapter fourteen in the "Animals as Agricultural Partners" section.

In summary, the chickens are utilized to manage pastures by grazing and fertilizing, and they provide eggs and meat that concentrate this trophically derived vegetable-based protein. Alternatively the chickens can be utilized in high-density locations to transform the landscape. Their incessant instinctual behavior to peck and scratch the earth clears zones of existing vegetation for a subsequent floral re-speciation.

Chicken Economics

Chicken eggs are not economically viable as a wholesale commodity for the small farmer. The price of organic feed is currently over twenty dollars for fifty pounds. While laying birds are sometimes available for around six dollars from commercial operations, it takes six to nine months from hatching for chickens to begin laying the first smallish eggs. During this period of adolescence the birds still require feeding and care. After nine months chickens may lay up to 80 percent of the days until they are two to five years old; then their production drops off dramatically. This short production life cycle marginalizes the economics of egg production. Commodity eggs are sold for as cheaply as three cents each. It has been estimated that the time, feed, and effort required for an average small farmer to raise eggs with fair wages would value eggs at fifty cents each.

We have never raised chickens strictly for meat. Roosters that are delivered in error with our hens are roasted or become soup once they attain an appropriate size or misbehave. While meat birds are a potential market for the small farmer, since humanely raised local meat is in high demand, such an industry has its challenges. Meat birds have been bred to grow at a tremendous rate and consume large rations of grain to feed their metabolism. Typically nine weeks is sufficient to raise a four- to six-pound dressed bird. After nine weeks the ravenous chicken's incredible metabolic growth slows, and the meat quality diminishes. I have witnessed a neighbor's meat birds that grew past nine weeks in confinement with unlimited grain until they were unable to support themselves with their legs and developed gangrene on their breasts from moisture and immobility.

Ducks

Our initial duck experiences were ill-advised forays in husbandry. We recruited a couple of Khaki Campbells for light weeding and postharvest cleanup tasks in our established garden beds. Unfortunately, while the ducks performed adequately in their daytime tractors, their overnight housing with the chickens was dysfunctional. In their instinctual efforts to bathe, the ducks splashed water from their dishes, and the chickens bullied the beleaguered ducks with their beaks. The bedding rapidly became a soupy mess, and the ducks were wounded by the encounters with the aggressive chickens. In our hasty search for alternatives we allowed the ducks into a zone occupied by the pigs. Though the pigs and ducks lived in harmony for a short period, this collaboration eventually ended in duck mortality.

Khaki Campbells were chosen because of their purported inclination to be mutually supporting contributors in a loosely organized ecological system. That being said, there are also many other varieties available. Ducks such as the Indian Runner, which we currently raise along with Khakis and Swedes, have been bred for centuries, to the point where they cannot fly because of disproportionately small wings. Other nondomesticated species have wings that must be clipped regularly or they will potentially fly away.

The thought was to allow these birds to be self-guided vigilantes of the garden. By allowing the Khaki Campbells to roam the landscape we had expectations that they could differentiate the beneficial insects and plants from the weeds and pests that we hoped they would consume. Unfortunately the birds are rather indiscriminate browsers who relish digging salad green seedlings as well as slugs. With their webbed feet they are not inclined

to scratch and therefore have little utility for soil cultivation. As our experimental observations of the ducks have progressed we have found that it is more appropriate to use these powerful collaborators in more discriminate and precise ways, rather than randomly throughout the landscape as browsers.

Ducks are a visual and aural addition to the farmscape. To truly appreciate the instincts that this animal has developed through eons of genetic development, it is necessary to provide them with proper habitat. The sight of ducks swimming amid a wetland, diving and gliding with grace as foragers in their natural habitat is a thing of beauty. The saying "like ducks taking to water" doesn't do justice to the observational understanding of this animal's ecological connection to its appropriate habitat. The ability of ducks to float and to shed water while maintaining their warmth is truly remarkable to witness.

Their raucous quacking is another element to consider. While they provide an aura of farm atmosphere for visitors, this additional noise can be a potential aggravation. In addition, male ducks dominate the females. While they are essential for reproduction, their malicious, violent dominion over the females can be disturbing to witness. The drakes, like roosters, will chase females until they can be subdued, then peck the back of the females' necks, often inflicting open wounds.

We currently utilize housing units for overnight predator protection as well as centralization of feeding and egg collection. We have had difficulty encouraging the ducks to utilize the housing as a dwelling, and they often seek laying locations outside. This has caused difficulties in time spent herding the birds into their shelters and searching for eggs. Herding the animals across a water body with long sticks is labor intensive and has resulted in several unintended baths in the duck pond. Fence height for ducks is dependent on the terrain and the animals' individual and species propensity for flight. Two-foot-high fencing is sometimes adequate to contain flightless birds such as the Runners.

Currently, we use ducks to generate fertility in areas that we are transforming into wetlands. Ponds and depressions that accumulate water are areas of immense biologic potential. Ecological niches where distinct resources and species habitat are combined are identified as "edges" in permaculture terminology. In this case the ducks serve to accumulate fertility along a designed edge, empowering a process of biologic perpetuation.

Within such a niche, ducks will efficiently and actively forage for amphibian and vegetative species. They will rapidly decimate the existing ecology and pollute water bodies with agitation and manure. Large bodies of water or rapid rotation procedures are necessary to prevent ducks from dominating the ecology.

We raise ducks for their meat, eggs, and aesthetic quality and as a means to translocate fossil fueled abundance to our waterscape. The grain with which we maintain these animals throughout the year is derived from the fossil fuel infrastructure of our epoch. In essence we are utilizing a fossil fueled delivery system to build the abundance of our landscape through the deposition of nutrients in the form of duck poop. The ducks also produce eggs, though primarily in the summer. The eggs are creamier in texture than chicken eggs and have a distinctive flavor.

The Grain Dilemma

In addition to predators the price of feed is a major concern when raising poultry in our region. Even though bulk and group grain purchases are possible directly from regional mills, in recent years the price has risen dramatically. While in the summer months pasture and forage rotational strategies can be utilized, alternatives must be considered for the other eight months of the year. To supplement their diet we experiment with storage crops such as beets, as well as household and restaurant scraps. To navigate the challenge of grain costs and availability into the future, we are attempting to maximize the potential benefits per bird. We are focusing on

hatching on-site and carefully considering flock size and whether a particular bird or breed has a greater capacity to endure this seasonal fluctuation in food. Further investigation into feeding poultry sprouts, aquatic plants grown in grey water, and fermented foods preserved from the growing season is necessary to develop a four-season diet for our birds.

Electric Fencing

Colonial-era agriculturalists were severely limited in their ability to contain livestock. Fencing was expensive, labor intensive, and unreliable. Stone walls were permanent installations and easily climbed by both hungry predators and intrepid livestock. Wooden rail fencing deteriorated rapidly and provided more of a visual barrier than adequate containment. In the 1800s, in an effort to tame the Wild West, the colonizers introduced barbed wire. This product provided years of reliable entrapment by wounding animals that attempted to evade confinement.

By the mid-1900s technology yielded electric fencing as a sophisticated mechanism to incarcerate large animals. The electric option provided long-range and lightweight portability. These fences provide an intermittent high voltage pulse of electricity that is painful though not lethal. As animals become aware of the fence's painful capacity it is generally sufficient to deter them from attempting to cross the line.

Electric fencing is our preferred choice for enclosing four-legged animals. The system consists of several elements, including the charger, ground rod, conductor, and insulators. The charger emits the electric voltage pulse. Chargers can be powered by solar, domestic house current, or a twelve-volt battery.

Proper setup is essential for successful fence operation. Instructions are provided with each type of charger, though generally two to three ground rods are recommended, spaced about ten feet from the charger. While six-foot half-inch diameter copper is the typical size grounding rod for a house we seek more inexpensive options. By clamping aluminum or copper wire to driven rods of four-foot half-inch rebar, adequate grounding can be accomplished.

We prefer the conductivity of solid steel wire, which is superior compared to the thin wire braided in nylon that is also available. The wire fencing can last decades and is transportable and reusable. The durable multipurpose wire is easily mended and extended. We use plastic conductors on fence posts and trees to string the wire around the perimeter of animal enclosures. The wire height is set to prevent grounding of the current and is dependent on the animal species and size and the season.

A fence tester can be purchased to gauge the electrical force of the fence. Monitoring the animals will also resolve questions of the fence's effectiveness. Effective fencing is painful and will elicit a yelp or sudden movement that corresponds to the cyclical zap.

Once the fence is set up and functioning, it is important to maintain conductivity. The wire must remain free of all noninsulated contact for the current to resonate effectively. Vegetation can grow up to touch the wire and ground the electrical current. When the current is grounded by an interruption the conductivity of the wire is diminished, and the fence becomes ineffectual.

Electric fence chargers cost from one hundred to two hundred dollars and have a three- to five-year lifetime. Chargers can be easily damaged in electrical storms. If possible buy a unit with serviceable internal components or an easily accessible and replaceable fuse. It is also possible to utilize solar and 12-volt battery powered units, especially long distances from grid power.

Electric fencing is a sizable investment, though we have found it very successful at containing the unruly foraging nature of pigs and goats. With proper set up and maintenance, this fencing is a fairly reliable means of animal containment.

Veterinarians and Animal Health

Domestic animals are survivalists that can maintain an existence in conditions of confinement and poor sanitation. These conditions may permit survival, but they do not encourage health. Animals need clean water, food, fresh air, sunshine, and the earth beneath their feet, as well as shelter and sanitary conditions to thrive.

Animals can develop ailments for which they need medical attention. Depending on the circumstance there are many medical procedures that can be performed by on-site farm personnel. While birthing, castration, and end-of-life care can eventually become skills in the repertoire of the typical farmer, it is helpful to have experienced advice and consultation during training and development. Even in conditions of an optimally healthy farm system it is prudent to maintain relations with a veterinarian in case of an accident or to seek a second opinion.

We continually monitor and evaluate animal health by noting characteristics of vibrant health such as bright eyes, healthy coats, group dynamics, and range of mobility. Signs of poor health can be visibly apparent and include slow and drowsy behavior or noticeable limps. If poor health persists, further steps must be taken to remediate.

For the animals we keep beyond a growing season it is important to develop a relationship that allows for regular grooming and health care. Animals should be taught to allow inspection and trimming of their feet and any additional regular attention.

CHAPTER TEN

BUILDINGS

We need functional structures to serve both as dwellings and as functional four-season space to fulfill the activities of the farm. However, as humanity entered the industrial era, large-scale toxic and resource-intensive construction methodology brought with it significant environmental consequences. Modern, modular styles of construction rely on manufacturing processes and materials that are polluting and energy intensive.

In this era of high-rises and sick-building syndromes it is important to remember our ancestral roots and our need for essential shelter. Buildings serve as a place for the necessities of food and water, as well as a location to develop friendship, music, and art. Our built environment serves as a crucial connection, a visible embodiment of humanity's response to environmental challenges that defines our role in relation to the physical space we call Planet Earth. We have demonstrated our ability to create structures of enormous scale, but these structures have had destructive consequences from the polluting and energy-intensive methodology of their construction.

What follows is the evolving praxis and philosophy that define our consideration, creation, and use of shelters. Although cognizant of modern building trends, we continue to redefine the concept and process of "shelter" using a regionally appropriate, sustainable, resource-conscious terminology. When we arrived at D Acres, we were blessed with the structures that my aunt and uncle had renovated and constructed during their tenure at the farm. While the original cape and barn were built in the 1830s, Delbert and Edith expanded the barn and built the garage shop, silo, and art studio. These additions served our needs and provided the basis for our initial operations. Since that time we have built many additional buildings to house our activities. From the tree houses and outdoor kitchen to the Community Building and G-Animal projects, we have learned lessons that we continue to apply to renovations and new constructions.

Design Build

We have explored and experimented with a wide range of construction methodologies. Our preferred choices are reused, recycled, salvaged, or natural biodegradable materials. Our aim is twofold: one, to construct permanent structures in which stability, durability, and longevity are essential; and two, to construct temporary structures defined by the time and materials available that can provide the adequate structural integrity necessary to meet short-term goals. Some buildings are constructed to be utilized for the foreseeable future, while other structures have a more limited life span based on the purpose and energy invested in construction. Our methodology is derived from local vernacular traditions that utilize regional materials and the skills base at hand. We also have drawn extensively from our experiences traveling the world to research building technology. Finding creative solutions to shelter represents a limitless exploration of how materials and the laws of physics can be applied for climatic control through designed constructions.

As we experiment and refine our principle-based design process, our construction style continues to evolve. The design process must enable, inspire, and empower versions of structures to be utilized for site-specific applications. Construction principles are based on observation and interaction with the intention of placing the structure as an element of the ecological system. Through design, a variety of yields are built into a structure, thus offsetting the investment of the construction. Our goal is to provide longevity and usability through design. Considering human usage patterns in relation to materials available and structural design provides a format for pragmatic shelter construction. Through our design we also attempt to mitigate the costs and maximize the benefits. Design is a process of creating microclimates and ecosystem mimicry. We are building shells from organic materials to endure the environmental conditions. These structures must be adaptable to needs by integrating intention into an evolving, goals-focused design.

The functional yields of a construction can provide living and working space, aesthetics, privacy, and security. Structures can provide a sense of place by sharing traditional architectural styles and vernacular. Traditional structures such as barns are an important component of regional identity and community and also provide an opportunity to capture renewable energy from hydro, solar, and wind. Renewable yields using rooftop gardening, solar hot water collectors, and rainwater catchment can be used to enhance productivity.

Longevity is defined as the durability of the construction against the damaging environmental consequences of climate and usage. To resist the perpetual climatic and gravitational challenges presented by our environment, a building must also be structurally prepared to endure and adaptable to the extremes of the climate. These events can include high and low temperatures, strong winds, and extreme rain and snowfall. Builders must be connected to the landscape of their constructions. Designers in Florida cannot evaluate structures to be built in the mountains of Colorado, and I should not design a structure to be built in the desert of Arizona.

The exterior and interior of the building should be considered as separate microclimates where the environmental conditions of solar incidence, temperature, humidity, and air quality are to be optimized through design. While these zones are distinctive the design dictates the degree of separation that the built environment provides. Structures can serve as an element to connect with the environment or can be constructed in a style that isolates the occupants from the exterior world.

Elements of design that should be considered include budget, time line, usage, accessibility, and aesthetic. Budget entails both the materials and labor that will be expended in construction efforts. Time line addresses the continuum of activities necessary to complete the various aspects of the project; time to completion is the standard by which all construction is measured. The benefits received must outweigh the costs associated with construction and maintenance. The primary purpose of the structure is generally the usage, which will be enhanced by the accessibility and aesthetic of the structure.

The construction process requires the means to communicate a plan and assess material needs. The drawings and vernacular of building terminology communicate the ideas for creating a structure. Common words such as "post and beam" are crucial to explain the design prior to construction. Drawings to scale help define the project specifically with regard to materials acquisition. Just as cooks rely on recipes, and garden maps define our food production process, we use drawings to refine and translate the building process.

Buildings require maintenance, regardless of their design and materials. The Taos pueblos in New Mexico, for example, are the oldest inhabited structures in North America, though the earthen construction materials still require regular maintenance. By anticipating maintenance needs before structural damage ensues, buildings can be utilized for generations. To ensure the process of perpetual maintenance, future generations must value the building, based on an analysis that weighs the benefits of functionality versus the costs of maintenance.

Building materials are evaluated on their performance and sale price, as well as the externalities associated with their production and a life cycle analysis of the materials, beginning with the sourcing of the materials and continuing beyond the viability of the material as a building component. To illustrate the distinction, while some conventional, industrially produced building components will remain toxic in the environment for thousands of years, a timber-framed house can easily be recomposed into the ecological system after its viability as a building material has been served.

The type of investment extended for construction should be considered in terms of the external costs of production that are incurred on the environment. In the bigger picture, choices such as the use of fossil fueled excavation equipment and polluting materials negatively affect the entire planet. By providing human-powered options constructed from locally sourced materials we can address the conventional building industry's external costs.

Modern construction has been formulated to reduce labor through standardized, prefabricated component assembly and is typified by industrially produced modular materials. While the fabrication of the materials is often polluting and resource intensive there are also additional resources expended in the vast distribution network necessary to deliver the materials. Modern construction often consists of modular squares stacked vertically, and human construction ingenuity remains confined to the parameters of the materials. The lack of square shapes in nature is indicative of its inherent weakness of form. Rounded shapes can deflect and absorb energetic forces such as wind, sun, and kinetic actions. Sharp edges are more easily broken; rounded edges are the common biologic form.

Ingenuity in the design-build process is only limited by the laws of nature. The most fundamental law that affects building integrity is the law of gravity. To effectively compensate for the gravitational pull of the earth, builders use the concept of level and plumb. Level indicates when a material is flat to the horizon. Plumb indicates the vertical nature of a material 90 degrees from level directly in line with the pull of the earth.

The weight of a structure is born downward by the perpetual force of gravity. To resist this dynamic it is necessary to structurally support weight that is incurred through the construction process lest the structure sink into the earth. In addition to the force of gravity, shear strength is engineered into a building design to resist the forces of Mother Nature, such as wind and seismic activity. Shear strength is obtained by connecting structural elements in various dimensions. This process provides rigidity to the construction that resists the twisting, pulling, and pushing of the earth's environment.

Water, too, is a powerful element that can undermine a structure's integrity. It is crucial to address this in the construction process, as water drains downhill, assisted by the pull of gravity. In our region we receive about forty inches of precipitation a year, which often arrives in the form of snow from October through April. Builders utilize constructed slopes in roofs and landforms to shed and direct the flow of this element.

Site Evaluation

Location is a crucial aspect of construction. Potential sites can be evaluated by many criteria. The elements of sun, wind, earth, and water define the location potential, as well as the food-growing capacity of the site. The influence of seasonal and storm water conditions will affect the suitability of the location. It is also important to consider the subsoil and identify potential issues such as rock outcroppings and ledges. Sites can also be judged by their proximity and accessibility to other infrastructure.

Site work is an essential element of a viable construction process. The degree of site work that is undertaken depends on the investment chosen to preserve the long-term utility of the structure. Site work is utilized to move or alter the existing landscape with the intention of improving the site for construction purposes. In essence, native landforms and terrain are altered to create idealized conditions.

Site work varies depending on the circumstance. The seasonal water table and water flows dramatically affect the maximization of the site. Other factors to be considered are usage and traffic patterns. Seasonal variations, including ground-freezing conditions, can be beneficial to strengthen the terrain during the winter. The freeze/thaw cycle, however, can also be a detriment to year-round success. The intention is to maximize the utility and longevity of the structure with a minimal investment in time and footprint. Site work can range from smaller manipulations of the landscape, such as clearing a trail to access a tree house, to major excavation in preparation for a full-foundation structure. The investment in site work can reduce future everyday nuisances such as muddy zones that develop in heavily traveled areas. Adequately planned and implemented site design can also prevent catastrophic structural failure caused by high moisture levels that rot wood or freeze/thaw conditions that heave the foundation.

Site work can be accomplished with hand tools such as shovels, picks, and rakes as well as with fossil fueled machinery such as excavators and skid steers. The surface of landforms can be graded and pitched to induce drainage. In addition the subsoil can be replaced with material of superior drainage and structural integrity, such as bank run gravel, riprap, or screened stone.

The Boots

The long-term viability of the structural component of a building begins with an adequate foundation. The foundation serves as the weight-bearing portion that maintains the stability and integrity of the aerial parts of the building.

Foundations should be built to be impermeable to rodents and moisture. Proper drainage surrounding a foundation is crucial to help protect moisture-sensitive aerial elements of the construction. By draining water away from the structure and elevating the height of the aerial elements from ground contact we design a response to common dilemmas. In the wet climate of New England poor drainage results in a wet cellar, which leads to rotting sills and eventually results in a compromised structure without a solid base.

Foundations must be sized appropriately to bear the weight of the structure. The weight must be distributed so the structure does not settle unevenly into the soil. If the structure begins to settle because the distribution of weight is inadequate, the building strength and original truth of level and plumb will be compromised. Once a building begins to shift because of movement in the foundation, the force of gravity will continue to work, pulling the structure to the ground, especially if the movement continues.

Foundations in cold climates must also contend with the effects of freezing ground conditions. As wet, saturated soils freeze, they expand. This can produce movement in the foundation structure. The freeze/thaw cycle in New Hampshire dictates that the foundation should be at least four feet below the ground surface. Although the frost rarely reaches four feet into the ground, extreme temperatures in snowless conditions can cause catastrophic damage to foundations that are not of sufficient depth. Driveways that are plowed or compacted, for example, can be more susceptible to frost than the snow-covered sides of a house because of the insulating quality of snow.

Foundations are typically constructed from durable, impermeable materials. Traditionally in New

England, nonindigenous settlers cut granite blocks to serve as the foundation material until the advent of concrete. Introduced during the Roman Empire, concrete has become the foundation material of choice in the modern era of construction. Concrete is a durable, mostly impermeable material composed of portland cement, sand, small three-quarter-inch aggregate stones, and reinforcement rods. These materials are mixed wet and poured into a form that then sets to a structural integrity within hours.

The original homestead at D Acres was constructed using the standard methodology of preconcrete construction. The builders dug a cellar so the consistent earth temperature would moderate the structure from the outdoor air temperature and provide food storage through the winter. Rocks pushed into position helped retain the walls of the perimeter onto which the granite slabs were laid. To cut granite slabs the builders drilled holes and allowed water to freeze over the winter, which cracked the stone into rectangles approximately two feet tall and one foot wide of various manageable lengths. The slabs were then laid out and the building timber framed on the impermeable, frost-resistant knee walls of granite.

The conventional structure that is built in New England today has a basement with concrete stem walls six to twelve inches in width built on a wide footer base two feet wide and one foot deep. The wall is then backfilled with well-drained materials, and a perimeter drainage system is installed. This system has the benefit of remaining relatively dry and structurally sound for hundreds of years.

Site work and frost-free foundations are an expensive undertaking in time, energy, and effort. This product can generally outlast the aerial components despite the best intentions of the designer. Under most conditions, organic materials typically used in the aerial parts are more prone to degradation than the concrete foundation. Hence it is important to consider carefully the future impact, reuse, and renovations necessary to extend the lifetime of the building.

Concrete is expensive and durable for lifetimes. Digging space for construction below the frost line and forming the space for the concrete to be poured requires adequate time, equipment, and knowledge. To ensure strength and integrity, steel reinforcement rods should also be added to the concrete.

But concrete has tremendous externalities that are unaccounted for in its selling cost. It is estimated that up to 10 percent of the annual CO_2 released into the atmosphere is a result of the fabrication of cement. It is also difficult to find appropriate means of disposal for the material, especially in a conventional waste-management system in which the material may be hauled with other construction and demolition materials to an incinerator or landfill. In rural environments we can generally find use for the material as fill or in road improvement projects.

There are alternative concrete foundation designs that use less material. These alternatives are designed to provide the structural basis and impermeability to moisture that concrete has provided to modern construction. We have experimented with Alaskan slab and rubble trench foundations designed to reduce the amount of concrete necessary for an adequate foundation.

The Alaskan slab concept originated in response to the permafrost conditions in Alaska. The extreme building conditions, in which the ground is frozen the majority of the year, are similar to the climate in New Hampshire. In these conditions, when the ground thaws in the spring, the entire soil structure becomes the consistency of soup. During these times traditional foundations not only sank but also became susceptible to refreeze conditions that undermined their structural integrity. There was also a limited time period each year when adequate excavation could be performed.

In response to these challenges of the landforms and climate, designers developed a floating foundation concept. A trough is dug along the perimeter of the structure to provide adequate bearing capacity and to help prevent penetration of moisture and frost horizontally. Then the entire base is poured so that a slab fills the interior. This slab is connected throughout with rebar. The reinforced slab then acts as a unit similar to a ship in the sea riding over the seasonal expansion and contraction of the surrounding landscape. This technique reduces the quantity of concrete poured and requires minimal excavation.

The rubble trench is a foundation design popularized by Frank Lloyd Wright's hotel construction in Japan. Wright's design utilized rubble pack as a primary base for the building to allow drainage. The idea was that the material would withstand the effects of an earthquake much better than a concrete foundation. While Wright's intention was to reduce the possible impact of seismic activity on the hotel, we have synthesized this style to reduce the amount of concrete we use and to encourage drainage.

Buildings we have constructed with these alternative foundation methods—the Ox Hovel with an Alaskan slab (2002) and the G-Animal with a rubble trench (2004)—are still standing successfully, with no signs of degradation thus far.

The Envelope

Walls are designed to protect the structure from outdoor conditions and provide for portals in the form of windows and doors. Walls increase rigidity against the downward pull of gravity and provide shear strength, which prevents wind or seismic activity from dismantling or toppling the structure. The load-bearing capacity of walls is dependent on whether the structure relies on individual bearing points, as in a post and beam design, or the walls are actively carrying the load of the roof, as in a conventional stick-framed 2×4 wall construction.

Walls and roofs also provide insulation, which moderates the extremes of the ambient outdoor temperatures and is an important component of the building envelope design. Overall performance depends on how well the envelope is sealed to reduce drafts. Large volumes of uncontrolled moving air flowing through the envelope will conduct the outdoor temperatures indoors. If drafts are not controlled, temperature efficiencies in the structure will be similar to building a good refrigerator and leaving the door open. Insulation can also reduce the transmission of sound through structures.

The goal of insulation is to maintain the preferred temperature range of the structure. In the case of human dwellings, that range is generally limited to between 50 and 80 degrees Fahrenheit. This is accomplished through deliberate use of materials and a design that creates a self-regulating, passive system of climate control. Outbuildings for animals are designed to prevent extreme heat or cold by retaining the warmth of the animals during the winter and by providing for adequate ventilation and shade in the summer.

Insulation is rated in terms of R-value, an engineered calculation used to assess the proficiency of different insulation materials. The R-value is based on the materials' ability to prevent the transfer of heat. The higher the rating per inch the more effective the material is at preventing the flow of energy. Insulation efficiency can be obtained by containing smaller pockets of air, which act as a barrier to moderate the flow of the thermal energy. This resistance to the flow of energy is what constitutes insulation value. To provide some examples, a single pane of ⅛-inch glass has a value of 1, while an 18-inch straw bale is rated up to a value of 32.

The relationship between moisture and insulation is also important to consider. Ambient humidity is as important as temperature with regard to building an envelope that meets the design goals of the structure. For instance, the hayloft of the Red Barn is designed to achieve hot, dry, and drafty conditions in the summertime, while the root cellar is designed to maintain cool, moist conditions. The conditions are produced by consideration of the environmental conditions and responding with design solutions.

It is necessary to control the moisture levels in livable structures. This can be accomplished through design that is responsive to the environmental conditions on the site. While the 100 percent humidity of a rainstorm must not be allowed to permeate and saturate the structure, humidity is also generated during the winter and must be allowed to breathe from the structure so that condensation does not develop on the interior of the envelope. Periods of rain tend to penetrate the building envelope, while in the winter the warmth of human activity can generate moisture, which must migrate out of the envelope. In essence humidity is resisted from the outdoors and is encouraged to exhaust when necessary.

Insulation techniques can be integrated into the building envelope depending on budget, philosophy, resources, and people power. The objective is to allow moisture to migrate out of the structure while shielding the building from external humidity. The insulation also traps the air in the structure, effectively isolating the building from the vagaries of the external temperature fluctuations. Organic materials such as wood, natural plasters, cob, and straw bale will conduct moisture through the envelope while restricting the flow of air. Plastic sheeting and closed cell foam can be used to shield or trap the flow of moisture, preventing migration in either direction.

The purpose of developing an envelope that provides a sufficient base of protection, while also accounting for draft, is to thermally isolate the interior of the structure from outside air temperatures. This protection provides the basis for the interior climate control system. The performance of the structure can be evaluated by its ability to retain or trap the internal temperature in resistance to the weather's environmental shifts.

Industrially produced, petroleum-based vapor barriers offer a conventional solution to the variability of daily and seasonal climatic shifts. These ubiquitous products with brand names such as Typar and Tyvek are laboratory tested to control the penetration of moisture while also allowing moisture to be expelled through the material. Tar paper has also been used for generations to provide a petroleum-based layer of protection as a barrier.

If properly applied these materials provide lasting protection from the environment, though they are not reusable and require industrialized processes of the fossil fueled economy to be produced.

There has been a progression of industrially produced products available to meet insulation needs since the popularization of fiberglass insulation in the post-World War II era. It was, and is, unsafe to breathe the dust and particulates generated by moving the fiberglass material. Fiberglass products are supposedly rodent resistant, though mice and red squirrels readily adapt to these housing conditions. The material is generally most effective in standardized stick-frame construction because the paper-backed or bagged product will fit readily in the framing cavities. The product is not suitable for wet conditions and can accentuate rot conditions in moist areas. The low cost and the effectiveness of the material in wall and roof insulation meant that the product was ubiquitous in conventional construction projects through the latter half of the twentieth century. Because of the health hazard posed by the material, however, alternatives have been sought to provide "greener" products for conventional insulation.

Foam is one such alternative product. It has been industrially produced in the post-World War II era and has become popular in our culture, with applications ranging from preformed coolers to drinking cups. For building applications the foam can be molded into panels for standardization of modern building requirements. There are also modular foundation forms and stress skin insulation products available to address structural elements of conventional construction.

While the foam is available preformed, there is also the option of spray foam, a chemical product that expands on application. The growth of the liquid when expanding to fill cavities, defies imagination, and thus is particularly effective at sealing drafts. There are several variations of this product, ranging from structural, closed cell urethanes to open cell foams with trade names such as Icynene.

Closed cell foam has been used extensively in the modern era as a structural building material. It has been used to build iglooish houses, swimming pools, and the oversize caricatures of cartoonish figures at amusement parks. This material is durable yet lightweight and is often chosen because it can be formed without the weight concerns of concrete. Closed cell foam is designed to be impermeable to water and is also used as a flotation device material.

Open cell foam is softer and has less structural integrity. This open cell product is designed to allow the migration of air and moisture. While it has a lower per-inch R-value than closed cell foam, it is superior for structural breathability.

These foam products are available from local hardware stores in aerosol cans and can be contractually delivered in fifty-five-gallon drums from trained installers. The products do require an industrial infrastructure for delivery, and it is difficult to imagine a plausible secondary reuse potential. Closed cell foam eventually will become saturated in wet conditions, reducing its R-value dramatically. While durable, both open and closed cell foam is affected by UV radiation and will degrade according to environmental conditions.

There are also foam products now available that are manufactured from plant-based material. These soy and corn derivatives are industrially produced and provide the same superior sound and insulation value of fossil fuel–derived foam. There are also industrially produced products that are made from recycled content. Cellulose insulation is made from the fibers of recycled newspaper. Loose cellulose offers insulation in smaller cavities, while a bagged product can be stapled into conventional framed walls. The loose cellulose will settle over time, diminishing its initial R-value.

Another alternative used in modern timber frames is exterior insulation packages constructed from structural insulated panels. These panels are produced industrially by sandwiching foam insulation between sheets of plywood. The panels are designed to provide insulation as well as structural capacity for both walls and roofs.

Our search for adequate natural insulation materials has been stymied by the natural systems at work in New England. The environment here features wide seasonal and annual variations in temperature and humidity. In addition, we have severe rodent pressure. During the winter months rodents seek the warmth of human dwellings to hibernate through the winter, raise their families, and deposit abundant excrement.

At times we have used plastic and foam postconsumer waste to insulate wall cavities in outbuildings on the property. Using the theory that once drafts are sealed the R-value is based on reducing the flow of air by creating small air pockets, we crammed packing foam and plastic bags snuggly into the space between framing members. This material is effective as insulation and resistant to moisture. Though it is not rodent resistant, it will defer recycling the postconsumer plastic until the next renovation.

Chip clay and straw clay are natural building material options available to provide insulation. Straw, wood chips, or a combination thereof are mixed with a slip clay, then poured into the walls. The mixture dries in the formed walls to ambient moisture levels. The clay slip coats the basic insulation materials, providing pest resistance, humidity control, and rigidity to the aggregated material. Once the clay has dried to achieve ambient moisture levels, it serves to moderate and neutralize moisture shifts by absorbing and releasing the ambient shifts without degradation. The clay naturally modulates and self regulates based on the relative humidity.

Combining Structure and Insulation

There are several criteria for choosing wall materials. In addition to cost and availability, we evaluate materials for their structural value: their ability to endure wind, fire, and earthquakes; hold up the weight of the roof even under heavy snow; and provide protection from critters. Materials also vary both in their aesthetic qualities and in how well they insulate from noise and temperature.

Cob

Cob is a natural building option that serves well as a wall material. While cob does not have the structural capacity of stone, it can be used as a load-bearing component. The material is superior to stone in R-value insulation capacity and presents unlimited sculptural possibilities. The material is nontoxic and reusable.

Cob is composed by mixing wet clay, sand, and straw. Its combination of 30 percent clay and 70 percent sand mimics the conditions found in the earth's crust. An ancient building material, this particular combination of ingredients is shaped wet, then left to dry hard and rigid like stone. This type of soil is blended in nature within the broad possibilities that silt, sand, stone, and clay provide building constituents. Natural cob can be recognized as the hard strata that resist erosion of stream banks, as well as the soil layer in which digging with hand tools is difficult. This earthen material has long provided *Homo sapiens* with a consistent and abundant opportunity for shelter construction.

The composition of soil will determine its potential to be utilized for cob building. Clay and silt can be confused when evaluating soil. While clay, sand, and silt form the basis of all soil, clay and sand are inert earth-based materials that are formed through geologic process. Silt, in contrast, is produced biologically and is largely composed of organic material. The organic nature of silt does not provide the stability and resistance to biological decomposition preferred in a building material. While clay and silt are both slippery to the fingers, soil with high concentrations of clay can be rolled into pliable strings, as if to construct a ceramic coil pot. For further evaluation of the soil composition a test can be performed by shaking a mixture of the soil and water, then observing the stratification that occurs as the soil particulates settle. Sand and rocks will drop to the bottom within seconds, while silt

and organic material will stratify in minutes, and clay particles can remain suspended, clouding the liquid for days. On a molecular level clay and sand combine to form a durably bonding, inorganic, natural building material. Clay provides an elastic glue that bonds the sand particles into a rigid framework. The combination can be compared to a stone wall, with the clay acting as the mortar and the sand behaving as the stone.

Straw is utilized in the cob mixture to help reinforce the structural capacity of the clay and sand combination. This dried organic material has low quantities of energetic nitrogen and degrades minimally in the walls of cob. This natural reinforcing rod provides tensile strength to the material while also improving the cob's insulation capacity.

Cob is made by thoroughly mixing the ingredients in the correct proportions, then directly applying the mixture to the structure. Walls can rise six to eight inches in a day, depending on the solar exposure, relative humidity, and wind, which dries the material so that further weight can be added. If too much cob is added to a structure the material will begin to loosen, sag, and slump. Once this slump and loosening of the materials has begun, it is difficult to rectify without time. In most cases this should be allowed to cure overnight and addressed after the cob has dried sufficiently to allow a machete to cut through the fibers of straw so that the bellied sag is removed.

While there are distinct advantages and considerations when working with earthen materials, keep the following considerations in mind regarding locating and acquiring the materials necessary. All are heavy. It is important to pay careful attention to how to transport and utilize the materials until their final placement on the structure. This requires planning to stage the materials in locations that are accessible yet allow space to maneuver on the job site.

In addition to traditionally defined techniques such cob, adobe, and wood framing, there are various ways that earthen materials can be utilized

D Acres hosted several alternative building workshops in 2004 during the construction of the G-Animal. This group is mixing a large batch of clay, sand, and water thoroughly before adding straw.

and integrated into wall systems. Stone, straw bale, cordwood, and cobwood are all techniques that can be adapted for load-bearing purposes and also utilized in timber frame designs as nonbearing wall elements.

Stone

Stone is a beautiful, durable, and weather-resistant material. Compared to other wall materials, stone has a relatively high thermal mass, though because of its convective nature it has a low R-value. Dry stacking techniques have been used throughout human history while mortared wall construction dates to the introduction of cement during the Roman Empire. While the material provides longevity, the actual construction requires heavy lifting, skill, and experience. A proficient mason can combine aesthetic design with structural integrity by proper placement of the variably shaped stones. In comparison to wood framing, wall building can be a slow, laborious process.

Helen and Scott Nearing documented a technique for stone construction in their book *Living the Good Life*. They utilized a slip-form construction in which stone and mortar is introduced into a form until it cures. Then the forms can be slid upward and the process repeated until the wall attains adequate height. The Nearings' process was designed so abundant stone could be utilized by relatively inexperienced builders to create shelters.

Joy Payton and Jacob Mentlick make the perfect cob burrito through the process of mixing the sand, straw, clay, and water to a malleable consistency, then folding it over using the large tarp. They are then ready to make cob balls for easy transport and application to the cob oven.

While stone limits the transfer of moisture, the material is not impervious. Stone and cement will wick moisture and sweat condensation. The Nearings' cabin in Maine has experienced issues with mold and mildew related to the stone style of construction. This effect can be mitigated by a design with adequate drainage and exposure to solar or air-induced drying through ventilation.

Cordwood and Cobwood

Cordwood is a traditional technique that has undergone a modern renaissance inspired by the author Rob Roy. His books, including *The Sauna*, describe in detail the strategy he has developed to construct the buildings located at the Heartwood Institute.

The technique is similar to stacking firewood except with cement mortar applied as bedding between the wood pieces. While cement is necessary for the process, the quantity is much less than that required for a wall of concrete and offers a superior insulation quality. We have limited experience with this technique at D Acres, but above the stonework that forms the base of the northern wall of the G-Animal structure, we chose cordwood because the lack of solar exposure made moisture resistance and insulation a high priority.

Planning for cordwood wall construction should begin months before the walls are erected. The wood should be debarked, split, and cut to length so that the wood can dry properly. The wood should be air-dried to the degree that it does not swell or shrink measurably with fluctuation in the ambient humidity. Removing the bark reduces the susceptibility of the wood to insect and moisture damage. The cordwood should be cut to a desired length so that structure and insulation capacity is realized. The wood typically starts between twelve inches and eighteen inches and can be cut so the wall width tapers to as little as six inches.

The cordwood technique is hampered by the expansion and contraction of the wood related to the ambient humidity. Unless the wood is kiln dried to less than 7 percent moisture, which permanently dehydrates and alters the cellular structure of the wood, the dimensions of the material will continue to fluctuate with the environment. Since the additional step of kiln drying is typically cost prohibitive, builders must confront seasonal shifts that shrink and expand the wood within the wall. As the wood shrinks it allows air to penetrate through the subsequent cavities. This can be alleviated by filling the joints with additional mortar. If the wood expands severely, it can cause cracking of the mortar and possibly subsequent structural damage. To help alleviate this dilemma wood is cut and split and stored under cover from rain approximately six months before it is used. Allowing the wood to dry to a moderate moisture level will avoid the

The G-Animal structure incorporates multiple wall systems to envelope the building. The lower half of the masonry wall is composed of large stones from the property. As the building increased in height, the move to cordwood sped up the construction process.

severe shrinking of fresh, green wood as well as the expansion of bone-dry wood as it is inundated with ambient humidity.

Cobwood is a technique that we began using through creativity born from necessity. During the fall of 2004 we were struggling to complete the G-Animal structure. We had worked diligently throughout the summer utilizing stone, cordwood, adobe, and cob methodology. As the days grew shorter the weather made mixing the near-freezing materials a physical hardship, and we became concerned that overnight freezing would undermine the structural capacity of the materials. Because of this weather constraint, we sought an alternative to accelerate the process to complete the roof before winter.

I had experience from Gaia in Argentina where we had inserted wood, trash, and even diapers within walls of cob. Since we had some extra cordwood available that we had dried for the lower sections of the wall, we began to insert the cordwood into the cob wall as we would have with cement mortar. This combination allowed us to rapidly accelerate the construction process in several ways. The walls were reinforced by the wood, so we were able to build more height within a day without the dreaded slump that is common when walls are overloaded with wet cob. We also used less of the cob material that was laboriously produced to meet our daily needs. The swift speed at which the earthen mortar dried between the cordwood conglomerate, in comparison to the slower drying solid walls of cob, allowed us further freedom to quicken the pace of construction.

Straw Bale

Straw bale construction is a relatively new natural building style, as baling equipment was first introduced to the plains of North America in the late 1800s. In straw bale technique the tightly compressed biomass tubes of plant matter are woven to shape and tied for the purpose of high-density storage. This innovation in building technology was ingenuity responding to the circumstance. The plains settlers were forced to adopt alternative techniques from the traditional wood or stone structures because of the lack of those materials. While the technique had been developed to provide for storage and transportation of animal feed, the bales' rectangular shape mimicked bricks commonly used in house construction. As an extension of the sod technology common in the region, early straw bale builders found these relatively lightweight, though proportionally larger, straw bricks capable of serving as a functional wall element to support the roof and provide the structure with insulation. The dry environment of the plains has helped preserve these initial experiments to survive for over a hundred years as usable structures.

This project was the first noninstructed, vastly unsupervised experience of my building career. The logistical challenges of real-life construction projects versus my prior experiences within the green academia became tangible for the first time. The problems became more complex as the scale and actual usability became apparent. For instance, when the 280 bales arrived from Canada via tractor trailer, they had to be stored to prevent moisture damage. Without an additional barn or warehouse, this influx of material created logistical challenges that took hours of shuffling, storing, and stacking to resolve. Had the process been sequenced and staged more efficiently, the time spent resolving this issue could have been better spent on actual construction progress.

Based on my experience there are several recommendations I would make regarding straw bale construction. In addition to low moisture content, bales should be relatively free of seeds and baled extremely tightly with wire. Because of the difficulty of cutting and rebanding bales, structures should be designed so the length of the bales corresponds to the placement of doors, windows and posts. If not, consider renting a baler to resize the bales as tightly as necessary. It is also important to consider preventive design elements to limit the avenues of entrance for insects and rodents. The introduction of borax and strategic placement of metal lathes can assist as pest deterrents.

The most important parameter of this style of construction is to protect the bales from moisture and rodents. The bales must be kept dry from condensate water while permitting water vapor to breathe from the interior. If the bales become moist their insulation capacity degrades and they begin to decompose rapidly. To protect the bales from the wet weather, the base of the walls should be elevated on an impermeable foundation, and roof overhangs should substantially shield the walls. Water vapor generated in the structure must be allowed to migrate to the outdoors through properly managed design. Rodents occupying a straw bale house are a grave concern. Walls should be sealed so that these pests do not jeopardize sanitation or the mental health of people residing in the structure.

Straw bale homes have an elegant, organic feel. The thick walls and sound-dampening capacity of the material provide a unique indoor experience. The plaster used to finish the walls can achieve curved and artistic sculptural designs and color possibilities.

Glazing

Glazing refers to building materials that transmit light, such as the glass panes of windows and doors. Glazing allows visibility between the outdoor and the indoor environment and connection to the landscape and nature. Glazing can also be utilized as a component of passive solar design. In addition to introducing solar energy into structures, glazing also has the ability to capture that energy through a process known as the greenhouse effect, similar to that of a car with windows rolled up on a hot, sunny day. The intensity of the sun is magnified and accumulated by the effects of the glazing's capacity to capture solar radiation.

Glazing can also be fixed or operable, offering an additional portal for relationships with the outside

environment. Opening doors and windows encourages the ambient noise of nature to penetrate the built environment. These operable portals can also aid in climatic control through the process of air convection and circulation.

The amount of glazing necessary to accomplish passive solar heating without being offset by the resultant heat loss from the poor R-value of the glass is a function of the design and the solar exposure on the site. There are many examples of housing designs with too much glazing that overheat in the summer and are difficult to heat during the winter nights.

The various materials used in wall construction can be utilized based on the orientation. As an example straw bales could be utilized on the northern walls of a structure, and glazing and cob could be employed on southern aspects. The northern straw bale walls would offer higher insulation value, while the southern walls would provide for passive solar potential with natural daylight and incorporation of the thermal mass. Within the same wall multiple materials can be utilized to augment the performance of the structure. As an example, the base of a wall can be constructed of stone, and other materials can be incorporated at higher elevations. The rationale for this combination would be to use the heavy, moisture-resistant materials in the lower portions while utilizing lighter materials with higher insulation capacity farther up the wall where there is greater protection from water.

The Hat

Structural capacity of the roof must be considered in terms of climate and usage. The roof should be functional to bear the weight of any potential snowfall. Other possible loads such as rooftop gardens or solar energy systems should be considered so the roof has sufficient structure to withstand the weight.

With regard to the pitch of the roof, it is once again necessary to consider the potential usage. The slope of the roof can be designed to maximize solar exposure for solar energy equipment. The angle of incidence that is chosen should be based on latitude and seasonal variances in usage. Slope is defined as the rise in elevation over a run of a horizontal length. A flat roof rises zero for every foot of horizontal plane, while a roof that rises twelve inches for every foot of horizontal forms a forty-five-degree angle roof in relation to the horizon. A shallow, nearly flat roof is ideal for rooftop gardening, but roofs that rise more than nine inches for every foot are difficult to work on without staging.

The distance that the roof overhangs the wall on the building is also important. This brim of the roofing system can be extended so the walls and foundation are protected from rainfall. This brim can also be extended to help provide shade from the solar heat in the summer months when the sky is high overhead. Conversely the brim should not be extended beyond its structural limitations and should not act as a curtain to prevent the low winter sun from fulfilling its potential to heat the structure.

Roofing systems can be considered metaphorically as the skin and bones of the building. The skin provides a membrane that prevents the transmission of moisture to the interior. The skin can also serve to block the sun's energy and shade the interior of the structure. The bones are the structural elements that sustain the weight of the roof. The structural capacity of the design is based on the usage of the roof space as well as potential regional environmental weather factors such as rain and snow.

Our favorite roofing material is metal that we've acquired from a prior utilization. Holes in the ridges seldom leak and can be tarred if sealing is necessary. Used roofing is a valued commodity and is particularly useful for outbuildings. Roofing can also be purchased and is available cut to length. Clear roofing is also available, which is particularly useful for daylighting and greenhouse applications. The fiberglass woven material is less expensive, though its fragility and opaqueness increases with solar exposure. There is also a UV-resistant, clear, pliable plastic product available, but at a significantly higher cost.

Climate Control

An indoor climate can be controlled without fossil fuels through intentional design methodology. The concept of passive solar construction provides the opportunity to integrate the building principles, concepts, and materials into a heating and cooling system powered by the sun. Glazing, thermal mass, convective loops, and insulation are all elements that can be coordinated through design so the heating and cooling goals are subsidized passively without inputs of daily energy.

A passive solar design is dependent on the supply of solar energy, which varies according to climate and latitudinal distance from the equator, which affects seasonal variability of the sun's declination. The intensity of the high summer sun can be overwhelming, while the low winter sun provides scant energy that is difficult to capture. These extremes are the parameters of the passive solar system of building heating and cooling that must be addressed.

Ventilation allows control of the airflow to different zones within the structure. This flow of air is designed to improve the temperature control of the structure. Ventilation can also be utilized to introduce or exhaust air with the intention of drying or moistening a zone or sector of the structure. Woodstove fires can aid in reducing indoor moisture throughout the year. The environmental conditions of wind direction and velocity can also be utilized in the building design to augment the heating and cooling goals for the structure.

Thermal mass is a term utilized to define a material's ability to retain heat. Some materials, such as metal, convect and transfer heat rapidly, while other materials, such as stone, soil, and water, will predictably retain thermal energy to various degrees. In the microclimates of our structures, the ability of a material to store thermal energy can be utilized to control the indoor temperature of a structure. On the regional scale thermal dynamics can be witnessed in nature through what is known as the lake effect. Because of the high thermal retention capacity of water, temperatures of landscapes situated in proximity to the ocean and other large bodies of water are moderated by the significant mass of the water.

Shade can be a component of structural climate control, whether provided by neighboring buildings, geologic formations, or vegetation. Structures built in forested areas are insulated from the extreme sunlight of summer, though the danger of falling trees and roots encroaching on the foundation can necessitate their removal. They are also susceptible to moisture problems because dry air cannot circulate to diminish the degradation caused by continuous high levels of humidity. Shade can perpetuate consistent moisture and contribute to structural decomposition. Shade that is welcomed in the summer can be a disservice during the winter months. Generally deciduous shade trees should be located due west of a structure so that their shade will be most effective in late afternoon on the hottest days of the year. For instance, there is a shade tree of that nature planted to the west of my aunt's art studio where I am at this moment writing and residing. The sugar maple was planted in 1971, the year I was born, to serve the purpose of shading the structure. The tree was planted less than ten feet due west of the building and is now approximately eighteen inches in diameter. Its roots are beginning to encroach on the minimal foundation of the structure, and the limbs have required pruning to prevent abrasive damage to the roof.

Passive solar design is also a function of the site's aspect and the structure's orientation in relation to the tilt of the earth and the variances produced by its annual orbit of the sun. The building site can be sloped in any direction. For maximum solar exposure in northern latitudes, south-facing slopes are ideal. This allows for the maximum daily allotment, as the terrain is angled to accept the solar radiation. For instance, a house constructed on a north-facing slope would still utilize a design with a southern orientation to maximize the solar energy gained from the southern direction. This process of structural orientation is obvious for builders who are connected to the landscape and the climatic seasonal fluctuations. Unfortunately, one of the legacies of the standardization and suburbanization of American housing is the disregard for structural orientation, which was replaced with the notion of orientation of housing based on visual curb appeal.

By integrating an active heating or cooling system into the structure, we are not completely dependent on environmental conditions to provide optimal indoor temperature conditions. Thermal mass concepts can be utilized with active heating and cooling elements. Masonry stoves and thermal mass rocket stoves provide efficient heating options. Water can also be used as a fluid, transferable conductor of thermal mass.

Green versus Natural Construction

The enormous, detrimental pollution to the environment caused by the conventional building industry has inspired a consumer-driven demand for alternatives. There are many building products and methods that are now being marketed as environmentally friendly. There are also many certification programs and ratings used to evaluate materials and entire structures.

That being said, the industry's response to consumers' concerns regarding building products has been similar to the tactics employed by other industries under pressure for their environmental perpetrations. Instead of revamping their production strategies and methods, companies have chosen simply to repackage the products with a greener image. There may be noble intentions with programs such as LEED certification; nonetheless, there are many shortcomings. The certification programs tend to focus on recycled materials regardless of the location and fabrication process of the material. While the buildings may be designed for performance standards in energy efficiency, there is no limitation for consumption. In fact, instead of conservation these standards encourage further uninhibited consumption. For example, a five-thousand-square-foot seasonal dwelling on a lakefront for a single family was certified to "green" standards in my town. These paid-participation programs sponsored by the building industry are designed to assuage consumer concerns while maintaining the status quo.

By using industrial high-technology building products, we are continuing along a path that abandons our roots and separates us from Mother Nature. By creating nests for ourselves and our families with toxic products, and relying on air exchangers to maintain adequate indoor air, we are losing our connection to nature. These ground-based spaceships shield us from the environment in a square, stifled atmosphere of sterility. By ignoring our role and capacity to create natural buildings in our environment we choose a path of isolation, exorcising our souls from our home.

Building Profiles

Each structure at D Acres was designed and constructed to meet specific needs that were prioritized at their respective junctures in organizational history and so reflect many styles and building methodologies. The methods employed were based on a combination of factors, including available materials, skills and number of personnel available, budget, knowledge, and experience, as well as our evolving building philosophy.

Community Building

The Community Building is to date the largest construction project we've undertaken and serves as the base of our operations. It is largely a conventional construction project, involving a high number of subcontractors and professional services, though with a unique design and purpose. The building is a foundational element of the farm system and provided the platform to begin the expansion of our community service operations and to continue experimentation with construction. Importantly, it serves as the public's main access to D Acres.

The Community Building was originally conceived during a two-week Design Build Course at Yestermorrow in Warren, Vermont. I attended the course to accompany my father as he pursued a design for his retirement home. Since I did not have a specific property or purpose for my design project, I chose an imaginary construction project that would be located in a village I had recently visited in Costa Rica.

The community building that I designed for the village in Costa Rica was intended to be a multi-purpose community center. When I had visited the small town as a volunteer for the abutting national park, I spent time playing soccer and observing the daily activities of the villagers. The villagers were mostly indigenous people who had been relocated to this concrete ghetto when their homeland was designated a national park. I noticed several areas where I felt the standard of living could be improved through collectivized infrastructure.

During the days the women spent large amounts of time doing laundry and preparing meals and had to travel downstream from their homes so they did not contaminate their drinking water with their laundry, cooking, and waste. At night, despite the proximity of a hydroelectric power plant, the village was barely illuminated except for the glow of the local bar, which attracted a majority of the men in town like moths to the light. Due to the design of the village the people were confined in a cycle of consumption without a design for production and efficiency.

The idea behind the community building was to economize the design by centralizing essential infrastructure so it could be utilized for the common good. The goal was to meet the diverse and collective needs of the villagers within a single structure. It was designed to be a hub of activity, with a particular emphasis on water usage. Water would be supplied to this central location, where it could be solar heated for cooking, bathing, and laundry. The greywater would then be treated on-site for reutilization in aquaponic and irrigation systems. The building would also house a central toilet facility in which toilet waste and kitchen compost would be aggregated to provide methane gas for the kitchen cooking appliances. By centralizing these facilities an improved standard of living would be achieved without the costs of installing a municipal sewer system. This facility would also provide a venue for meetings and production of cottage goods.

The design of this structure also included solar water panels, a commercial-scale kitchen, water treatment facilities, and a playground. The goal was to offer immediate, remedial measures necessary to slow water pollution while also providing an investment in the village's future. Based on concepts of collectivism, this center could provide the forum for the harmonization of the village and the creation of a sustainable economy.

When we moved to Dorchester in 1997 the facility we initially occupied as our center of operations was the Red Barn. This structure was originally constructed in the early 1800s, though Uncle Delbert had built additions, including a silo to augment its agricultural capacity. This post and beam structure served as a stable for horses, oxen, pigs, and a tractor. The upstairs still contained ample evidence of the generations of hay stored in the loft.

The original barn walls were packed with sawdust. The sawdust had settled down over the years, so the top section of the wall cavity was empty. The middle of the walls had become a haven for nesting mice. The mice created an odor and a nuisance that did not induce rest or relaxation. The sawdust on the bottom of the walls was susceptible to high humidity pressures that led to mold and mildew.

When we arrived we began the process of cleaning and evaluating the structure. We hired a neighbor to help us rewire the building for safety and residential convenience. We reconfigured some interior walls and corrals so the stable area became a kitchen/living area with computer and phone service. We installed a woodstove and insulated as the structure allowed. Along the woodline, 165 feet from the back door, we built an outhouse, which was our first tree house.

We soon discovered the inadequacies of the Red Barn as a community center. Water was supplied via a pitcher pump connected to a hand-dug well. The pipe froze solid and limited availability during the winter months. When water was available we cleaned dishes by heating a basin of water on the stove and rinsing the hot, soapy water off by spraying the dishes individually with a ketchup bottle filled with clean water. The uneven floor and terrain provided poor accessibility to the structure, which made visitation difficult, particularly for seniors. The indigenous rodent population prevented proper sanitation conditions in food preparation areas. The walk to the outhouse could be perilous, especially during the icy winter months. We purchased dowels of green white birch from the paint stirrer factory in Plymouth and huddled by the meager, hissing fire that the wet wood produced.

As we endured our first winter in Dorchester, we began envisioning a structure designed to meet the evolving needs of the organization as well as serve the inhabitants of the farm. The origin of the structure and subsequent decisions were a series of compromises made by those who invested in the project.

My initial plan was to build a timber-framed, gambrel-style building. The envelope was to be an integration of materials, with a southern-exposed earthen Trombe wall and a northern wall comprised of straw bales. This design would represent an experiment in natural building and result in a farm clubhouse of sorts with kitchen, work areas, storage, and sleeping quarters.

As the idea fermented, other stakeholders appeared with distinct visions and desires for the structure. My father became interested in the possibility of a woodworking shop for usage in his retirement. My mother expressed interest in overnight accommodations for family and friend visitors. I chose to be dependent on my parents as the landowners and financiers of the building project. D Acres became the general contractor for what was designed to be an investment in farm infrastructure. While I used revenue generated

The roof rafters and sheathing are being assembled in anticipation of the metal roofing material.

by the construction project to finance the development of the D Acres organization, the intention was to receive fair compensation for the work that the building construction entailed. This partnership required continual dialogue regarding the design, materials, aesthetic, personnel, budget, and timetables.

My role in the construction process was primarily as the general contractor. I was the principal foreman and carpenter of the framing crew that built the structure. I worked and learned daily, assisting subcontractors and attending to specific details, thus coordinating the project to completion. My objective was to attain the highest quality while controlling costs. My construction experience had been primarily on hotels, condominium complexes, and larger family dwellings, so the scale of the Community Building project did not initially hinder my enthusiasm. But as we overbuilt the structure, spending considerable time framing it in the

shipshape format that I had been taught, the duration and expense of the project grew, along with my personal anxiety and frustration. While I may have intellectually understood the potential difficulties of a project of this scale, I was unprepared for the logistical challenges of orchestrating subcontractors and work crews with varying levels of experience and motivation.

My parents also demanded certain specifications for their investment. They chose to hire a friend as an architect. We also contacted a distant relative, a builder in the area, to be the principal on-site oversight for the project. As other parties became involved it was clear that it would be difficult to arbitrate between the financial and aesthetic considerations of my parents and my own preferences. For instance, my mother vocalized to the architect that she did not want a structure that resembled a barn, whereas I had no opposition to that style. As the on-site participant who was investing with sweat

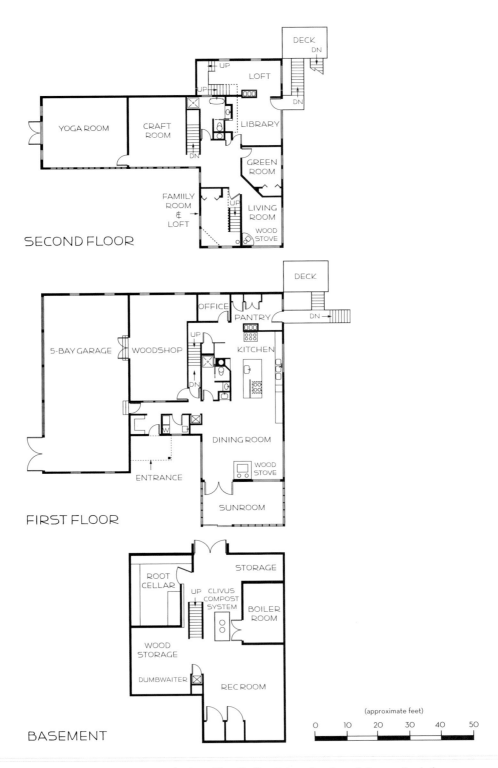

SECOND FLOOR

FIRST FLOOR

BASEMENT

(approximate feet)

The floor plan of the Community Building was designed by Shelley Pripstein, though improvised changes were made by the onsite construction team.

equity, my views were often in conflict with the goals of the financial investors.

As we proceeded with planning, the structure of the building grew conceptually. A library, commercial kitchen, root cellar, and storage area were conceived to meet the many needs that were evolving at the farm. A blacksmith area and screened porch were initially proposed, though these zones were discarded as being impractical to the design. The architect provided blueprints that offered elevations for all four sides and top views for all three floors that were scaled to indicate window and door openings.

From the moment the initial footprint was marked in the field there were questions regarding the scale of the project. For a small family the structure would be a proverbial McMansion. During the three years spent on the building site, the immensity of the structure haunted me. I debated whether the structure would be utilized to its full potential. Would it sit idle, with usage limited to a select few? Was the building just a reflection of the excesses of its era? The notion that the building was a "white elephant" made me question the legitimacy of my efforts. I was not relieved until the building began to be utilized for its organizational purposes.

Cost/Benefit Analysis

Our success in completing the Community Building was due to financial resources, perseverance, determination, and able, organized assistance. The building construction costs were immense, both in terms of cash and energetic footprint measured in materials, equipment, and people power. After three years Betty and Bill spent over five hundred thousand dollars on the project. Large portions of this money were spent on subcontractors and materials; the D Acres LLC administered the funds as a general contractor. A portion of this was used to support our burgeoning educational programs and operational costs as we struggled to build the structure and develop a farm system.

In addition to the dollars that were invested, the physical and mental costs of this project were enormous. The daily grind of twelve-plus-hour workdays, combined with the inexperience of our crew, led to increased anxiety. Uncertainty regarding the future of the project also affected our positive engagement with the structure. The farm crew's enthusiasm was stifled by usage of conventional, toxic materials and doubts regarding whether the structure would be used for public or private family purposes. The crew was reluctant to compromise the pure vision of a sustainable, benign construction that would be shared for collective utility. The worksite also produced several interpersonal and professional conflicts among subcontractors, volunteers, and hired hands.

But the benefits of this infrastructure have been substantial. The Community Building has become a symbol and is regionally recognized. The building has inspired thoughtful conversation and designed implementation based on its model. By articulating both its successes and failures through open houses and group assessment, we have provided an example for future deviations and variations. Through its designed accessibility and openness, the multiplex serves the community. While the street façade of garage doors lacks curb appeal, this ready access provides an open portal to the flow of people. The ample doorways and halls provide the necessary width for two-way traffic.

The space serves the on-site residents as well as the broader community as a hub that concentrates opportunities to access knowledge but also as a rural support and social networking mechanism. The building promotes aspects of the camaraderie offered historically by rural Grange societies and other civic organizations such as the YMCA, Masonic Lodge, or Rotary.

For the on-site residents the Community Building provides all the luxuries and conveniences of the modern era in a shared space. The kitchen and bathroom are spacious and sanitary. There is electricity and running water, as well as phone and Internet access. The building meets all of our personal and business needs without excessive duplication of infrastructure. Instead of private bathrooms and individual phone lines these elements are shared so the overall energetic footprint is reduced. This design depends on the acceptance that sharing requires negotiated communication for all parties to remain satisfied by the arrangement. By sharing the benefits of the infrastructure we are also committing to perpetuating it for the future.

The Community Building is also the hub for our food operations. We store our seeds in the dry, climate-controlled closet upstairs and start our first plants of the year in the basement with the soil stored from the prior fall. As the season progresses we use the kitchen to prepare and preserve the harvest. The root cellar, basement, upstairs closet, and attic space are designated food storage areas. Meals throughout all four seasons are prepared daily for the residents and visitors in this zone of interaction.

The value-adding made possible through the integrated functionality of the building provides for our livelihood. The woodshop and the food production infrastructure reduce overhead expenses while also generating income. These systems help transform low-value natural resources into products that can be sold for a higher price than can be realized in the commodity market. By raising the value of the commodity the farm system derives greater income in relation to the resources expended.

The upstairs craft studio and library are both examples of ongoing proliferation and productivity

The library is a stable and growing resource for visitors, residents, and community members. We have over twenty-five hundred volumes on subjects from permaculture to politics, spirituality to slaughtering farm animals.

provided by public space. The resources for painting and sewing that have accumulated through thrifty frugality and generous donations offer endless opportunities for creativity in the craft room. The library has grown through gifts and diligent acquisition, from the books I initially arrived with to its current collection of over twenty-five hundred volumes, which are catalogued by community volunteers.

The building provides an activity nucleus for our farm activities. By being the hub of food and facilities, the building serves to encourage interaction among the residents and visitors of the farm. The structure has responded well to the heavy traffic and use patterns and is cleanable and durable with minimal maintenance requirements.

The Community Building was constructed and designed to allow us to both demonstrate and live our values. Its roof pitch allows us to mount solar panels and have easy access to the electrical utilities. Rain falling on the roof accumulates and drips from the eaves in large volumes and is then channeled and collected in ponds. The height of the foundation walls protects the building from the splash of rain drips and osmotic wicking of seasonal humidity, thus preventing moisture penetration to the sills. The roof overhangs on the second floor allow three-season sun access while preventing midsummer blazing from the glazing. While curtains and operable frames are necessary to maintain climate control year-round in the building, the system is readily manageable. The building's orientation on

In 2006 the NE Permaculture Gathering utilized the many spaces available on-site, especially around the house. While there is a maximum capacity of the numbers the farm can host, we can welcome large crowds and have seen up to 250 people on-site for various special events.

the east-west axis allows for the collection of solar energy along its southern side. The overall window plan provides a maximization of sun from the southern and eastern exposed directions. The lack of glazing on the northern and western elevation protects the structure from the heat of midsummer's sun while minimizing heat loss during the winter months. The west face of the structure collects the wind through the garage doors on the first and second levels, which introduces fresh air ventilation during the warm summer months.

Now fourteen years old, the space continues to evolve. The yoga room, which was originally desig-nated for storage, has become a model for space adaptation. By carpeting the floor and painting the walls with an attractive mural, we made the space inviting for a wide variety of usages. We have hosted presentations in it for education, music, and entertainment, and it is regularly used by overnight guests ranging from budget travelers to Boy Scout troops. Practitioners of yoga, Reiki, and massage frequent the room, and it is also utilized as an over-flow space for children during community events.

We have plans to expand the capacity of the struc-ture in the future by identifying the microclimates that the structure can provide, including an icehouse in the shade of the northern side of the building. On the southern side we have implemented plans for a long-awaited solarium, utilizing the thermal mass of the rock wall and the domestic heating system to augment our growing season and passive solar heat-ing potential. We also have plans to utilize the build-ing's greywater to supply a living machine designed to clean our water through a series of sediment and biological filtration processes.

The basement was designed to be an accessible working space, so we constructed a bulkhead that opens to grade on the northern side of the structure. A wide stairwell with handrails was constructed to access the first floor. The basement quickly became a zone for starting plants in the spring. Another section has become devoted to musical equipment and recreational games such as foosball and Ping-Pong.

Recently a bicycle-powered washing machine and a silk-screen printing station have been established. The basement also houses the root cellar, heating plant, and wood storage. There is a chute to load wood via the garage and a dumbwaiter that can hoist wood to the first and second floors.

Cob Oven (2000)

As the construction of the Community Building drew to a close, we were excited to begin experiment-ing more with alternative materials and designs. The first project we undertook was the cob oven, which shifted our focus to locally sourced, natural materials. The building of the oven coincided with educational workshops, meaning that the labor was provided by nonprofessionals instead of subcontrac-tors. Our ideas and knowledge were enhanced by experiences with the Gaia Association Ecovillage in Navarro, Argentina, where we were instructed in methodologies of cob construction and provided with designs for cob ovens.

The hands-on experience at Gaia immersed us in cob technique. The residents had simplified and improved the construction process with several innovations. They utilized the subsoil directly from a pond dug on-site, which had sufficient quantities of clay and sand to make an adequate composi-tion for construction. To consolidate the materials we operated a walk-behind rototiller on a round concrete slab approximately twenty feet in diame-ter. By adding the loads of subsoil onto the slab to a depth of four to eight inches and spreading the hay on top, we utilized the mechanization to mix the native field hay into the composite. While water accelerated the process of thoroughly mixing the materials, a sloppy mix would not be stiff enough for the purpose of wall construction. Dry materials were more difficult to adequately mix and lacked cohesion. Through experimentation the percentages and order in which water, hay, and subsoil were combined were adjusted to maximize the speed and fluidity of the process, while also ensuring the qual-ity and functionality of the material.

Josh demonstrates proper cob application technique while constructing the sidecar to the cob oven.

Though our work was geared toward the construction of a building, Gaia also provided in-depth exposure to other cob constructions, including a wood-fired cob oven and a cooktop. The oven design integrates both metal and earthen materials. A fifty-five-gallon drum is transformed into an oven by placing the drum on its side, fashioning shelves, and fabricating a door on one end. The drum is then attached to a skeleton frame that elevates the drum above the firebox and allows the combustion gases to rise around the drum and out the chimney. This structure provides a form so the combustion energy can be retained within a shell of cob mass that retains thermal energy. A shield is inserted above the fire to help deflect heat to both sides of the oven and mitigate unequal heat distribution on the bottom of the unit.

The cooktop is similar to cooking on an open campfire except that it has two principal advantages: efficiency and air quality. The cob oven cooktop is more efficient than metal stoves and open flames because the cookstove accumulates thermal energy in the construction material, which is then released throughout the cooking process. Instead of heat that is rapidly conducted away via metal or air, the thermal energy is conserved in the cob's mass. The cookstove design is also superior to open fires in terms of air quality because the combustion gases can be directed away via a chimney.

The cooktop is devised by forming the cob material into a shape that allows for pots, skillets, or woks to be inserted directly above the firebox. The flames from the burning material lick the bottom of the cooking vessels as the combustion gases make their way out the chimney. Leaping flames create high, boiling stir-fry temperatures, and coals distribute simmering warmth. It is important to note that the outside of pots heated by direct wood flames will become extremely difficult to clean. We designate several pots to the cob cook stove and simply clean the inner surface of the pots to avoid the significant challenge of cleaning the outside of the smoked pots.

Cookstove design should consider versatility, longevity, efficiency, and usability. The size of the firebox and corresponding cookware is important, as are the working heights for loading the stove and stirring the pots. Particular care and attention should be focused on a means to lower pots into the flames while supporting the weight of heavy ingredients. For our design we inserted rebar across the firebox at six inches to support the pots' weight. Because of the overall depth of the ten-inch firebox, flames can circulate around the pot, and adequate cob thickness of three inches can be attained on the top of the firebox. To form the top side of the cookstove we cut old wire fencing to within one-inch accuracy of the width of the primary cookware we intended to use. We covered the wire with cob to a depth of about three inches and formed the top side directly to the primary pots. The cookware to which the cob is formed should be coated with vegetable oil or a layer of newspaper so it is easier to remove once the cob dries.

Open-Sided Building (2001)

The Open-Sided Building was created in response to our evolving storage needs and was the first construction project to incorporate both forest and recycled materials. The garage of the Community Building was packed with materials from the construction process. This mess of materials threatened to disrupt the usability of the garage area. We were also facing increasing pressure to meet the hay storage needs of the oxen.

To address this storage crisis we chose to use part of the problem to create a solution. A large quantity of the building materials in the garage was pieces of steel that had been mistakenly ordered during the deck construction process. The heavy, awkward steel pieces provided the basis for an open-sided storage building design.

The shed was designed based on the length and quantity of steel beams, and the sixteen-foot lengths of wood that are commonly available. We wanted to divert minimal resources to provide a simple, durable, covered space.

We took the open-sided approach because of the opportunity provided by the structural integrity of the bearing posts and the weather resiliency of the extended roof system. The post and beam system allowed us to avoid walls so we could maintain easy access to the storable items, particularly long lengths of wood.

Over time the pressure for wood storage has decreased and the convenience of this building for other materials such as hay and appliances has increased. As the building purpose has evolved, more walls have been added, which increases storage capacity for hay by allowing more bales to be packed into the space for dry storage.

In addition, the warm southern side of the structure has been modified with a greenhouse expansion we refer to as the "Big Cold Frame." The goal of the Big Cold Frame is to increase food production while lowering the fossil fueled inputs necessary to derive that production. The southern wall has a three–post and beam system that bears the weight of the roof, and reclaimed sliding glass doors inserted to maximize solar gain. The roofline of the addition was framed using long roundwood rafters. The roofing material is a clear, supple, UV-resistant material with superior light transmission and durability compared to the

Since enclosing the Open-Sided Building, we have been able to increase our hay storage capacity. The building continues to be used to store wood and other weather-sensitive items such as woodstoves and puppets.

cheaper, fading, brittle fiberglass roofing product, which rapidly grows opaque. The solarium-type space allows ample sunlight, protection from heavy winds and rain, and additional warmth for seedlings in the spring. This space is particularly useful because of its proximity to the main house as we transition away from the energy-intensive fluorescent light seed-starting option. In addition to housing hundreds of seedlings as a spring nursery, the space is also an adequate work area. Although it primarily serves as a potting shed in the spring, the sunlight and warmth generated in the space can be useful for various activities and

projects year-round, such as drying beans, hardening squash, storage, or accommodating a cabin-fever-inspired group meeting.

The expansive roof of the Open-Sided Building provides a tremendous opportunity for water catchment and solar energy. The square footage collects a tremendous amount of water during storms. During heavy downpours the water cascading from the gutters can form a column 1½ to 2 inches in diameter. It is necessary to design a diversion to retain this concentration of water so it can be maximized in the system. The rooftops have maximum exposure to sunlight, and solar panels require minimal

Beth works with volunteers to pot up seedlings in the Big Cold Frame on the south side of the Open-Sided Building.

maintenance, with the exception of snow removal. By placing future solar panels on the rooftops we save precious footage on the ground.

Ox Hovel (2002)

The Ox Hovel was created to answer an urgent infrastructure challenge. When the oxen were offered to us I had just purchased a ticket to leave for Argentina. We had less than two months to build adequate housing for the animals' needs, so we sought the help of our neighbor, Jay Legg.

Jay brought his bulldozer and cleared the site of the topsoil. Then we imported sand to level the terrain. The sand was watered until it settled, and

we began forming an Alaskan slab foundation. An interior retainer was added, and we dug a two-foot-wide channel one foot below the sand base. We then formed an exterior perimeter so that concrete would fill the channel and a floor to a depth of six inches. A grid of rebar was tied into the forms using sixteen-inch spacing.

This Alaskan slab foundation has not degraded structurally. The perimeter continues to bear the weight of the building, and there is no noticeable deterioration. The block wall installed along the length of the building has prevented the sills and structural elements from exposure to direct contact with water.

The Ox Hovel is a great example of sturdy boots and protective hat. Snow and water shed dramatically off the building. The window glazing on the south side of the building is at a height that reduces solar gain in the heat of summer, yet allows the low-horizon winter sun to filter in nicely.

Once the footprint of the Hovel was established, we built two courses of block along the southern and northern walls. Stucco was applied to these knee walls for aesthetics, strength, durability, and moisture resistance. The wood for the shelter was then sourced via Jay's new sawmill, which he had recently purchased. He also provided us with the framing materials that we needed for the project. With the help of a novice builder interning at the time, I framed the structure in two weeks. After we finished the siding and roofing we were able to welcome the oxen to their new home on schedule. During the winter the first-floor walls were insulated with plastic bags and containers.

The roof style was a gambrel with a shed extension on the north side. This shed was designed for wood chip storage so the materials could be readily unloaded. The side doors of the structure allow the dry chips to be forked into the animals' bedding area. The upper area of the structure is used for hay storage. The extended roof in the front of the building matches the aesthetics of both the Red Barn and the Community Building and provides a perch for the hay pulley system. This prow offers protection from the rainfall along the gable end of the building, creating a sheltered niche for the structure, as well as an opportunity for storm spectating.

The footprint of the main gambrel structure is approximately 14×20 feet, though the second floor of the structure was extended to increase the usable space and protect the foundation. The hayloft is uninsulated and stores approximately 150 bales.

The windows placed high along the southern wall are designed to prevent the oxen from accidentally breaking them with their horns, as well as to maximize the solar potential. The windows' placement and roof overhang combine to encourage the sun to penetrate the building during winter when it is low in the sky, while shading the glazing during summer when the sun is high.

During the winter the building is noticeably warmer than the outdoors. Protection from the wind combined with heat generated by the oxen creates an environment that is generally 10 to 30 degrees warmer than the outside temperature. During the summer the well-ventilated and shaded building provides a cool environment for the oxen.

In designing the structure we sought to maximize the viability of the structure beyond usage with the oxen. In the event that Henri and Auguste would not be suitable and we chose a mechanized route, the rear doors are also accessible to wheeled equipment.

G-Animal (2004)

The conception of this multipurpose facility sprang from the realities of our climate and was undertaken with education and experimentation in mind. Greenhouses are essential to extending the growing season at this latitude. Faced with the windy, slippery, and frozen landscape that prevails half the year at our site, we pondered the possibilities for a greenhouse that integrates animal husbandry and plant food production. By maximizing resources we hoped to conserve human labor hours and improve the quality of life for humans, animals, and plants. Many traditional homes in New England are a combination of house, garage, and barn all under one roof. Morphing that idea, we designed G-Animal to be beneficial to both plants and animals that required daily attention.

The energetic cycles of heat, air, water, electricity, biology, art, and construction materials are all integrated into the design and performance of the structure. The sun provides energy to the structure through the glazing's salvaged glass, which extends the lifetime of the embodied energy invested in its original production. The massive earthen walls of the structure provide insulation as well as thermal mass, which moderates the extremes of temperature. Animal manure is composted and reused in the garden beds. Even the air is recycled within this system, as the oxygen that the animals breathe is then exhaled as carbon dioxide, which feeds the plants, who in turn produce more oxygen.

The structure is designed to collect and conserve the energy required to meet our goals. By stacking functionality we acquired multiple benefits from the investment. We designed G-Animal so that these needs are met adequately, using redundancy where appropriate to strengthen and augment the overall system. As an example, heat captured by the greenhouse is complemented by the body heat generated by the animals.

Our goal was to site the building in a location that would maximize its utility. We decided that geographic proximity and solar exposure were vital, though we did not want to detract from the existing landscape. We chose not to connect G-Animal to the existing building because of the potential fire hazard, although placing the structure close to the Community Building ensured that it would be easy to visit daily. The space to the south and east of the Community Building was well drained on a gentle north-facing slope. By considering the incidence of the setting sun during the summer solstice we sited the building so the greenhouse is not obstructed by the shadow of the Community Building.

The southern wall of the animal husbandry section is a half moon that collects the sun's energy throughout the day. The greenhouse roof sweeps away from this wall in the three southern façades of this semicircle. We chose to excavate below grade so the overall height would be minimized while still providing two floors for animal husbandry. The two levels attain a minimal though adequate floor-to-ceiling height, which is comfortable for work and conserves heat.

By observation we concluded that, while the site's soil was well drained, during intense rain or spring snowmelt the water would flow across the ground toward the building site. As a precaution we addressed these concerns during design and implementation to keep the structure dry. Before construction began, we dug a trench swale to mitigate runoff and push it around the site. This site work was necessary to prevent the potential saturation of the building foundation envelope.

We chose to build two distinct foundations for the structure. The foundation of the inner animal husbandry area was constructed as a slab below grade. The floor height was excavated to maximize the consistent ground temperature and to minimize the overall height of the building. While this interior slab mirrored common traditional methods to mitigate materials and excavation costs of footings, the perimeter foundation was a more severe departure from conventional technique.

The perimeter foundation of the greenhouse was poured onto a rubble trench foundation. We dug a

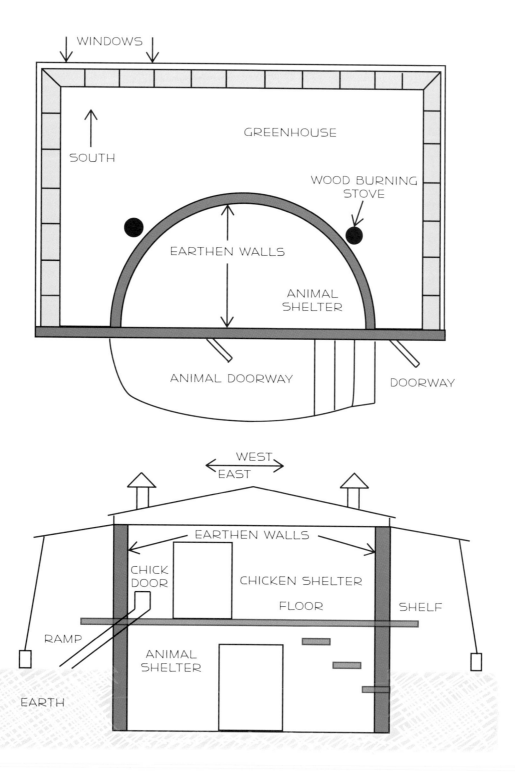

The original digital sketch of the G-Animal structure, while very basic, allowed for the vision to be thoroughly discussed for size, siting, and materials.

twelve-inch-wide trench approximately three feet deep around the perimeter so the water would flow around and downhill from the building. To ensure the flow of water we placed a six-inch perforated pipe into the trench, then packed the trench with rubble consisting of squared chunks of stone between four and eight inches in diameter. The rubble-and-pipe combination prevents water from becoming trapped in the trench.

On top of the rubble trench we formed a stem wall eight inches in width that fluctuated between eight and twenty-four inches in height based on attaining a level grade. We then poured concrete up to a nominal height. This created a surface-level beam around the perimeter of the greenhouse. The concrete provided a dry and rot-resistant base for the greenhouse construction. All the concrete was reinforced with steel rods to maintain its integrity through freeze and thaw cycles.

Part of our intention with this project was to experiment with building techniques so we could determine options for future projects. After we poured the foundation of the animal husbandry section we began building the walls from stone and mortar. This impermeable base rose about a meter in height and was backfilled with soil in the greenhouse section. As the walls grew past the height that would be vulnerable to rain splash and ground moisture, we began experimenting with two styles of alternative construction up to the first-floor ceiling. On the north wall we chose a cordwood style because of its insulating capacity and resistance to moisture, while the southern wall would be cob. Cob's thermal mass made it ideal for this location

Adobe bricks were formed to build the bulk of the G-Animal walls. They proved to be labor and material intensive and dependent on weather for drying, yet they became solid building blocks.

Recycled glass windows and roundwood sourced on-site help to complete the greenhouse element of this multifaceted structure.

because of the solar exposure. After obtaining sufficient height we began to install the floor of the second level. After the floor was installed we began construction of the second-floor walls. We utilized adobe blocks that we had dried during the first phase of construction. Though these blocks stacked quickly, our supplies were limited, and the cold of winter was approaching.

While ice and snowstorms delay every construction project, these weather conditions are particularly severe for earthen construction. The mixing and the drying of the material require temperatures above freezing. Also, earthen construction that is left exposed to heavy winter rains is salvageable though it is severely eroded by the experience. Consequently,

The interior garden beds allow for season extension of vegetable production. A late fall crop of lettuces and kale is cared for, even with extra protection from frosts, by the use of a movable cold frame.

the immediacy of finishing the walls became apparent as the cob began to freeze in our hands.

As the time-sensitive nature of the project increased our urgency, we searched for a hybridized solution to speed wall construction. The idea to use cobwood was born from this urgency. We found this technique sped the process with the lowest embodied energy compared to cob, adobe, and cordwood.

G-Animal's roof is important for passive solar and drainage considerations. Its angle of incidence determines solar energetic gain. It also consolidates precipitation deposited immediately at the drip edge or via melting, falling snow. The roof overhangs the building strategically to keep the water away from the walls and foundation, shedding snow and water to the perimeter, especially on the north side. The vertical greenhouse glass maximizes the winter sun's trajectory, while the horizontal greenhouse roofing captures the summer sun. The metal roofing of the animal unit is pitched to the north. To minimize the height of the structure while preserving headroom in the second floor, the pitch of the roof is minimal. The roof was also extended six feet on the back side to provide for a covered stairwell and further protection from precipitation.

Our intention for G-Animal as an educational site was not only to create a functional building but to evaluate different building materials and methods and to practice their associated techniques. The varying amounts of labor, cash outflow, and natural resource use required for each were recorded, and we are still in the process of evaluating their performance. Evaluation requires time to observe how materials insulate and serve as thermal mass, as well as how they endure. There are many unresolved questions. Will the time and expense saved by not digging a full foundation be worth it when compared to the longevity of a monolithic foundation? How long will clear plastic roofing last compared to metal? Will there be shrinkage and air gaps around the cordwood installed in the cob? Will the cob be affected by severe changes in temperature and moisture levels? Thus far the

animals love the warmth of the building, but there have been high moisture levels in the second-floor chicken area, possibly as a result of recent warm, wet winter weather.

We built this structure with the intention of its being usable without major renovations for at least fifty years. We expect minor maintenance will be required annually. Some predict the concrete in the foundation will remain intact for at least four hundred years. We hope that the building can be maintained for the lifetime of the foundation.

Cost/Benefit Analysis

Building and utilizing G-Animal confirmed some common sense expectations. Commercial materials such as concrete, rough sawn lumber, and metal roofing are conventional because they save time. These materials require capital and unaccounted natural resources to produce but conserve physical labor on-site. Earthen construction requires significantly higher resources of time and sweat equity but is accessible for less cost and a reduced resource footprint.

The costs and benefits of the project can also be measured in social terms. While the project provided a practical, hands-on experience for the crew of interns living on the farm that summer, pushing to meet the time frame for construction completion tested their volunteer spirit. The constant effort required to meet the constraints imposed by the weather was difficult to maintain for a temporary crew with little investment. The slowness of the cob walls' growth and the repetition of the building process drained the building crew's energy, while the rapid progress on the wood framing and roofing provided immediate gratification.

If the volunteers had been paid an hourly rate the costs would have been immense. A total of 1,602.25 hours were spent on the project. We spent 291 human hours digging the foundations and building the concrete forms. Including acquisition of the stone, we spent 214 hours on the masonry for the base of the walls. The time spent dancing in the cob, building adobe blocks, and raising the earthen walls amounted to 652 hours of moving mud. The wood framing, finishing the details of the wood-stoves, and framing of doors and windows ended up taking 392 hours. The actual roof took less than 50 hours of labor. And there are always more details. Experienced construction workers might have built the structure more quickly than students and volunteers, but the cost of wages would have been higher.

There was also a capital outlay for this structure. We spent almost four thousand dollars on materials, including concrete, masonry cement, reinforcing bar, cob materials, the cob mixing drill, foundation materials, milling of the lumber used for framing (sourced on-site), and roofing materials. About 75 percent of the materials' cost was for the foundation because we could not arrive at an alternative to concrete. But the high investment for this project was deemed worthwhile because it served as a hands-on construction learning tool for over fifty workshop participants, nineteen interns, and four staff persons at the farm that summer.

Subsequent years of use have proven its worth, as G-Animal has extended our growing season, conserved energy, and enhanced our lives through its design. The thick earthen walls moderate the temperature in the structure, preventing extreme spikes, particularly during the beginning and end of the growing season. It has also proved to be a fantastic location for seedlings in the spring. Plants that

The G-Animal in springtime is bursting with seedlings ready to be exposed to their first natural sunlight. Protected from any extreme elements, they gain strength before being potted up.

are started inside are brought into the greenhouse to harden and adapt to the natural sunlight. The structure provides the intermediary space as plants are transferred from the lights of the basement to the field plantings. On a typical March day the G-Animal building provides a tremendous atmosphere to launch into the work of the growing season.

Overall, the aesthetic of G-Animal's curving walls and round wood structure integrated with plants, people, and animals is wonderful. We cut salad greens until Christmas. The sounds of the pigs, ducks, and chickens, mixed with the aroma and feel of fresh soil between fingers is divine. The moderated temperature and ample sunlight soothe the senses. The three walls of glazing provide a vista to the outdoors while serving as an oasis from the wind, heavy rain, and snow that shorten the New Hampshire growing season.

Summer Kitchen (2008–2012)

The Summer Kitchen was developed by integrating three distinct, complementary components, beginning with the greenhouse, then the kitchen, and ultimately the vegetable wash station. Our focus with this structure was on stacking functionality and maximizing the use of reclaimed materials.

The greenhouse portion of the structure was donated to us by a considerate carpenter who had been hired to deconstruct the prefabricated unit. The cedar wood and glazing had been sold as a kit during the passive solar boom of the 1970s as a house addition. After thirty years the weather sealing had deteriorated to the point that the homeowners commissioned the dismantling of the structure. Fortunately, the carpenter who was hired was familiar with our organization and offered to gently remove the structure so we could retrieve the fabricated parts and revive the structure to its solar glory.

When we picked up the trusses and panels of glass, the difficulties of storing this unit quickly became evident. The glass panels were fragile and the curved roof trusses proved challenging to store because of their size, weight, and awkward shape. Luckily, we had just completed the Open-Sided Building, so there was covered space for storage. Nonetheless, this storage space was quickly filled up, and the greenhouse parts soon became a liability.

After several minor moves it became apparent we needed to use or lose this reclaimed treasure, and we began the process of siting a location and designing the overall structure. The kit was designed to allow for precipitation to be shed while capturing the maximum solar gain. To compensate for the north-facing slope of our terrain, we chose to excavate the southern side of the structure. This removal of soil also served effectively to reduce the height necessarily attained so that water would continue to be shed from the structure. For the purpose of simplifying and expediting the process, we chose a two–post and beam system to carry the trusses' aerial weight.

After a summer spent deliberating the design of the kitchen addition based on our goals and the materials available, we began the kitchen construction in 2009. Our goal was to provide an open-air cooking and dining location that would be accessible to both the residents and the guests. The structure now services the farm as a space for large-scale cooking, food preservation efforts, and kitchen workshops, as well as providing overflow for multiple parties of guests.

The deck was elevated by pouring several columns of concrete. This elevation was necessary to maintain a level floor, as the grade dropped away to the north, and it also created additional storage space underneath the floor. Once the deck was installed we began assembling the aerial components of the structure. Posts that had been salvaged from a house fire and a renovation were utilized to support the roundwood beams and rafters sourced from the thinning of our own woodlot. Plank wood milled on just two sides was chosen as triangulating bracing for its aesthetic curvature and structural value.

After reused metal roofing was installed we began debating the floor plan. Eventually the sinks and

The installation of countertops, sink, and dish storage has allowed the summer kitchen to be a comfortable place for cooking in spring, summer, and fall.

The PDC class enjoys lunch in the Summer Kitchen.

stoves were placed on the east side of the structure, and the kitchen was divided by a dining and food prep counter in the center. Additional counter space was included, and half walls were installed to reduce the effects of constant wind, while still maintaining the open-air tropical character of the structure.

Subsequently we installed a water system for irrigation and kitchen utility as well, including a batch hot water heater to assist the cleaning of dishes. To treat the greywater we installed a rudimentary filtration system consisting of several bathtubs in which the water cascades through aggregates of stone, charcoal, and sand before being discharged into a pondscape.

After further evaluating the structure and our growing organizational needs, it became apparent that the north side of the structure would be ideal for a vegetable washing station. Washing vegetables in the Community Building brought dirt into the indoor kitchen space, which was already constrained by various daily operations. The existing shade, centralized location, and available water infrastructure of the Summer Kitchen made it a better and a natural fit for the work. By obtaining a used stainless steel two-bay sink and constructing a stand, we transformed the location into an essential, multifunctional element of our operations during the harvests.

In general the structure has surpassed our expectations by providing low-cost, viable alternatives that add to the value of the experience at D Acres. The infrastructure not only provides essential,

functional space for residents and guests but also an outdoor aesthetic. While it provides for shade and cool breezes during the hot summer days, the building also provides an open-air tropical ambience into the fall season. The building provides an additional outdoor space to further connect with nature, along with the comfort and functionality of the shelter.

The Summer Kitchen is an example of utilizing creativity and available resources to meet our needs. The investments in infrastructure such as shelter, water delivery, and treatment provide a basis for further ancillary design with cumulative benefits. This concept of stacking and incorporating multifunctionality into design is a major principle of permaculture processes.

The north side of the Summer Kitchen provided space enough for the installation of an outdoor washing station in 2012. Summer intern Sydney Bennett washes large batches of greens, salad mix, beets, and carrots.

Greenhouses

One town to the north of Dorchester, there is a double-layered hoop house where hydroponic tomatoes are grown through the winter under grow lights. This commercial-scale facility has been operating for several years; on winter nights the lights provide a visual spectacle to travelers on the state highway. This demonstrates the region's capacity to grow vegetables year-round in this climate, but the question remains: Does that tomato embody less energy than a tomato grown in sunshine and shipped from Florida? We have demonstrated it is possible to grow vegetables in Antarctica and outer space, but what is the real cost of this food? At D Acres we use greenhouses to help us reduce our carbon footprint rather than increase it to replicate the diet of another climate. Our greenhouses are seldom heated, and our food production depends on growing season-appropriate foods, including super cold-tolerant greens such as mache and spinach between December and March.

For maximum effectiveness at this latitude the thermal mass of the earth must be maintained above freezing. This can be accomplished through liquid or air heat exchange that brings heat energy down into the soil. Steve Whitman, a permaculture teacher from Plymouth, New Hampshire, has designed an excellent example of a homestead-scale air exchange in which a fan and pipe system thermostatically propels air heated during the day down into the soil, to be released throughout the evening. Unheated greenhouses can be effective for midwinter food production of specific cold crops when kept at temperatures appropriate for harvesting. As temperatures plummet into the negatives for successive days, the ground freezes in the interiors of even large hoop houses.

The lack of sunlight at this latitude also prevents any significant growth during this period. While this photosynthetic deficit is significant, it is the freezing temperatures that ultimately prevent plant growth. As indicated by the frost-to-frost dates of the growing season, most vegetables must be maintained above

Early on, movable cold frames were used to warm the soil in the Lower Hoop House to increase growing capacity in the fall and spring. Micki is pleased to see a patch of lettuce and claytonia, even with snow on the ground outside.

freezing temperatures to survive. When winter temperatures drop to –20 degrees Fahrenheit the cold permeates both the ground and air. In addition, winds leach heat, pulling the warmth inexorably to colder areas, equilibrating the atmospheric temperatures. This relentless atmospheric pressure requires tremendous designed capacity and resources to resist. Generated heat competes with universal entropy to be dispersed into the prevailing climate. While mulch and floating row covers can insulate the ground from freezing, if the ground does freeze and stays frozen long enough, these measures tend to retain the cold even as the temperatures rise.

Greenhouses are also affected negatively by rodents, ranging from mice and voles to rats, that enjoy the temperatures and ample food these spaces provide. While traps are effective, diligence and consistency is necessary to minimize their impact.

We have two greenhouses of typical hoop house style, which utilize tubular steel and framed end walls to provide support for translucent plastic. This plastic can also be double layered, which necessitates a fan to separate the plastic and maintain the insulation air channel. The initial 12 × 20-foot Lower Hoop House was constructed during our first

spring in Dorchester but was crushed by a snowstorm that dumped twenty inches, after which the roof was not shoveled. When the structure was rebuilt we installed additional wooden post and beam members to help support the frame in the advent of heavy snow. This minihouse has provided hundreds of pounds of produce in three rows of beds sized for sliding-glass-door cold frames. The large 30 × 90-foot Upper Hoop House has six beds that span the distance of the structure.

For the most part the construction process for greenhouses is basic, provided there is flat terrain and the soil is free of stones. A savvy crew of three to four people should be able to piece together the metal framework in a day, framing the end walls

Situated at the top of the Lower Garden behind the barn, the Lower Hoop House is a prime example of a simple season-extending structure, heating up quickly in the spring. It's large enough to produce early spring salads and convenient for growing hot crops such as peppers and tomatoes.

and applying the plastic on the second day. Once the greenhouse is built, short-term maintenance should be limited to soil health and plastic replacement.

In greenhouses fertilized with animal manures, salts can accumulate in the soil and inhibit plant growth, because of the lack of drenching rains that nature provides throughout the year to unsheltered gardens. Hence, it is advisable to remove the plastic every three to five years for a winter or summer and allow the rains to percolate the salts from the soil. This purging period can coincide with the replacement of the plastic.

The replacement of the plastic also should occur every three to five years, as it can become torn, and its opacity increases through exposure to sunlight.

To install the plastic, calm weather is helpful. High winds create dangerous situations, as the plastic can become a powerful sail. The plastic is difficult to unroll and pull into place, but there are two methods that have proved effective here at the farm. We have unrolled the plastic on top of the ridgeline along the length of the structure, then unfolded the two sides. This method is more effective if personnel are comfortable with heights. We have also unrolled it along the length on the ground beside the greenhouse, then tied rocks to the plastic with ropes along one edge. The ropes are then tossed over the ridge and the plastic pulled up over the structure.

The plastic is sized so that it drapes over the sides and end walls, where it is affixed. The end walls are

Matt Palo preps the beds in the spring, aerating the unfertilized, never-worked soil. While outside the snow still covers the ground and hugs the high tunnel sides, it is sweltering inside. The beds are measured to allow space for a truck to drive through the center if needed.

pulled tight and permanently affixed while the sides are rolled onto a bar that can be opened to vent the structure. The sides typically roll until the first curve of the frame at approximately three feet from the grade. The height can be customized to allow additional ventilation during extreme heat.

Tree Houses

Tree houses are an essential aspect of communal living at D Acres. These humble abodes are designed to have a minimal fossil fuel footprint while providing their residents with privacy and connection to nature. Because necessities are shared in the Community Building, individual dwellings remain rustic, frugal, and calm. Tree house residents awake to the sounds of birds and running water from the creeks and have witnessed bears, coyotes, and moose approach over the years while they remained safely higher up. For one-thousandth of the cost of a typical thirty-year mortgage, people can construct superior dwellings with minimal tools, time, and experience using the natural resources at hand and repurposed materials.

During the first winter we car camped and shared sleeping quarters in the barn. The space in the barn was crowded and noisy with its office, kitchen, and living space. Craving warmth, my sister set up a futon in the crawl space where the chimney exited from the woodstove. Needless to say, rest, reflection, and personal space were difficult to obtain.

Before settling on tree houses we considered options including pop-up trailers, yurts, and school buses. Yurts are expensive investments. Although they are popular in the Pacific Northwest, we were concerned that their insulation and roof strength would be inadequate for our cold, snowy winters. Pop-up trailers are cramped, cold, and uncomfortable, with persistent leaks and mold and mildew issues. School buses are ideal, reused and mobile, but they are difficult to insulate and often rot, to become more permanent fixtures of the landscape. At the time we were also concerned about tax implications for new structures. Because wheeled vehicles and greenhouses without fixed foundation were exempt, we were optimistic that our tree houses, without fixed ground foundations, would also be excluded from additional tax assessment.

We had come from the West Coast with visions of tree houses high in the evergreen canopy, inspired by the partially fulfilled constructs of our youth, as well as the popular books by Pete Nelson. Nelson's books detail grandiose, palatial structures constructed in magnificent Douglas firs and sequoias, which set a high standard for our aspirations in Dorchester.

Our visions were tempered by the practical realities of our situation, however, including Eastern tree species and our (at the time) limited building experience. Trees in the Northeast are very different in age, habit, and shape from those in the Pacific Northwest. The forest landscape of New England does not offer many trees with three-foot-diameter foundations to which you can affix a sizable structure. The dense, vigorous forests of our property are mostly packed with competing pioneer species such as aspen and paper birch. While larger hardwoods, as well as older pine and hemlocks, occur sporadically, they do not dominate the landscape. The limited number of large single-tree possibilities led us to design tree houses attached to multiple trees.

Also, although enthusiastic, we did not have the experience or equipment to construct at the heights attained in the picture books. While they inspired our efforts, we had limited hands-on experience with dropping trees and moving them into position. We also lacked knowledge on how to safely elevate heavy beams for durable attachment to a living foundation. The snow, budget restrictions, remote location, and our evolving building philosophy all challenged us to refine our technique and strategy.

We began the first tree house project in January of 1998. We sited the structure just beyond the wood line and began cutting trees for beams and joists. The heavy snow eased the challenge of using the come-along to skid logs into place beside the trees. We drawknifed the bark from the beams and attached

The Creaker is a great example of lessons learned in tree house building, affixed to the trees in such a way that the entire structure creaks when the wind blows.

them to the trees using a generator-powered drill and threaded rod. The work was slow and laborious, though we were enthusiastic and excited. We constructed a platform onto the initial beams using roundwood joints and dimensional lumber floorboards sourced as seconds from a local mill. Once the platform was complete a metal shed roof system was attached to the trees using two beams and roundwood rafters. This served as a tent platform for several weeks until the walls were framed in.

Our tree house construction process was refined by this first experiment in the snow. We were essentially exploring inexpensive alternatives in natural resources and recycled housing. The first D Acres tree house, known as the Creaker, represented a crucial stepping-stone for the development of our

entire building system, teaching us valuable lessons regarding tools, materials, and architecture.

Our basic tool belt contained a tape measure, a speed square, and a marking pencil. We used the plumb bob and water levels to evaluate our horizontal and vertical alignments. A majority of the wood for the frame was provided using the chain saw and drawknife. We purchased a small Honda generator to allow portable power essential for drilling the hole for the connecting threaded rod. We used cordless drills and screws extensively to fasten the structure. This simple array of tools was mobile and scalable for our remote building sites, as well as intuitive to use and suitable for introducing building techniques to novices. We sourced reused building materials such as windows and hinges from the stockpiles my uncle had collected, and we began to learn the distinct attributes of different species' wood, both the living trees and the debarked structural components.

The Creaker also presented structural challenges. The design connected the roof system to the four support trees ten feet in the air. When the wind swayed the trees in a rhythmic fashion, it sounded like a ship at sea. Moisture issues also challenged our design. It is difficult to prevent rain pouring down the trees from seeping into the wooden structural connection and causing rot. The trees' continual movement in the wind, as well as growth in girth, made implementing a suitable rain shield difficult.

After the Creaker was occupied in early spring, we proceeded to build two additional tent platforms as satisfactory housing for the summer until we were able to revisit construction projects in the fall. At that juncture we constructed the dwellings known as Eastside and Irie House on the existing platforms. They were both constructed with a roundwood post and beam system and included sleeping lofts and screened, operable windows.

Building projects resumed again the following spring with the construction of the Crow's Nest, the Sanctu, and the Skinny Shack. The Crow's Nest's substantial height was made structurally possible by using 45-degree-angled siding to reinforce the tall walls and many windows. The Sanctu was situated in the ravine of a hemlock forest, perched above a beautiful section of stream. We utilized dimensional lumber to frame the Sanctu, which is our squarest tree house to date. Farther up the creek on the edge of the ravine, the Skinny Shack structure was constructed. It represented the culmination of our tree house design and construction techniques of this era. Its design maximized the use of roundwood rafters to create a sweeping gabled roof and featured casement windows left over from my uncle's renovation work, as well as carpeted floors, a woodstove, a covered porch, and eventually solar power for lights and music.

Reinforced on an I beam and chained to the tree, the Crow's Nest received an extension of its life in 2011.

This construction frenzy in our first two years resulted in six personnel dwellings. The majority of wood for the tree houses was sourced from the land, and the hardware, doors, and windows came from renovations. They provided inexpensive yet extremely satisfactory shelters for the residents. The average tree house investment was approximately one hundred to two hundred dollars for new materials (roofing, screws, bolts).

But unfortunately, because of several factors, these tree houses have proved difficult to maintain. The severe climate and our construction naïveté contrived to bring several houses down from their perches. We had made many mistakes in construction because of inexperience, choosing large-diameter pioneer trees such as poplar and white birch as living posts. These trees were already at the end of their life cycle, and damage from the drilled holes accelerated their demise. But juvenile, coppicing red maple was also problematic because the trees were competitive among one another. The beam connection zone became the breaking point of numerous living posts toppled by the wind.

Our selection of trees for use as beams was also dubious. Aspen and beech are particularly susceptible to mycelium, which in conjunction with moisture rapidly depletes the wood of all structural value. Especially in areas of high moisture, these mushrooms contribute to a rot that can decimate heavy hardwoods within five years, to the point that a pocketknife can be buried up to its handle.

The Crow's Nest was the first tree house to fall victim to beam rot and pioneer red maple competition. Luckily, the tree house remained wedged in the air. The shape and structural integrity of the structure was preserved by the 45-degree-angled siding, which strengthened the building. Since the initial deelevation, we have been able to chain and reinforce the structure in place.

The Irie House was evicted from the landscape by development. My parents chose the tree house site as the location for their retirement castle. Eastside was the next tree house requiring renovation as it completely toppled to the ground eight years after construction. Despite the fall of over six feet the timber frame structure remained intact. Using come-alongs and car jacks we were able to level the floor and rebuild the foundation on the ground. The Skinny Shack's threaded rods were bent by the weight of winter snow, and brackets were lag bolt–mounted to ensure safety from the rod's possible failure.

The structural failures of the Crow's Nest and Eastside forced us to evaluate the safety of all our tree house dwellings. While we enjoyed the tree houses, we were unwilling to chance the catastrophic fall of an occupied dwelling. We systematically reviewed the possible dangers and implemented safety measures on all tree houses. These further structural reinforcements included additional posts and steel beams.

Nearly a decade after the initial creation of the Creaker, we also undertook to construct an updated tree house that addressed our prior design oversights. The Lighthouse was constructed among three youthful and vigorous hardwoods, each between eight and ten inches in diameter. The platform was framed between the ash, sugar maple, and beech using eight-inch steel I beams instead of wood. This impervious foundation increases the longevity of the structure from ten to twenty years to an estimated twenty to forty years. We also improved our design aesthetic by utilizing immense panels of glass and an expansive roof design. A loft/storage area augmented the usable living space. Timber lock fasteners were used extensively to accelerate and simplify the attachments between framing members.

Our practice in tree house construction has taught many lessons. Appropriate tree species are now selected based on their vigor and normal life cycles. Steel beams have become my preference as structural supports for their longevity and can be sometimes salvaged from buildings and bridges. Hemlock is also a great choice for a wood beam, though potentially susceptible to sunlight damage. White pine is another favorite, as is red oak, which though heavier is more rot resistant. For the roundwood joists and rafters I prefer the lightweight, straight woods of spruce, pine, and fir. Hemlock and hardwoods are more difficult to attach with fasteners and are heavier to work with. Debarked ash and red maple roundwood framing material has a beautiful aesthetic.

Living in the trees can be risky. An intern once fell while relieving himself in the night and required twelve stitches above the eye. I am also acutely aware of the fire danger posed by tree houses. On one occasion I woke to a candle burning into the wood shelving of a tree house. This experience reminded me of the limited fire suppression services available in our region. In general, precaution should be taken to provide railings and teach fire safety and suppression techniques.

There are also limitations concerning insulating these structures. The nonstandardized construction makes using conventional materials such as fiber-glass or foam board difficult. We have experimented with alternatives such as a straw clay mixture and recycled plastic trash and packing materials. Technically it is difficult to insulate a floor that is exposed to the air underneath, seal it from moisture penetration and mold, or guard it from rodents. We have discussed insulating the floors by building an insulated floor on top of the original subfloor.

Renovation Projects

New construction has been substantial at D Acres since its inception, which has tended to overshadow a principal component of our philosophy: valuing the existing historical buildings. While often overlooked, renovation will be the primary building activity at D

Acres for many years to come. As buildings age and their usages evolve, repairs and upgrades become necessary. Renovation also provides experience for future new construction by pointing out deficiencies in prior design.

Delbert and Edith improved the original structures over the years, although during my lifetime because of advancing age they struggled to maintain the property. When we arrived in 1997 the buildings were beginning to show significant signs of decay. The two-car garage, the barn, the silo, and the residence at 241 Streeter have since all undergone significant renovations during this era of farm rejuvenation.

The first objective of these renovations has been to locate potential problems in a building's roof, wall, or foundation. A roof can be repaired or replaced

The 1830s cape in 2006, with the Grays' front porch, sunroom, and back-house studio space additions.

and structurally improved, insulated, and extended. With regard to walls, we focus on structural integrity, insulation, and glazing. Drainage and integrity are the primary foundation considerations.

The problems that the older structures endured tended to result from water. Where water had either leaked through the roof or splashed and permeated the base of the wall, the wood rotted. Water also constantly expanded and contracted for decades in freeze/thaw cycles, causing the various elements of the noncontiguous foundations to shift. These shifts in the foundations undermined the intentions of the designed load distribution and pulled the buildings apart.

The barn is a hand-hewn post and beam timber frame whose roof structure comprises three trusses connected in parallel by purlins, then planks, from which the steel roofing is attached. When we first arrived, the floors in the animal stalls had heaved and broken the concrete extensively throughout the southern side of the structure. The large volumes of water that fell from the drip edge had migrated into the structure, causing severe damage to the concrete floor and stem wall.

After we replaced the obvious essential flooring and beams, the building languished for many years. Folks were not impressed by the damp, dirty scene in the Red Barn. During the summer of 2007 we were encouraged to apply for a grant to assess the structure with the New Hampshire Preservation Alliance. The assessor provided a plan to renovate the southern wall, which had deteriorated to the point that the roof was sagging noticeably. We then hired the assessor as a contractor to support that

The Red Barn and Silo prior to renovation

segment of the building so the wall and foundation could be replaced. We built the stem wall up higher with stone based on a contiguous footing, insulated it inside and out with closed cell foam, and rebuilt the wood framing with modern operable windows with screens for enhanced building performance.

The structural improvements have noticeably improved the environment in the structure. The porous, cracked, and rotting wall has been replaced by a stable, moisture-resistant, and insulated construct. By extending the overhang of the roof, precipitation is pushed farther away from the building, and while the building is shaded in the summer, the winter sun can still shine in. There has been a drastic reduction in humidity inside the perimeter of the southern wall, especially during the spring thaw.

The building is also easier to heat. By eliminating significant air passages and filling the wall cavities with clean plastic refuse, the R-value provided by the walls has improved. Operable windows have also enhanced the indoor environment by allowing in the fresh air and sounds of the outdoors.

The silo is another example of enhanced benefits through renovation. It was constructed in the 1950s, and stamps on the wood indicate that it was delivered as a kit to Rumney, New Hampshire, via rail. When we arrived the roof leaked and daylight was visible through the holes in the rotten wooden base. By 2008 the structure was tilting perceptibly, so we hired an outside contractor to support the building. Once the building was supported we cut back the rot almost thirty inches and built a stone

A new insulated wall with airtight windows improves the barn living space.

The silo is raised and ready for a masonry foundation to rebuild the footing to last many more years.

Regrounded, with new roofing and fresh paint, the silo is ready to be inhabited. The interior is set with two floors, for either storage or sleeping.

wall onto which the structure was lowered. The silo's roof asphalt was replaced with metal, though the weathervane horse continues to accurately predict inclement conditions. Additional windows in the loft provide ample light on the top level, and the three floors triple the space available for storage. The roof extenders and elevated impermeable stem walls have radically improved the silo's weather resistance. The interior framing has also served to reinforce the structure.

The Red House

The Red House was the original homestead on the property. I have fond memories of arriving late at night from distant travels to be greeted by the light on its porch. Edith once conducted a title search to determine the history of the land's ownership and discovered that the original proprietors were the Eliot family, who were buried in Cheever Cemetery on Hearse House Road. From their tombstones the life span of the original deeded tenants could be determined. Some years later we removed several large sugar maples from the cemetery and utilized the firewood grown from the Eliots' remains to heat the house they had constructed.

When my aunt and uncle purchased the property in 1946, the Red House and barn were the only structures. The Grays arrived to a house that had no indoor plumbing or electricity and had been through many years of neglect. The roof leaked severely and was causing rot throughout the building. Edith and her daughter, Pat, were skeptical that they could reside in these conditions. But with the advent of the rural electrification program they were convinced to make the move to reside in Dorchester. After that Delbert, Edith, and Pat spent their lives fine-tuning the dwelling to their preference.

Delbert spent years building cabinetry and carpentering the furnishings. He worked intensely to excavate and reinforce the basement, transforming a crawl space into a half cellar. A front porch was constructed from huge windows salvaged from the mills of New Bedford by my great-grandfather.

In the early 1970s the Grays also built an addition, connected to the back of the house with a solarium-style transom, which served as Edith's art studio. As he was over seventy years of age at the time, the studio was the last major construction project undertaken by Delbert. As he eased into a more sedentary existence before his death in 1989, Delbert relaxed his regime of annual maintenance on the property's structures.

While the house was a masterpiece of traditional New England aesthetic and charm, several factors undermined its long-term viability. Over years of renovations and remodeling, a succession of electrical talents haphazardly hung wires throughout the basement, creating a shocking experience on the wet floors. Insulation was nonexistent in parts of the house, including the roof, which contributed to conditions where icicles formed on the ceiling. The foundation, once excavated, had proceeded to collapse in on itself as water and soil pushed into the basement hole. Before my arrival in 1997 my parents consulted with an assessor to determine the viability of the structure. The determination was that the building was unsalvageable and should be allowed to deteriorate until it was condemned.

This assessment affected my parents' plans for their retirement. Up to this point I believe Edith's assumption was that when Betty and Bill moved from North Carolina and retired they would reside in that building. She even offered to move into a trailer for that to occur. As Betty and Bill began to consider the alternatives the Red House continued its descent

The dug basement was small, dark, and wet.

into disrepair. As Edith aged beyond ninety years, we worked hard not to disturb her with changes such as a renovation of her dwelling. During the last year of Edith's life Betty and Bill moved into the upstairs and assisted Edith with her daily needs. After Edith passed away they made minor cosmetic alterations by replacing carpet in the studio and the linoleum flooring in the kitchen.

By this point Betty and Bill had begun constructing a separate retirement home. When this structure was completed they relocated, and D Acres proceeded to rent the Red House to a couple. While the income derived from renting the structure was helpful, the foundation's water problems and lack of insulation continued to go unresolved. After several years the couple chose to vacate the lease, partly due to the mold and indoor air quality precipitated by these circumstances.

After our tenants had left the property Regina and I lived in the building for several winters. We migrated toward the almost contemporary insulation of the studio, which supplied the electrical potential and heat retention desirable during the cold, dark winters. Over time the main house was

deemed unlivable. Several short-term residents attempted to use the facility, though consistently the mold and air quality were perceived as inhospitable.

Consequently, we were forced to make decisions regarding the structure. While it was designed primarily as a residence for a small nuclear family, we currently did not have that constituency on-site. Regardless, the structure was unsuitable for occupancy in its current state. So we explored discussion of our other needs. Additional indoor overnight accommodations and retail space were determined to be high priorities for the organization. The hostel was booked to capacity on holiday weekends, and our garage was not sufficient as an inviting farm store. In addition we were seeking additional classroom, meeting, and workshop space that could be handicap accessible.

To fill those needs we debated the merits of renovating the Red House versus building anew. There were many vocal proponents for tearing down and starting from scratch. Part of the renovation would entail the removal of the large brick-and-mortar fireplace, which was connected to a fire marshal's

Almost emptied of furniture, the Red House is ready for demolition. This is a view of the central fireplace in the front room.

A view of the same front room. Without walls and a drop ceiling, the beauty of the interior emerges. The hand-hewn beams are a testament to an era of skill and survival.

Bob Guyotte and his crew have already spent many hours strategically placing the house on wooden cribs so they can completely dig out the old foundation stones and excess dirt to work with a clean slate.

nightmare: a labyrinth of flues connected on all three floors. Renovation is dirty, methodical, and tedious work that does not arouse the same enthusiasm as new construction projects. In addition, it was concluded that the time and money spent on renovating the building would be greater than to build a new, similar structure. The site was problematic because of a huge rock on the southwest corner that was assumed to be a part of the rock ledge. It was suspected that the ledge would seep water into the basement in any event, despite our best intentions. There were also concerns about the effectiveness of any particular course of action taken to rectify the site location's limitations.

As the decision was debated we considered partial renovation of one corner of the building. Judging that to be another Band-Aid attempt, we finally chose to invest in a full renovation. This would require lifting the building up to install a conventional concrete foundation. By raising the building on a secure base with proper drainage we hoped to eliminate the problems that had beleaguered the building for nearly two hundred years.

We chose renovation over new construction for several reasons. Although problematic because of the rock ledge, the location's proximity to the road is ideal for its visibility and accessibility, which create a welcoming atmosphere. To conform to town zoning regulations we could not build a new structure at the same location. We were also motivated to renovate because disassembling and reconstructing allows for maximum utility of

The large and open basement invites the possibility for more food storage space.

salvageable materials. Ultimately, we believed in the approach because it valued the land's history and the continual spirit of investment in that place. By opening the structure to the public upon completion, we hope that visitors will be able to witness the style and techniques of construction that predate the era of the sawmill.

In May of 2012 we began gutting the interior. After the last of the furnishings was removed, we took out the partition walls, cabinetry, and appliances, stripping the building bare. The chimney was disassembled and removed from the top down with chisels and hammers. During this process we discovered evidence of several potentially catastrophic fires that had somehow been averted in the form of charred wood hidden in the walls. As

we removed the ceiling we were pleased to uncover the long, hand-hewn beams that support the second floor. These remarkable beams span the width of the building and are capped by a three-truss system that supports the roof with an exceptional display of roundwood purlins. The gorgeous wood frame that was uncovered more than paid for the gamble of undertaking the renovation.

Bob Guyotte arrived in June with another contractor from Ashland who specialized in wood frame renovations. The building was excavated, then supported by the frame and elevated so that further excavation could occur around the perimeter and the new foundation could be poured into place. Once the building was stabilized we rebuilt the floor system and have begun the process of

The Red House in the fall of 2013, closed in around the sills, with brand-new windows and weather-tight doors. The basement walls have been insulated, and the electrical wiring has been set. With a new chimney cap and weathervane, the house is approaching reoccupation.

replacing windows and the electrical service, as well as the front porch and back solarium transom.

The structures at D Acres, whether new or renovated, continue to evolve based on our needs and as our techniques are refined. Our attempt is to continue to develop a building vernacular of place that is both functional and aesthetic. The goal is always for construction to be local and of this place, in terms of both its materials and its builders.

CHAPTER ELEVEN

RENEWABLES

There are several ways to address energy usage. But first we have to ask, what is the goal we wish to achieve? We must decide the form and amount of energy that we wish to harvest and garner, and the time in which we would like the task accomplished. For example, to generate hot water for dishes do we have the time, materials, and enthusiasm to build a fire, or would we prefer the convenience of propane if accessible?

There are slow and incremental ways to accomplish our everyday tasks. We are looking for solutions that are based on the reliable fundamental laws of nature, including gravity and thermodynamics. A clothesline, for example, is a cost-effective alternative to an electric drying machine. To meet our goals we also often choose redundant and integrated means to accomplish the task at hand. Consequently, we are attempting to integrate kinetic activities so that a hydro system, for example, could be augmented or replaced seasonally with wind or people power. If one system fails or needs repair, then another option can be relied on.

Renewable technologies are often site specific and require conscious placement for successful utilization. While the ergonomics of a clothesline is one factor that encourages usage, we also need to consider factors such as wind patterns, proximity and access, and solar exposure. Likewise a hydropower system would depend on installing the power generator in a location that maximizes the head and the flow of the water.

These technologies have been installed based on economics and a rationalized course of action, as well as twists of fate and circumstance. While we prioritize renewable strategies based on their cost effectiveness, we often have pursued specific projects based on coincidences or the interests of the personnel. We are also interested in experimenting with technologies that are particularly applicable to other farming operations. While we have experimented with many technologies thus far, we will continue to refine the existing systems and explore further alternatives.

While some of the renewable strategies involve industrial technologies such as evacuated tubes not readily fabricated on a community scale, other strategies are similar to the types of inventions on Gilligan's Island. Our pursuit of these alternatives is based on the philosophical credo of exploring options that are reliable yet readily accessible to the general populace. We are particularly fascinated with the concept of reusing and upcycling items that would otherwise be thrown away or melted to be downcycled into disposable items.

Maximizing these various sources requires the ability to directly use or store the available energy when it is abundant. Storage can exist in many forms: potential physical energy, such as water stored uphill or amassing electrons in batteries; energy can be even stored within materials through thermal mass and the caloric intake of animals, both to be consumed as protein or to invigorate physical strength, endurance, and mental capital.

The purpose of these technologies is to develop feasible alternatives to our uninhibited consumption of the world's natural resources and the consequent environmental destruction. The ideas for many of these projects were cultivated by exposure to the information during my time at Solar Energy International (SEI), as well as time spent observing innovative alternatives on islands and other resource-scarce situations throughout the world.

In addition to the various infrastructure projects that we build, our program also depends on a systematic conservation effort. Every year we assess ways in which we can reduce the size of our fossil fuel and resource footprint. Toward this end we have evolved a version of the traditional New Year's resolution by enacting an annual D Acres conservation resolution.

Initially our efforts focused on obvious appliance choices, such as removal of a microwave and toaster. These tools of convenience drained electric energy while providing negative contributions to our health and home safety. Recently, we sold our clothes dryer and dishwashing machine on craigslist. At other times we have resolved to change our behavior or protocols to conserve energy. One winter we vowed to utilize the root cellar in place of electric refrigeration from November to April. On another occasion we schemed to institute a power-down day, during which we avoided any and all energy usage. This process has been a gradual and effective method to systematically reflect as a group, then institutionalize ways to reduce our organizational energy consumption.

This chapter discusses the following renewable energy infrastructure we employ: domestic hot water technologies, photovoltaic uses, solar dehydration, solar cookers, cob oven, root cellar, bike power, and methane digestion. As of this writing technologies such as hydro, wind, and living machines have been discussed but not yet brought to the design and implementation phase.

Domestic Hot Water

Hot water is crucial for daily household and farm operations, including cleaning, laundry, and bathing, as well as energy delivery and storage. Heating water with the sun is a common practice. When the sun is available, it is our primary source of energy for the purpose of space heating, bathing, and washing. To accommodate times when the sun's energy is not readily available we developed an integrated water system in which alternatives to the sun are readily attainable. Our primary alternative is the stored solar energy available in the form of wood.

The energy required to heat one gram of water 1 degree Celsius is termed a calorie. The energy necessary to heat a certain amount of water to a certain temperature is therefore fairly predictable. Water drawn from the earth typically is about 55 degrees Fahrenheit. For water to feel warm it must generally be at our body temperature of 98.6 degrees or above. A hot tub temperature is quite warm at 105 degrees, while dishwashing water is suitable at 120. Hence, the caloric energy necessary to increase water temperatures from well water into a usable form can be incrementally achieved. Water intended to be heated from 60 to 120 degrees can be preheated in the sun to 100 degrees, which would save two-thirds of the energy required to attain 120 degrees.

With this in mind it is important to understand the typical components in a solar domestic hot water heating system. These include the energy collection mechanism, a vascular system to transfer and utilize the liquid, and storage if necessary. These component parts are common to other renewable energy systems as well, such as wind, hydro, and photovoltaic.

The energy collection device amplifies and concentrates the waves produced by the sun. These heat waves are then embodied in the liquid, which can be transferred and stored. The energy of the sun is collected and amplified through various means. Glazing is used to capture the energy through the

We utilize two solar hot water collectors to supply hot water to the Solar Shower, located on the far right.

greenhouse effect. While lighter colors reflect the sun's energy, painting the collector black can help maximize the absorption of the sun's radiation. The collector can utilize amplifiers such as mirrors to further intensify the solar capacity. In addition geometrically produced parabolic reflectors have the capacity to project the energy toward a focal point. Metal is utilized as a construction material because of its ability to transfer the energy and to withstand extreme heat.

The collection device is designed to maximize its capacity to capture the heat waves generated by the sun. The greatest possible amount of sunlight must be allowed to enter the device, whereby the energy is trapped in a form that can then be utilized. The

The large silver tank in the center is the main thousand-gallon heat exchanger. The two brown tanks beside it are connected to the solar hot water collectors located on the roof of the Community Building.

capacity of flat glass to absorb sunlight depends on its position relative to the sun's celestial journey. When the glass is perpendicular to the sun it absorbs its maximum; as the glazing exposure shifts away from a 90-degree incidence, more of the energy is reflected than is absorbed.

The vascular element of the system transfers the energy to storage or to its point of usage. When there is a storage mechanism that incorporates a heat exchange element, or when water is used directly in a heating system, petrol-glycol is added. While water is functional in many circumstances, such additives prevent freezing and elevate the boiling temperature. This antifreeze ensures that pipes do not freeze and rupture while also allowing

additional energy to become embodied during periods of high heat. The vascular system must be able to endure this thermal intensity and should also be insulated as much as possible to maximize the efficiency of the liquid in its course toward its utility.

The storage element is crucial to accumulate energy that can be used as needed. During dark nights and cloudy weather, stored solar energy can still meet our needs. An insulated thousand-gallon water storage tank serves this purpose here.

We have two distinct styles of solar hot water collection devices on the Community Building: flat panel and evacuated tube. The flat panel design has been common since the Carter administration. The unit consists of a black box in which copper pipes

Taylor "Sunweaver" Mauck, a long-time advocate of solar power and solar system installations, installs the water pipe components that connect from the evacuated tubes into the water tanks in the basement. Notice also in the background the large satellite dish, used as a parabolic cooker.

zigzag along the bottom. The top is glazed, and the unit is painted black and constructed primarily of extruded aluminum, while insulation on the bottom further increases efficiency. Each unit has an inlet and an outlet, which is plumbed into a heat exchange unit in our boiler room. The unit is low tech and reliable. As long as the copper pipes do not leak, the unit should be operational for many years. The safety glass is fairly rugged, though replacement of a pane broken at my parents' house proved impossible to repair.

The evacuated tube style consists of a rack of twenty three-foot tubes, which are connected along the upper edge to a manifold. This manifold runs the length of the rack. The tubes collect the energy of the sun and focus that heat on the manifold. In the manifold is a vascular system of copper pipes that encase the focused heat to absorb and transfer the heat energy. The manifold also has an inlet and an outlet, which connects to the heat exchanger in the basement.

The evacuated tube style produces a higher output, particularly in the nonsummer months. The design generates heat in cooler outdoor air temperatures, and the circular shape of the collector has more surface area for greater solar exposure. They are more expensive, however, and necessitate the mining of rare earth metals to fabricate. The vacuum of the tube seal is also a potential weak link that could undermine the long-term performance of these units. The evacuated tubes are also more delicate and can be shattered relatively easily. The flat panel construction is more rudimentary and can be constructed and fabricated in a garage shop.

Both flat panel and evacuated tube units were installed parallel to the pitch of the roof. The roof pitch is a rise over run of approximately 4:12, which means it rises four inches for every twelve inches of length. Expressed in degrees from the plane of the horizon, a 12:12 roof produces a 45-degree incline. The 4:12 roof pitch converts to less than a 20-degree angle from the horizon. This low angle prevents the units from maximum efficacy during the winter

months. Because the sun is nearly directly overhead in the summer months these units are most effective during the hot months when we choose not to use the wood boiler to meet our Community Building needs.

The wood boiler is the final element of our domestic hot water system at the farm. Firewood is readily available in this region, and the boiler makes this energy available for domestic hot water and space heating needs. Our unit is a Tarm, which is produced in Denmark and distributed in this region by BioHeat USA. These units are renowned internationally for their long-term performance and efficiency. Our boiler is a wood-gasification unit in which wood burns in a primary firebox while the combustion gases burn in a secondary chamber. By burning all the possible elements of the wood at a high temperature, the system achieves maximum efficiency.

A strong, efficient wood fire generates heat with temperatures that range from 350 to 400 degrees. Contrary to a steam system, a boiler must find a way to transfer that energy or it is forced to shut itself down. Instead of dampening itself down every time boiler temperatures rise above 212 degrees, the boiler is able to continue burning at maximum efficiency for the duration of the fire. This is possible because the wood boiler is augmented by the storage capacity of the thousand-gallon heat exchanger. The rising temperature in the boiler is pumped via pipes to the tank across the room so the fire can continue efficient energetic combustion with minimal air contamination. By allowing the fire to burn at maximum efficiency the particulate emissions from the fire are minimized and the chimney does not develop creosote.

This insulated tank serves as a battery to store the energy generated by the Tarm. The heat generated by the wood boiler is pumped through a long copper coil of pipes in the thousand-gallon tank. The design allows the pipes to radiate the heat energy into the thermal mass of the tank. We draw from this reservoir to meet our needs for domestic hot water, as well as for central heat, via similar copper coils that draw the energy from the tank and through our flooring.

We've experienced a lot of difficulty in integrating the various components. The installation was the largest and most sophisticated that the original contractor had then attempted. In the years since there have been many contractors who have attempted to improve the efficiency and operations of the components. At this juncture we have deprogrammed any computer element to the system and rely on humans to control the boiler room operations. We are proud that we have not utilized the backup propane boiler in many years.

We also have constructed several simple batch heater designs to provide hot water for our outdoor kitchen and showers. To construct these heaters we salvaged old chest freezers from the dump, then found metal tanks that could be inserted into these insulated boxes. We then plumbed pipes as an inlet and outlet for the water, and painted the tank and the inside of the box black. The complete setup was then covered with an old sliding glass door. The final steps were to tilt the unit so it could receive the maximum solar incidence and fill the pipes with a source of pressurized water.

We have found these units to be functional during the frost-free months of the year. We have used several old electric hot water heaters whose internal heaters had failed as solar water tanks in our system. By removing the insulation of the electric heater, the sun can transfer its energy directly via the convective metal. The inside of the insulated box heats up in the sun much like a car with the windows rolled up.

To efficiently use this heated energy requires planning of an insulated vascular system that can readily transport the energy to be utilized or stored. Energy is invariably lost through transmission within the vascular system. It is important that the insulated plumbing from the heater to the insulated storage unit is as short as possible so excess heat is not lost in transmission.

The solar hot water system for the Summer Kitchen is the same as the Solar Shower, a freezer (as the insulated box) turned on its side with a water tank painted black. The water is plumbed from the Community Building.

Photovoltaic Uses

Photovoltaic is the terminology for creating electric energy directly from the sun. Of the energy that impacts the earth from the sun at any given moment, we are limited in the percentage of energy that we can convert into electricity by both technology and cost. We generally can convert less than 20 percent of the solar radiation into kilowatts. The kilowatts produced correlate to the costs of production generated by the industrial manufacturing and resources necessary to produce the panels. To create this green source of energy requires substantial investment in industrial production, pollution, and resource extraction. Cost per kilowatt and efficiency in collection per area of panel ratios continue to improve the cost effectiveness of this technology.

Once constructed and installed, however, the technology is extremely reliable. While the electronics of an inverter may require maintenance within fifteen years, the panels can be expected to perform up to twenty-five years. This reliable life

The array being installed with the help of Sunweaver and PAREI (Plymouth Area Renewable Energy Initiative) volunteers in 2005

In 2011 we added eight more panels to our grid-tied system.

span provides the economic incentive to purchase these panels. While the economics of solar may be marginal at this juncture, the investment today is based on the assumption that the costs of energy will continue to escalate.

We have two systems that supply a portion of the electrical needs in the community building. One is a 1.7 kW array mounted on a pole by the driveway in which each of the fifteen panels in the array is rated for 110 watts. The array is designed to track the path of the sun so that the incidence of the radiation is maximized on the surface of the panel. The tracker can also be manually adjusted to adjust the angle at which the rack is mounted toward the south.

We also have eight panels fixed on the roof. These panels offer 2.2kW of additional electricity through the 275-watt rating. The upper roof pitch is 9:12, which is approximately 35 degrees, suitable for year-round collection.

Roof-mounted systems have the advantage of being in close proximity to where the energy will be used. By reducing the length of the vascular system, the costs of construction materials are mitigated and the efficiencies increase. The roof systems occupy space that otherwise would not be utilized and do not take up ground space. Pole- and ground-mounted systems depend on an aerial or buried conduit to transmit the energy to their desired location, while the roof system is accessible to readily penetrate the building envelope. Roof systems depend on the pitch and orientation that is functional for the performance of the units; otherwise additional racks must be constructed to set up and orient the collectors. Maintaining a weatherproof envelope and addressing the dangers posed by snow and wind are important in the planning and implementation of the installation.

Solar panels produce direct voltage, such as is utilized in vehicles and typical RV and marine applications. However, the power that arrives from the utility company is an alternating current. To power a conventional electrical system the solar system relies on an electrical inverter. This device changes

A full view of the system after completion

A small solar panel is hooked up to two car batteries to provide the living space with a small light and enough power to operate a CD player.

the form of electricity so we can connect directly to the existing electrical grid.

We also use solar panels directly or in conjunction with batteries to power remote equipment on the farm. In these efforts we have experimented with water pumping. Solar water pumping is most effective when utilized to move large amounts of water over a period of time at low pressure. Direct current pumps can push water continuously uphill, though they lack the ability to push or suck large amounts of water at high pressure and volume, as do gas- or electric-powered pumps. Because of this fact, solar water pumping is less than ideal for direct irrigation and immediate usage purposes, but rather it is more suitable for slow and steady incremental acts, such as pumping uphill to fill reservoirs over the course of a day.

Additionally, solar-powered equipment operates our electric fence chargers. These systems often provide the only plausible way to maintain vigorous foragers in remote locations without the convenience of grid-tied electricity. We have also used solar panels to provide twelve-volt electricity to

tree houses and private dwellings on the property. This arrangement is dependent on sufficient light to charge a battery. Once charged, the battery powers twelve-volt lights and music equipment salvaged from automobiles, boats, and recreational vehicles designed for direct current applications.

Solar Dehydrator

The solar dehydrator is designed to provide a nonelectric option for the preservation of food. To function, the dehydrator depends on its capacity to achieve and sustain the elevated temperatures and adequate ventilation necessary to induce drying, and in conducive meteorological conditions. Hot, dry, windy days reduce the amount of time necessary to dry effectively, while damp, cool weather inhibits the process. Drying time is affected by the dryer design, the weather, and the specific plant material to be dried.

The dehydrator comprises various components designed to capture and transmit the sun's energy to the contained plant material. Our fabricated

Glass front panels of the solar dehydrator absorb the sun's heat. Removable screen frames allow for access to drying plant material.

unit collector consists of a metal box that has been painted black and covered by a transparent sheet of glass. This 1½ by 3 foot box is approximately six inches deep and is attached to the unit such that the angle from the horizon is approximately 35 degrees. The unit faces due south to capture the sun's radiation throughout its daily journey. This box has ventilation holes at the top and the bottom to enhance the forces of convection and to propel the heated air up through the plant storage element of the collector.

The plant container device consists of stacked shelves with plastic mesh screen bottoms to allow the air to flow through the material. It is important not to utilize galvanized metal screening for these shelves because the coating will contaminate the food. These

A close-up view of the fire box

shelves are on racks to allow them to be emptied or refilled inside to avoid windy or wet weather.

The design of our dehydrator includes a wood-fired heating element. Under the storage unit is a firebox that provides heat directly to the chamber above. This augmentation is particularly effective during the shorter, cooler days in the fall. The smoke from the fire is channeled through the end of the box and out a chimney on the top so the material is not tainted by the smoke.

Interestingly, our barn has similar qualities, and we utilize it as a giant dehydrator. During the summer months the loft of the barn can become unbearably warm. The aluminum roof is painted black and is inclined toward the sun to collect the full day's solar journey. Inside the loft there is ample vertical height to hang leafy vegetables and root crops such as garlic from the ceiling to dry.

Solar Cookers

Solar cookers are designed to reduce our reliance on wood, electricity, and fossil fuels to prepare our meals. There are two main genres of solar cookers: those that function as ovens by augmenting the greenhouse effect and those that operate as a cook-top by using parabolic geometry to create a specific point of heat.

The oven must be movable so the glazing collector can directly face the sun as it moves across the sky. A crucial element in the design of these structures is the capacity for the hot cooking ingredients to be maintained upright in their vessel as the cooker tracks with the path of the sun. The construction must be adjustable to the varying incidence of the sun throughout the year and be durable enough to withstand the wind. Some solar ovens are designed for the outdoors in fixed, permanent locations, while others are less weather-proof and more readily storable. For all of the cooking equipment it is important to consider that hot and heavy cooking equipment is difficult to transport and monitor. Design should consider

Sun ovens are an easy and practical way to cook beans, grains, and stews.

the safety and comfort of the cooks with regard to the ergonomics of these appliances.

The solar ovens concentrate the energy of the sun into an insulated box. This box accumulates heat to temperatures necessary for baking, relying on the greenhouse effect to concentrate the energy of the sun into the enclosure. While different styles abound, all units consist of an oven portion in which the food cooks, along with a means to amplify and accentuate the energy from the sun. When I was at SEI, for example, we constructed an effective oven using the simple materials of cardboard, newspaper, plastic wrap, aluminum foil, and masking tape. We also created an oven with glazing from an old sliding door that could cook eight dozen cookies at a time.

We have two manufactured sun ovens that are commercially available for around two hundred dollars. They consist of an insulated box with a glass seal in which a gallon pot can reasonably fit; we have cooked whole chickens as well as loaves of bread in these units. Attached to the box is a system of mirrors arranged to focus the sun's energy into the oven. Black enamel pots with lids are more effective conductors of the heat than reflective stainless steel or aluminum pots. It is important not to use materials that will release toxic gases when heated. At times moisture can build up on the interior glass and obscure the box, limiting the capacity of the unit. Wind can also slow the process by chilling the device, or tipping the lightweight units over, spilling the contents. Still, these units are reliable on clear days throughout the year, except when temperatures are less than 32 degrees.

I enjoy cooking with solar ovens because, unlike with gas or wood, it is difficult to burn food to a black and unpalatable state. Because there is no direct source of heat, that focal point for combustion does not exist. While some items may dry out, I have not ruined food because of inattention. If the solar cooker is left unattended the sun's path progresses and the cooking ingredients cool, which protects them from being overcooked.

To minimize time spent adjusting the unit, I estimate the perfect setup for the unit in relation to the sun's location thirty minutes into the future. Therefore for the next thirty minutes the temperatures will rise, and I can make another adjustment in an hour to further intensify the heat if need be. The hours between 10:00 a.m. and 2:00 p.m. offer the maximum solar intensity, although I have successfully started food earlier and later in midsummer. In full sun on a 70-degree day in the summer, a pot of rice or beans will cook in two hours or less.

The second style of solar cooking is commonly referred to as parabolic cooking. This cooker utilizes the geometry of a parabolic shape to locate a focal point—where the sun's rays concentrate to produce intense heat, similar to that supplied by a cooktop and effective for frying pans as well as pots.

When I was at SEI, for example, we used a unit constructed at a German technical college for worldwide distribution. This unit's reflective panels and tubular steel created a highly practical cooker. The cooking surface was elevated directly in the focal point of its umbrella shape so that sunlight was focused from all directions when properly aligned.

Later during my travels in Argentina, I encountered this design again at Gaia. While we have been unable to replicate the versatility of either the SEI or Argentinean units here, we have constructed a parabolic cooker from a salvaged satellite dish. This huge unit was difficult to adjust and fix in place, though it was extremely powerful. The power of the unit forced the cooks using the equipment to wear welding goggles and cover their skin for protection from the intensity.

Root Cellar

The construction of the root cellar was part of the Community Building project. Our intention was to preserve food without electrical inputs. For our root cellar we chose a location in the basement inside the northern perimeter of the foundation. Instead of the slab concrete that was poured into the rest of the basement the root cellar floor remained sand. The sand floor has allowed the cool yet consistent temperature of the earth below the frost line to permeate into the space. The walls of the room are insulated to isolate the room from the ambient heat generated by the appliances in the basement. The floor of the root cellar is level with the top of the footings of the foundation wall below the frost line. This creates a space that is chilled by the freezing temperatures of the surface of the earth yet maintains resistance to that cold because of the consistent temperature of the earth below the frost's reach.

An active and a passive ventilation system enhance our capacity to control the temperatures by taking advantage of outdoor ambient conditions, composed of a six-inch pipe with a cap as well as a louvered fan. In the fall we can lower the temperature in the room by inviting the cool night air into the space. If the space becomes too cold we open the door to the rest of the basement, allowing the heat of the building to enter.

Humidity in the root cellar is preferred for crop storage. Unfortunately, high humidity negatively affects common natural building materials, so we choose moisture- and mold-resistant drywall similar to what is utilized in bathrooms for the fire protection on the ceiling and close-cell, urethane foam as insulation. These compromises were made to insure the longevity of the structure. Typically we do not proactively augment the humidity in the root cellar and are satisfied by the level supplied by the natural conditions. We prefer to sacrifice maximizing the storage conditions in favor of preserving the structure.

The root cellar is extremely effective for storage, particularly in the winter months. From November to April we are able to safely store food items at temperatures consistently below 42 degrees, enabling us to limit use of the refrigerator to the summer months. Instead of standing in front of the refrigerator to decide what to eat, we bring a basket into the walk-in cellar and choose from the selections available.

We have considered building an icehouse to augment this system. While this work has traditionally been done on lakes by cutting large blocks, we aim to freeze containers of ice and store them in coolers in a shelter constructed in the earth on the north side of the Community Building.

Bike Power

We are interested in bicycle- and human muscle–driven devices, which help illustrate the reality of the fossil fuel derived energy we take for granted. The amount of work accomplished by fossil fuel energy on a daily basis is unimaginable. The difficulty of supplying just one horsepower of energy continuously through human power with engineered mechanical advantages is immense. Imagine how many humans it would take to pull the weight a horse can.

However, bike power is helpful but is not a remedy for the lifestyle to which we have become accustomed. For example, a typical washing machine has up to a one-horsepower motor. This kinetic potential is required to agitate and spin heavy wet fabric

in the confines of the machine. For a human to exert the energetic force necessary to power this process requires an equal investment in energy. Even with mechanical help, providing that energy is difficult. Our inspiration to create bicycle-powered devices came from the Maya Pedal videos available online, which illustrate the use of these devices in rural Central America.

We have relied heavily on cheap electrical capacity to accomplish our daily tasks. But instead of electricity, we can accomplish some motor-driven tasks with bicycle power. The most basic design is a direct drive from the pedals to the implement. Increased efficiencies are obtainable through gearing and the use of a flywheel. The gears allow the user to adjust the power-per-resistance ratios while the flywheel builds the potential embodied energy.

A flywheel is a heavy round object that can be spun on an axis. The concept is that as the flywheel gains momentum it becomes difficult to stop, as with rolling down a hill. The user can start the wheel in a low gear and then shift up as it accelerates and inertia decreases. By constructing the axis using a freewheel the user can occasionally rest as the flywheel continues to spin. The weight of the flywheel then drives the implement.

We also are designing equipment that is interchangeable so the bike and flywheel become a power station that can operate several different devices. We hope to create a power station in which one or more riders can power different devices depending on their needs. By combining people power we can overcome the energy quotient necessary to complete chores we've become accustomed to accomplishing with fossil fuels.

To achieve maximum utilization of bike-powered tools they must not only be effective at replacing the convenience of an electric device, they must also be accessible and fun. Designing our bike-powered tools to allow purposeful, intense workouts as well as social and recreational sensibilities will improve the frequency and enjoyment with which these tools are utilized.

The rear wheel of the bike belt drives a freewheel flywheel that powers a vintage washing machine.

Methane Digestion

We have experimented on a limited scale with methane digestion. This process uses gases generated through anaerobic digestion as a fuel source. Similar to the way our stomachs produce flammable gas, the methane digester uses microorganisms to produce energy.

The concept of methane digestion is well known globally. In our region there are large-scale electricity-generating facilities in Vermont that depend on five hundred–plus cow dairies and landfill-capped systems that generate electricity from the buried trash. Our intention is to create a system with small-scale applications. During my time at SEI as an intern after college, I studied with a designer who installed a greenhouse digestion complex in Tennessee. We constructed a fifty-five-gallon experimental batch system that filled an inner tube with combustible gas. I also studied the balloon systems that were being designed and installed in Central America to provide cooking fuel in response to the deforestation pressures.

The methane gas is generated in a liquid of decomposing biologic materials that has available nitrogen, more or less a manure soup. The containers must be able to endure the caustic chemistry of these potent ingredients. As the gases are generated they are either utilized directly or stored for future consumption. To perform optimally this decomposition requires a critical mass similar to that of an exothermic aerobic compost. Once a critical mass is attained, the activity of the microorganism generates heat sufficient to induce further activity, creating a chain reaction of heat production. This chain reaction creates enduring biologic activity that produces the flammable gas in consistent, usable quantities.

We have been stymied in our experiments by the ambient temperatures in New England. Even during the summer the digestion is inhibited by cool temperatures, while during the winter it would require protection from the cold and supplemental heat to induce the decomposition process. After further research to determine the viability at this latitude, we would consider the possibilities of burying a unit similar to the common Chinese model or building a unit that incorporated a passive solar heating element to augment the heat generated in decomposition.

CHAPTER TWELVE

WATER SYSTEMS

We receive approximately forty inches of rain per year from the natural energetic cycles of evapo-transpiration. This abundance fosters a flourish of biologic activity. Following the spring runoff of melted snow, aquifers are charged with a fresh, clean ingredient for agriculture. The water that falls uphill from the farm sheds downstream with gravity to join the South Branch of the Baker River, which then joins the Pemigewasset, a tributary of the Merrimack that eventually pours into the Atlantic Ocean. This valuable resource carries nutrients from the earth's crust down into the coastal estuaries, where this oxygenated and nutrient-rich water feeds the life of the estuary ecology.

Agricultural success depends on successful utilization of water. While the volume of water in our region is adequate, the propensity for extremes requires proper planning. There are periods of rainy weather and high-precipitation storms, including fall hurricanes and the famed nor'easters, as well as stretches of dry weather in which rain may not fall for thirty days or more. These variances create challenges that water management systems must be designed to handle.

Conservation is crucial in all energetic cycles from which we seek to derive benefits. Water is no exception, and our designs must reflect that reality. When the water resource is maximized, the resource is efficiently utilized to address our needs without waste. The process requires perpetual observation

and analysis of the prioritization and functionality of the system and its components.

Bodies of water also serve as habitat for a multitude of aquatic species. These species range from native amphibians and birds to introduced aquatic food crops such as fish and rice. The edges of water bodies provide an ecology edge niche where land and water meet, allowing for species native to both to intermingle and the diversity of species to increase. This is true in the microcosm along the edge of the stream as well as in the macrocosm in estuaries where fresh- and saltwater are mixed along the coast. By encouraging water to take the longest route possible we increase the amount of available edge. A winding river has more edge and available habitat than a straight channel.

Water can be utilized within the farm system for a multitude of purposes. Water that falls with the force of gravity can be used as a power source. Bodies of water are useful to draw from for fire suppression and irrigation, as well as recreationally for swimming. Water falling on the landscape provides aural and visual serenity. The tranquillity of this experience is as primal as gazing into a campfire, and it rewards the soul. By retaining and capturing the energetic potential of water falling from the sky within the farm system we enable a multitude of energetic yields to be harvested.

It is crucial that water be utilized to its greatest potential within the farm system, and to capitalize on the existing energy embodied in the global hydro-logic cycle. This potential must be harnessed with an awareness of the strength that this resource can manifest with potentially negative consequences. Too much water or too little water is not good: It is important for water levels to have continuity and permeate the ecological system over time, versus the consequence of extreme events.

Water in excess can be counterproductive to the farm system. Surface runoff creates erosion. As water accumulates and drains down toward the sea the strength of its current impacts the land. Within the soil, light organic materials float away, while smaller sand particulates are washed downhill. As the volume and speed of the water increases, so does the erosive potential. Other water-related effects we need to contend with include the freeze-thaw cycle and the weight of snow. Heavy snow can cause a structure to collapse or pose dangers to people and animals if it slides off a roof. The destructive potential of water must be recognized and factored into the design.

Snow affects the climatic conditions at this latitude by amplifying and extending cold conditions. The white landscape reflects solar radiation that would otherwise be absorbed by the darker colors of the earth. Snow coverage chills the air and radiates throughout the landscape, particularly between sunset and sunrise. Weather conditions such as fog and rainfall similarly affect not only the humidity but also the temperature.

As winter recedes and the days grow in length, the frozen reserves melt into the aquifers and recharge the water table with the resources required for spring vegetative growth. Water retention measures aim to amplify the capacity of this abundance so that it can be used throughout the growing season. While the inconsistent summer rains do provide some relief, attention must be paid to providing perpetuating sources of water for irrigation throughout the growing season.

The water table fluctuates based not only on seasonal shifts but also on the variations of depth and reserves. When you dig a hole most anywhere on the property during spring thaw or following heavy rains, water will spring forth. Through the summer heat and drought, the water table shrinks back down into the earth until it is refreshed by the heavy precipitation in the fall. With the exceptions of some notably well-drained deposits of glacial sand, our mountainside at D Acres is composed of granite, soil, and water and features bellies and bowls in which springs and surface hydrology combine to provide wetlands and small bodies of water.

Surface permeability dramatically affects the rate of absorption for water. Consider the landscape as a massive catchment system designed to enhance the biologic capacity of water for the area. Impervious surfaces such as concrete, conventional roofing, and pavement shed nearly 100 percent of the water that falls on them. This lack of permeability means that adjacent ground must absorb the excess or it will become surface water, causing flooding and erosion. In conventional agriculture, for example, the plowpan caused by the repetitive shearing of cultivating equipment on the subsoil creates an impregnable layer that resists absorption of water. The result is stagnant water that sits in the spring fields until it is evaporated into the atmosphere.

The power of water can be quantified predictably and formulaically based on reliable gravitational and friction-resistance coefficients. The power is an expression of the factors of head, flow, and resistance. The head or height of the water exerts a gravitational pressure. This pressure (pounds per square inch or PSI) is consistently expressed regardless of the volume. Thus one gallon of water one inch deep exerts the same pressure on a drain pipe as one hundred thousand gallons of water one inch deep. There is a linear progression of pressure based on the height of the water at 4.2 lbs/per foot. The flow is the volume of water that is energetically involved in the process. The flow is often dependent on the size of the opening through which it moves. The principles of hydrologic physics dictate that a smaller opening offers

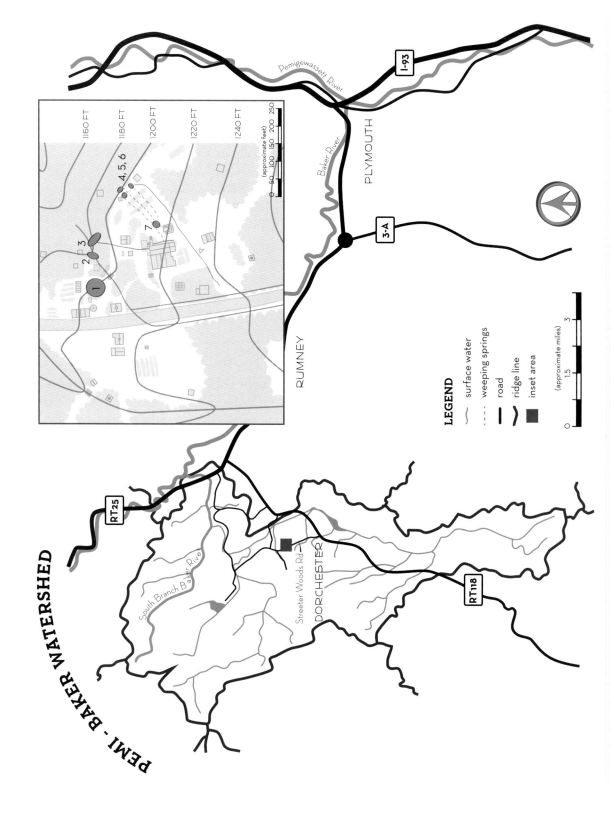

This map illustrates the source and movement of the region's tremendous water resource as it migrates across the landscape at D Acres. The Baker River Valley is also illustrated to contextualize D Acres in the larger scheme of the watershed. Illustration by Marylena Sevigney.

substantially more resistance to the flow of the liquid through a portal.

Above all, water seeks company. The physics of its properties at a molecular level encourage cohesion. By recognizing these physical properties, the utility of water can be designed for and integrated into the landscape. All phases of the water cycle should be considered, from collection to usage and purification.

Watershed Design

Hydrologic construction has been practiced for millennia by beavers. These industrious communities of animals provide an important ecological service; by constructing dams across the waterways they slow and spread the water. As the velocity of the water is slowed, the organics and minerals that the water was transporting downhill are concentrated and become a resource for biological activity. The dams serve as strainers that trap the silt and increase fertility. Eventually, the prodigious beavers will clear an area of their food supply or the pond will silt in, at which point they migrate. Once the beavers vacate, ecological succession continues as species move in to occupy the zone and compete for the fertility and solar access. This ecological succession transforms the landscape as the dams form terraces through the watershed.

Though not as single-minded as beavers, we attempt to create the perpetual availability of water in the D Acres landscape through taking advantage of natural rainfall, with conservation measures such as mulch and water retention, and through landscape features such as swales. As humans our understanding and relationship to water in the landscape begins as children by playing in puddles. With our fingertips we start to understand the physics of water flow and hydrology. Through our initial efforts to channel and concentrate the flow of this substance its fundamental principles are imbued. The properties and dynamics observed on a small scale and through physical experimentation can be extrapolated to help the design of a whole landscape.

Our goal is to reduce erosion and build fertility through the availability of water on the landscape. A fundamental caveat of watershed manipulation is that our human endeavors are designed to enhance the energetic capacity at this elevation without causing negative consequences downstream. Our intention is not to deplete the flow that feeds the valley below but rather to augment the watershed potential. The invigoration of our water table has positive effects for the surrounding areas. By retaining water throughout the watershed, we mollify seasonal fluctuations.

A properly designed system allows for overflow in the event of extreme volumes of water. The location for the collection overflow should be designed to mitigate damages this flow can cause. In situations where the water table becomes saturated all precipitation becomes runoff, so purification systems, such as septic and greywater, must be designed to prevent pollution and contamination of the surface water. In locations where water has been retained, it is essential to plan for overflow. When water overflows a water retention structure, the flow of water can damage the landscape downstream as well as the structure itself. The force of water freeing itself from the restrictive measures can be destructive. Often at the lowest point of least resistance water can be spread evenly, forced into a sluiceway, or diverted around the water retention structure to avoid damage.

Along the north-sloping landscape the watershed is split by the Community Building. South of the building (upslope), water travels from the parking lot and oxen house pasture to the east and into a series of ponds by the resource pile depot. This tributary is connected to the Community Building ground gutter, which overflows downhill into Pond Number five, then into Pond Number seven. North of the Community Building the water is also aggregated through a delivery and storage system. The original pond is fed by a groundwater spring, while Pond Number 2 is fed by number 1 overflow, greywater from the solar shower, and the ditch that brings water from the roadway and farther uphill. This flows into Pond Number 3, which is augmented by the perimeter drain of the Community Building and the greywater of the outdoor kitchen. At the bottom, below Pond Number 3, the two microwatersheds join, creating the possibility of further expansion of water potential.

Southern Microwatershed

The history of this system begins in 2004, when we undertook the construction of a parking area to the southwest of the Community Building. Cutting into the forest and removing its protective canopy changed the water dynamics in this area dramatically. The exposed surface area provided the space for precipitation to accumulate and migrate via runoff directed toward the Community Building. The southern garden bed and roadways became progressively muddier, and the humidity in the basement increased. The ramifications of our convenient parking included a muddy walk to the homestead.

To rectify this situation we decided to divert the water via a ditch-type diversion system, which captures water accumulated in the parking lot and channels it eastward uphill on the roadway. A four-inch perforated pipe runs along the bottom of the entire length of a three-foot-deep trench dug from the parking lots nearly four hundred feet until it empties into a small pond. We covered the pipe with clean, well-draining

From the parking lot two pathways diverge. To the right is a drivable road that leads to the G-Animal, animal areas, access to the ponds, and field space.

The road leading to the G-Animal along the south side of the house from the parking lot is a ditch line on the contour. Within the ditch is a buried perforated pipe bringing water to a pond. Above the ditch is a hedgerow of native trees and shrubs, including serviceberry and wild blueberry. Over time we have begun to stabilize this ditch-line hedgerow by transplanting other perennial and fruit-bearing trees, shrubs, and herb plants, such as Nanking cherry, medlar, gooseberry, mullein, and walking onion.

sand. A shallow surface ditch follows the course of the pipe above its route, creating redundancy. This integrated ditch captures the surface runoff as well as water that migrates through the first three feet of the soil horizon. During most months of the year the flow is continuous and is substantially higher during spring runoff and intense rainfall events.

Northern Microwatershed

This system has a longer history, beginning with Pond Number 1, the original pond across the street from the Red House, which was dug in a location that the cows always avoided because of its muddy conditions. My uncle Delbert, with the help of a Korean War vet who had experience with heavy equipment, obtained an excavator and dug out the hole. Although they were stymied by a large rock in the center of the location, the muddy area had indicated a spring, and the hole soon filled with clear, cold water. As a youngster I swam in the pond and sunned on the rock. I also spent countless hours observing the golden fish my uncle had stocked into the pond. By the time D Acres was created the pond had filled in substantially with

The original Pond Number 1 behind the barn in 2004. The Lower Garden was already well established.

silt and cattails, and we began to evaluate the possibilities for further earthworks aimed at augmenting our available water resource.

In 2010 we undertook a major excavation project to enhance the water resource. Our goal was twofold: (1) to dredge the existing pond; (2) to construct earthworks to capture other sources of abundant seasonal water that flowed across the property. While the dredging was fairly straightforward, our attempts to capture and clean the water flowing from the roadside ditch were more difficult. The steep terrain necessitated a series of ponds to slow and capture the energy and nutrients carried by the seasonal and storm water flow.

While there is always a concern for fragile wetland ecology, the primary governmental restrictions placed on the construction of earthworks limit the dimensions of dams. Dams are structures that impede the natural flow of water. It is illegal to construct a dam four feet in height above the preexisting grade of a site. This height provision is intended to prevent liability downstream in the event of dam failure. Essentially, the government mandates a property owner to dig deep to construct a water body rather than construct a tall dam. Interruption of a flowing stream or waterway is also severely restricted, and activities that encourage erosion are subject to regulation.

In pond construction the objective is to dig deep rather than rely primarily on a retention structure. Shallow, wide ponds heat water more rapidly,

Pond Number 1 in 2010, after all excavation was complete and the conservation seed mix had already begun to grow

The first step in the process was to remove the trees in the area.

A roadway was built along the topside of the series of ponds.

The completed waterworks provide a multiuse platform for aquatic systems development.

A view of Pond Number 1 from the north one year later.

Taken at the height of summer 2011, one year after the excavation project

May of 2012 brings the cattails back to the edge of the pond. The pondscape is beginning to flourish, with a diversity of native and transplanted plant species.

though more is lost to evaporation. Six-foot-deep ponds are necessary for trout, as they require cold, aerated water and a zone that remains unfrozen throughout the winter. While depth is important, a significant portion of the water's edge should be shallow so that any animal, such as a cow or moose, that falls into the pond can escape.

All ponds and water bodies must be constructed to withstand flood conditions. In conditions of extreme volume, it is important to consider multiple solutions for overaggregation of water. In Pond Number 3, for example, there is a pipe underdrain that maintains the water level below earthworks that retain the water. If this pipe clogs or is overwhelmed by volume, the earthworks were graded level so

that the water will seep over along the length rather than cut into one spot, which would result in severe erosion. Spillways can also be designed to circumvent the direct course downhill, ensuring that the erosion of a flood event does not compromise the water retention structures.

Ponds and other bodies of water can serve as a resource for fire suppression and irrigation. For fire suppression it is important to develop four-season contingency plans. During the earthworks accomplished in 2010 we installed a dry hydrant that extends from the roadside down into the pond below the frost line. The hydrant is designed so the fire department can quickly couple onto the pipe and pump filtered water rapidly from the pond.

To dredge the pond adequately we chose to pump out the pond and install a pump shelf in close proximity to the spring. This pump shelf was installed by digging a deep hole, then inserting a twelve-inch pipe vertically. The submerged portion of the pipe was perforated to allow water to enter easily. This pipe was then backfilled with riprap, four- to six-inch irregular stone that has maximum structural and water permeability capacity. A pump can then be dropped down into the pipe so that water can be pumped from the pond without sucking up excessive silt.

Watershed Protection

Weather conditions are an important consideration when undertaking earthworks. Dry conditions will speed efforts and improve the quality of the work accomplished, especially when using mechanized equipment. Slippery, wet conditions make the process messier and can result in additional time spent with equipment damaged or stuck in the mud, and it becomes necessary to bail a hole before you can adequately proceed with any further digging. Consequently, it is important to build earthworks during dry seasons if possible.

After a soil disturbance we will seed or mulch to protect the wound of earthworks from further erosion. Our choice of seeds depends on the time of year and our goals for the site. In the spring or fall we generally plant rye, which is inexpensive, cold tolerant, and fast growing. During the warmer months we plant more clover and alfalfa, particularly in areas where animals may be pastured. If dry weather persists it is crucial to use methods such as irrigation to ensure that the seeds germinate and thrive.

Plant roots, mulch, and foliage are helpful to prevent erosion. The impact of rain and the possible effects of runoffs are more severe if the soil is not covered by vegetation or mulch. Their crisscrossed architecture acts to diminish the catastrophic effects that occur when water overwhelms the structural integrity of the soil system. The filaments and fibers of the vegetative material create a structure resistant to erosive forces.

In particular we have found cattails and willow to be vigorous and productive in our aquatic environments. Both these species can be readily propagated, willow by cuttings and the cattail through division. The cattails are edible, and the mat of their roots filters water and retains soil. The willows grow rapidly to provide an expansive root system and can also be coppiced for basketry.

A variety of materials can be utilized for mulch. Covering the soil with a layer of organic material preserves moisture and builds fertility while also suppressing weeds. We have utilized seaweed, leaves, grass clippings, hay, straw, barnyard manure, cardboard, newspaper, compost, and combinations thereof to create mulches.

Swales are intended to slow the progression of water as it flows across the landscape, while releasing the water over time to maximize the biological capacity of the area, and are constructed by piling biomass in correlation to the contour of a slope. They are typically smaller than terrace structures, and the land is not leveled to the swales' height. The biomass acts as a surface sponge resisting the downward flow of water, absorbing the available water, then distributing it slowly down the slope. Often swales are designed so that vegetative plantings can get the most out of the accessible water.

Partnering with animals within the waterscape requires conscious management to attain our identified goals. Animal density and their duration at a given location are primary considerations. While the fertility of the pond systems can be enhanced by waterfowl, ducks, for example, can annihilate any existing amphibian populations. Animals can cause damage by destroying preferred plants while ignoring the competitive undesirable species. While pigs are very effective at compacting the soil of a pond so that seepage is minimized, they also may dig or otherwise destroy earthworks structures. The predation and turbidity of the water caused by confining waterfowl affects the native species as well as any fish that have been introduced. While their urine and manure may be beneficial to the long-term fertility of the site, concentrated dosages in a short time frame can result in an algae bloom that will kill any fish.

Woods Roads: Access and Erosion

When we first moved to the property, we began to realize the extent of the damage from the mechanized logging that had occurred several years before. The loggers had often chosen the most direct routes down the hillside. The process of dragging whole trees down the slope with a four-wheel-drive skidder loosened and exposed the soil. Once the loggers left the woods, the water relentlessly eroded the exposed roadbeds. The skidding paths had not been protected with erosion control practices, and years of neglect created a network of tributaries that propelled water and soil down from the mountainside. In several places water had cut three-foot-deep channels into these old access routes. The deep gullies and exposed rocks prevented access to a large portion of the upland section of the property. At that point our principal intention was to reduce the speed and volume of water traveling down these logging scars to reduce its erosive force.

After inspecting the damage to the forest, our neighbor Jay Legg agreed to work with us to rectify the situation before it escalated further. We applied for funding that was available for storm mitigation on forestlands. After funds had been secured Jay brought his bulldozer, and we began working up the roads, replacing the soil and installing erosion-control earthworks. To channel water off the roadway we pitched the grade of the roadbed so the water shed to the side. We also periodically installed water bars across the roadway, depending on the steepness of the slope, to mitigate erosive flow and velocity of the water.

Well Options

Traditionally wells in this area were constructed by lining holes with laid stones. Such holes were less than a yard wide, required careful construction, and were generally less than thirty feet deep. The drinkability of these hand-dug wells could be undermined by an animal's feces or accidental drowning. Modern, hand-dug wells are built with round concrete tiles to capture water and provide safe accessibility.

To provide our drinking water we utilize a drilled well. This energy-intensive and expensive process is similar to that of drilling for oil or fracking. Large derrick equipment is required to drill into the earth, then insert a submersible pump. These pumps access water in the bands of aquifers and subterranean water available through the strata that has been drilled. Our pump was plunged five hundred feet down the hole. The volume of water available is determined by the depth of the water table and the recharge capacity that feeds the well. The deep subsurface wells provide ample, clean water; however, the initial cash expenditure is high, and the system depends on the substantial infrastructure of pump and wiring, as well as the electricity required for operation.

Irrigation

Our food crops thrive when they have consistent availability of water. While there are aquatic species that thrive in standing water, most plants prefer sufficient drainage and an adequate water supply.

The active delivery of water to our plants relies primarily on a low-pressure drip irrigation system. We have chosen a drip system for its efficiency at delivering water directly to the plants with limited evaporation and incidence of weeds and plant disease. Sprinkler systems in which water is sprayed into the air are inefficient because of the large amount of water that evaporates; they also encourage fungal diseases to spawn.

We purchase most of our supplies from DripWorks, a California retailer that focuses on gardening in arid climates. There are three primary options for delivery of the water: soaker, emitters, or inline.

The Upper Field is the largest garden space to be outfitted with a gravity-fed drip irrigation system. There are over forty beds, all complete with two separate lines running down the length. The piping is connected to a tank that sits at the top of the hill above the field. Water is pumped from the ravine below, then sent through the lines to irrigate specific zones.

The soaker oozes water from its many miniature holes. The soaker hose is typically buried so the water is delivered directly to the roots. Although the roots cannot clog the delivery process these hoses require sufficient pressure to force the water from the pores and permit delivery. To utilize these hoses it is often necessary to bury them before planting, making them less appropriate for annual cultivation, although we have been successful growing potatoes by laying the pipe as we plant. Pulling the pipe up during harvest also aids access to the spuds. Soaker hoses can be easily damaged by cultivating tools.

Their soft plastic is easy to puncture and difficult to repair. If a hose is punctured during the middle of the growing season the section must be dug up and replaced.

Emitters release water in different quantities, adjusted to release either a spray or a variable drip. With a simple puncturing tool, these units can be inserted into the pipe at exact locations for precise delivery to plants. These emitters can clog, and prices vary depending on their capacity. Emitters are not capable of adjusting flow in relation to pressure variables, so flow tends to concentrate unevenly.

The tubing is laid out to irrigate in the vicinity of seeded or transplanted crops. While drip irrigation allows the water to penetrate the soil slowly over time, direct-sown seeds still require some initial care with hand watering.

Inline piping is our preferred choice in terms of economics and reliability. This system is more difficult to clog and will operate with low pressure, delivering equal volumes of water evenly the entire length of the pipe. The pipe can be purchased with different options for the spacing of the inline emission. In general we choose eight-inch or twelve-inch spacing for usage in a variety of circumstances.

The capillary capacity of the living soil is immense. The web of life formed by the microscopic strands of mycorrhizae and mycelium in the living soil provides a network for the absorption and horizontal transfer of water. When you dig below a surface emitter you can get a sense of the broad, horizontal absorption in the soil profile. The water travels horizontally in the profile captured by the osmotic and capillary potential of this biologically connected underground universe.

We use significant footage of the half-inch black poly pipe, as well as the associated connectors and shutoffs of that size. The components are designed to allow installation with minimal tools and can be adjusted in the field as necessary.

To deliver the water we receive assistance from gravity. Because of the low-pressure requirements of our drip system, we generally will fill a storage tank so it can drain into the garden of its own accord. The head or height of the water above the garden is proportional to the pressure of the water that can be obtained in the system. Filtration of the water is essential to prevent the various emitters and pipes from being clogged. Pumps and tubing must be drained seasonally to avoid damage from freezing conditions.

By designing our gardens in response to the slope of the land we are constantly searching for ways to garden with the contours of the terrain. By following the contours, the consistent level line of elevation across a slope, we are spreading the resource rather than accelerating its departure. In our field conditions the no-till mulched beds absorb precipitation, allowing the water to sink into the landscape rather than being evaporated or carried down to the Atlantic. Every row and bed lies across the course of the water flow, acting as a sponge and sucking the water up until saturation.

Roof Rainwater Catchment

People are amazed at the number of gallons of rainwater that can be harvested from rooftop runoff: One inch of rain will produce seventy-five gallons of water on a ten- by twelve-foot roof. In the simplest scenarios buckets are placed under the drip line of the building. We also utilize gutters—either salvaged or fabricated on-site—to concentrate the volume of water to specific locations. Gutters function by catching the water as it drips from the roof and channeling the accumulating flow to a downspout device. Rather than having water cascade and splash down from the gutter, a piped downspout or chain of links can be utilized to help direct and control the water flow.

To collect the water we can utilize a tank or a drum. A fifty-five-gallon drum will fill rapidly even with a small roof, and systems need to consider overflows. Water can also be piped downhill to other locations such as ponds or animal water tanks. Rainwater is potable, although roofing material can erode into the liquid, and there are concerns regarding mercury and other atmospheric contaminants.

On the east side of the Community Building, water presented us with a challenge. The expansive roofs concentrated a tremendous flow of the water, which caused severe erosion to the garden beds we were attempting to establish. The height of the roof diminished our enthusiasm for a removable gutter system, and the winter snow would certainly destroy any four-season gutter installation. In response to this dilemma we introduced an innovative ground gutter system. For this system we cut some twelve-inch pipe lengthwise, then laid the pipe on the ground so that it was pitched from north to south along the drip edge. Where the water

A simple gutter system that easily fills two large storage tanks. These tanks are connected to drip irrigation, and they gravity-feed water to the Mandala Garden.

arrived at the southeast corner of the building a pipe was laid crosswise and pitched down along the eastern slope. We backfilled around the pipe, then inserted a perforated pipe, which was then covered with half-inch stone. At the end of the pipe we dug a hole with a bulldozer and raised pigs in the area for a season. The pigs compacted the edges of the minipond to reduce seepage, and the pond has held water since.

Wastewater Management

We refer to water that drains from sinks, laundry, and showers as greywater; urine is classified as liquid gold; and any water contaminated by fecal matter is deemed blackwater. Urine contains high concentrations of nitrogen and other available minerals that can provide fertility to the landscape. The blackwater and fecal matter is extremely nutrient rich, yet is dangerous because of pathogens such as *E. coli* that can have serious health effects for humans. All three of these streams require treatment and remediation for safe handling, reuse, and maximization of the resource.

When considering options for treatment of wastewater it is important to consider zoning statutes and regulations. The intention of traditional septic systems is to spread all of this liquid underground so that both the smell and any dangerous pathogens are trapped in the soil and filtered over time. Our

Once the G-Animal was built, pigs were able to inhabit the area. In 2007 a ditch was dug to bury perforated pipe to more effectively move the water into the pond.

systems for wastewater treatment intend to clean and reuse the water versus relying on a system of burying our waste.

Water can be biologically and chemically contaminated. Biologic hazards include bacteria and microorganisms, which are organic. These contaminates can be mitigated through filtration and chemical treatments that kill the organisms or reduce the composition into a benign form. Chemical contamination results from nonorganic compounds and can be mitigated by dilution or distillation. Without these cleansing processes the contamination of the water will persist.

Greywater

To clean and maximize the utility of our domestic wastewater stream, we have built several rudimentary greywater systems through the years. These experimental designs are used for outdoor applications, such as showers and the sinks at the outdoor kitchen area. Despite the lack of fecal material this liquid can become septic if not treated properly and in a timely fashion. Greywater that is not moved rapidly through a purification process can rapidly develop pathogens, undermining the efficacy of the system. Since this water was not initially septic or blackwater filled with fecal pathogens, the goal is to filter the water and reuse as much as possible for irrigating the landscape.

In a typical greywater system we try to maintain the motion and flow of the water through several different stages. By ensuring that the water is perpetually moving through the system we reduce conditions conducive to contaminating pathogens. In the first stage we screen-filter the material to remove the large debris, such as plate scrapings. After the fluid is screened, it passes into a tank, typically a salvaged bathtub, which consists of layered one-inch stone, sand, and charcoal. The tanks are connected so the water flows with gravity through the system. The initial stage allows rapid filtration and maximum permeability with higher proportions of stone. The second stage is similar, although the

sand and charcoal are utilized in a higher concentration, to physically and biologically cleanse the water. In the third stage we release the water into a container with a soil medium in which we grow water-loving plants with filtrating roots, such as mints or cattails. We also rely on the diluting force of rainfall, which regularly showers the unit and helps to rinse the entire system.

The greywater system that I observed at Gaia was a direct discharge system. The greywater flowed through a series of perforated pipes that were submerged in a lagoon with a multitude of water-loving plants. The lagoon was lined with plastic, and the pipes were covered with sufficient sand, then soil, so the water pulsed through the roots and through the plants without saturating the soil medium and becoming visible on the surface. The size of the system was able to withstand the volume of subterranean greywater such that the odor and sanitation of the system was not an issue. The plants were able to absorb and purify the quantity of water discharged throughout the relatively warm Argentina winters.

At D Acres we have plans to enhance our greywater management by constructing a "living machine." The idea is to utilize the greywater from the Community Building as the basis for a biological system that would clean, reuse, and amplify the potential of the wastewater. The design of the living machine has several components. Water that drains from the building would be initially filtered for organic particulate with sand, charcoal, and stone. After the initial treatment the water would be utilized to grow plants. Depending on the complexity of the system the water could then be further utilized as habitat for fish. At this point, while we have established a site and some preliminary designs, we have resisted this construction for several reasons. While purifying our wastewater through this process would be an improvement, we have an existing septic system that we were required to build as part of obtaining a commercial kitchen license, so there is a deficient yet existing system on which we currently rely.

This is the greywater system utilized when the Red Barn was the primary community center. The used water was strained into the blue container at the top left. The water then filtered through two containers filled with gravel and sand, then to a third that contained soil and a moisture-loving plant such as mint. And finally into a larger holding barrel from which a watering can could be filled for use in the garden.

While we are optimistic that the warm water exiting the building into a glass-enclosed structure will provide for rampant biologic activity, we are also uncertain about the sizing and design of each of the stages. Because of the economics involved and the experimental nature of this project, thus far it has been deprioritized.

Humanure

Conventional flush toilets waste tremendous amounts of clean water. Water that could be utilized for irrigation, bathing, washing, or drinking is polluted through the standard conventions of flush and forget. By seeking alternatives to the standard sewage systems of conventional society, we aim to maximize fertility while minimizing energy consumption and human health safety issues.

The standard sewage system uses water to convey and discharge dirt, grease, and human waste. If the dwelling is not connected to a centralized municipal system, it is typically addressed by the conventional septic system. In the first phase of the process the sewage flows to an underground tank. This concrete or plastic septic tank allows solids to settle out as the liquid migrates to a leach field. The leach field filters and dissipates the liquid into the groundwater.

Leach fields depend on the ability of the soil to percolate the liquid downward. If the leach field is not situated in a location that can absorb the quantity of water generated in the facility, then the liquid will seep to the surface. This water has a noticeably unpleasant odor and can present a health hazard.

Our desire to create wastewater and humanure alternatives is based on the observation that the conventional system is impractical and imperfect to the point of counterproductivity. Municipal systems pump liquid from every household to a treatment plant. The rural system depends on a buried tank and ground filtration. In both these systems, by mixing the liquid from all usages, water is needlessly contaminated with fecal foulness. The septic tank system also depends on occasional pumping and hauling to a centralized chemical treatment facility.

The study of wastewater management has been advanced by the implementation of alternatives documented and designed by several noteworthy authors. Nancy Jack Todd and John Todd provided blueprints for alternatives with their book *From Eco-Cities to Living Machines: Principles of Ecological Design*. While this book focused on a wide range of wastewater issues, other authors focused more exclusively on the issues of human waste. Joseph Jenkins's *The Humanure Handbook* serves as an indictment of modern septic and sewage systems while also providing practical alternatives for the conventional flush systems. *Liquid Gold: The Lore and Logic of Using Urine to Grow Plants* by Carol Steinfeld continues the vein of effective waste management while offering an additional urine separation and reutilization regime.

There are many alternative options—after all, the conventional toilet is relatively new to human history. Western civilization has known since the plagues of the Middle Ages that sewage must be properly managed to avoid disease. Over millennia we have developed many alternatives for fecal disposal that do not rely on polluting our water resources.

Composting toilets offer an alternative to the standard system. The number one benefit is that they separate the fecal material from the wastewater treatment. Isolating the fecal material guarantees that it can no longer contaminate the water. If the wastewater is not contaminated, then it can be more readily filtered and reused for other purposes, such as irrigation.

We operate a Clivus Multrum in the Community Building. This unit was chosen during the construction phase because we wanted a dependable, tested alternative to the standard flush toilet. The Clivus is imported from Sweden and was developed to meet the needs of northerly locations where septic systems were not feasible or not cost effective. The Clivus system begins in the bathroom with a toilet seat that resembles the standard porcelain throne, except that the Clivus seat lacks a flush mechanism. The seat sits directly above the collection unit in the basement, connected by a sixteen-inch-diameter tube. During a standard sitting the vast majority of the material follows the laws of gravity and lands in the reinforced plastic unit. After completion of the excretion the toilet paper and two cups of wood chips are tossed down the chute. The tubes and the collection mechanism are fabricated in sections from a heavy-duty reinforced plastic material.

The Clivus also offers reliable odor control thanks to a fan and pipe system that creates suction, pulling the aroma from the unit through the exhaust pipe installed through the rooftop. This unit does not have a urine separator at the toilet. Instead, after any liquid effluent percolates through the pile it is pumped into a separate tank. This dark liquid can then be poured onto compost piles so that its nutrients are retained in the agricultural system.

While this unit may have been effective for its original purpose of occasional usage at alpine lake cottages in Scandinavia or even the needs of a nuclear family, the capacity of the unit has been maximized at D Acres because of the volume of material inputted. During summer months from five to fifteen people plus guests can regularly use the

The entire Clivus unit sits in the center of the basement. The white circulation fan is on the top.

equipment, straining the ability of the unit to decompose the material before it needs to be extricated because of spatial limitations. When the collection unit fills and the material creeps back up the tubes, the fan becomes ineffective, and the aroma of the unit permeates the structure.

The unit is difficult to stir and mix; in the cold basement the conditions necessary for aerobic hot composting are challenging to create. As a result the decomposition is generally a colder, slower moldering process versus the faster, hot aerobic process. The waste material is also somewhat awkward to remove from the composter. Even with the luxury of a wheelbarrow placed besides the unit, the process can be messy, tedious, and odoriferous.

Given these challenges, we regularly add a microbial supplement to fuel the microscopic biology of decomposition and also introduce earthworms. As part of regular maintenance every week we rake the tops of the piles so that the material is spread through the collection unit rather than forming pyramids under the tubes. The process helps aerate and mix the material on the surface of the piles so that decomposition can proceed. When the unit is full, we shovel out the collection unit and finish composting the material in specific designated outdoor piles. Our strategy is to always leave sufficient material in the unit so that it remains inoculated with the proper decomposing microscopic biology and earthworms. The material's moisture content is also important to maximize decomposition potential. Dry material lacks vital moisture, while a soupy pile can become anaerobic, which slows decomposition and elevates the foul aroma, so it is important to maintain a balance of thorough saturation.

There are other alternatives for fabricated compost toilet units. Some units possess an agitator to assist with aerobic decomposition. While this may serve in the short term, the mechanism can jam over time from the unruly nature of the material. There are also units with thermal elements that heat the material, thus enabling accelerated thermophilic decomposition.

Looking beyond fabricated units to the more traditional outhouse concept, one common design is known as the two-seater. I have seen many examples of this design modified for practical applications throughout North America. The concept of the two-seater allows for two toilet locations in the same outhouse above two separate collection bins. Generally, underneath the toilet seats are two fifty-five-gallon drums or concrete enclosures. One collection bin is filled to capacity before the other seat is used, until it too is maximized. At that point the material in the process of decomposition can then be removed from the first collector, and the process can begin anew. The goal of the design is to allow a period of decomposition so that removal of fresh material is not necessary.

In all of these units it is important to consider the ideal finished product when mixing the ingredients. The goal is to provide a medium in which the odor and flying insect activity is limited. We also want systems that ensure health through proper handling and transport. In addition our systems intend to accelerate decomposition so the fertility can be digested by the landscape.

The humanure process requires a bulking agent to assist decomposition. This bulking agent acts to provide aeration to the pile and to balance the high percentage of nitrogen in the manure. By adding a bulking agent with a high carbon content and

We have two conveniently located outhouses that use a five-gallon-bucket system. This five-gallon-bucket outhouse is a three-sided structure, giving the occupant a view of the nearby stream.

surface area we can accelerate the decomposition of the manure. This bulking agent also shields the manure from flying insects and helps trap odors. At D Acres our preference is hardwood lathe or planer shavings, although we've also utilized leaves, hay, wood chips, and sawdust, as well as combinations thereof. The long, thin strands of shavings can absorb moisture and are conducive to aeration and decomposition. Hay, leaves, and chips can be messy and do not always suppress the odor or absorb the moisture. Sawdust works fine, though the decomposition can be aerobically impeded if the material becomes too wet and compacted.

Generally a soup or slurry of human manure has a pungent odor and is difficult to transport. The odor of urine is particularly intense. With the exception of the Clivus, which has an effluent pump, we generally separate urine from our human feces. Urine can be directly applied to compost piles or accumulated in five-gallon buckets. For nasal relief baking soda and vinegar can be added to help diffuse the odor.

It is important to preload a compost toilet with sufficient coverage of bulking agent. This technique helps prevent the humanure from sticking to the bottom of buckets so that it can be easily removed. If maggots or other flying insects become a nuisance, lime and wood ash can also be added.

Our first outhouse was a tree house in which the seat was set above a fifty-five-gallon drum. To empty the unit we would attach the lid and roll the bucket to the appropriate pile. While this action helped mix the ingredients for further decomposition, the actual dumping of the drum into the pile was awkward and messy.

The five-gallon-bucket system helps facilitate the movement of smaller batches of humanure. We currently have two outhouses in which a five-gallon bucket is set under a toilet seat. Between each seat and bucket is another bucket without a bottom, so that a degree of separation exists between the user and the vessel. When the buckets are filled they can be lidded and easily transported to appropriate piles.

CHAPTER THIRTEEN

FOOD SYSTEMS

In the summer of 1997, as my friends and I were preparing to embark on our journey to what would become D Acres, I helped out informally on a harvest down the Doe Bay Peninsula on Orcas Island, Washington. As I clumsily pulled salad greens up by the roots and plucked their leaves haphazardly one by one, a resident from Texas named Emily asked me in a sweet yet pointed manner, "Are you certain that you want to be farmer?" I responded with why community and farming made sense to me intellectually. After enduring my extrapolations politely for some time, she looked at me and said, "That's all well and good, but farming is more than talk and thought." Then she picked up her bushel full of greens and headed for the washing station, leaving me to slowly continue my awkward harvest in the field. While I was ranting about the importance of agriculture, Emily had focused on providing food rather than talk.

Prior to Orcas Island, I had very little experience growing and preparing food. Although I had started and assisted several gardens before arriving at D Acres, I'd never faced the responsibility of full-season planning or implementation of seed-to-plate farming. My youth was spent in studies and leisure. I had not grown up doing substantial daily field chores or farm-related tasks, and my hands were uncoordinated in the subtle and refined operations of plant propagation. While I was fit for exercise, I did not understand the physical and mental demands of year-round farming, which I would come to learn is a seasonal marathon, not a sprint.

In addition to my inadequacies in the field, my culinary artistry had been stifled by the process I learned from the restaurant industry. Creating a full meal from raw, seasonal, homegrown ingredients was beyond my ability. I was accustomed to preparing preportioned food without being a part of all the stages of the process. I could go to the grocery and get staples such as carrots, apples, eggs, or meat and prepare them. I did not possess the competency to butcher and preserve an animal, nor did I know how to can, ferment, root cellar, or dry plants for food. I did not understand the necessity of grounding local foods systems in a diet that cycles with the seasons.

Through D Acres I developed my understanding of a regional food network based on local climate, geography, and low-energy infrastructure, as well as the skills and energy required for such a localized system to succeed. While the industrial food production model calls for the increasing specialization of skills, sustainable food systems require a multidimensional skill set and a fluid workflow structure that allows for functional continuity through the fluctuating activities of the seasons.

D Acres' food system evolves in response to both our dietary and environmental needs. Describing its components is a lot like describing an elephant piece by piece. You can describe separately its skin, feet, tail, ears, or tusks, but to understand an elephant, all of these parts must be taken as a whole. As the individual elements of the system evolve, so do their interrelationships, resulting in the metamorphosis of the overall system from year to year. To begin the discussion of D Acres' food system, we can examine how the broad categories of people and place help to define a food network.

With people the aptitudes and attitudes of the participants are central to the long-term health and vitality of the system, as are the economic conditions that impact both the individual's and the community's food decisions. To be nourished by an equitable food system each individual must provide input for the production. Beyond the skills, knowledge, and physical capability that people bring, individuals involved in a farm food system must share a common food philosophy, which provides an ethical basis that instills enthusiasm, confidence, and continuity toward the system's goals.

Sustainable food systems are also largely based on place. Resources such as water, soil, and sun provide both potential opportunities and limitations that affect their development. Built infrastructures ranging from roads to buildings such as greenhouses are necessary components of a functioning food system. Smaller though crucial constructions such as root cellars, dehydrators, and smokers—infrastructure influenced by geography and climate—will affect and determine the food system possibilities.

By weaving together the complementary elements of people and place, we provide the basis for a food system suitable to the climate of New England that is resilient and perpetuating, one that is designed for minimal imports of fossil fueled energetic inputs and for long-term maximization of benefits. This process is organic and fluid yet requires discipline, planning, and conscious decision making. By recognizing the various elements of the system and addressing inconsistencies or dysfunctions that affect the efficiency of the overall operations, we then address issues and seek alternatives for improvement.

Food Philosophy

The food philosophy in practice at D Acres stems from our perception of the global food system's inadequacies. We recognize that agriculture provided a bounty of storable crops that propelled humanity's ability to achieve a stable existence. This stability, however, is currently jeopardized by the globalization of a fossil fuel–driven food system, which gives the illusion of choice, freedom, and security. The reality is that access and availability to nutrition are diminishing as corporate agriculture values food for profit rather than as the basis of our civilization. Instead of a diverse subsistence economy prompting long-term sustainability, farms are forced to generate a commodity, thereby reducing their ability to implement a holistic approach to farm operations.

On the consumer end of the industrial food model, the path of personal and cultural dietary preference is wrought with explanations and justifications. Food choices can be deeply personal and even a source of conflict. For myself, as I learned about the inequities and inefficiencies of the contemporary Western diet's imbalance toward animal protein, and of the inhumane conditions and suffering of the animals, I began to question my consumption of meat. This intellectual and spiritual revelation led to my becoming a vegetarian during my first year of college.

I made this dietary choice as a result of interpreting the biotrophic pyramid of life. Humanity was and is carelessly abusing water, land, and food resources to support the extravagance of an animal-based diet. Animals given a grain diet in feedlot conditions lose energetic potential. As they grow, their metabolism requires tremendous energy that is not maintained in their flesh. Over the course of their lives, the energy that is consumed at each trophic level is reduced by a power of ten. Ten pounds of grain can provide food for ten people or it can be fed to a cow, which will yield one pound of equivalent protein.[1]

In less than a hundred years the United States' per capita consumption of meat more than doubled, to over 195 pounds per person per year.[2] The preference for this luxurious diet has been made possible by the energetic potential of fossil fuels as well as policy-driven human exploitation of resources. Our cheap, animal-based diet is dependent on a Concentrated Animal Feeding Operation (CAFO)–style agriculture that is resource intensive and polluting. This system is also heavily dependent on exploited migrant labor to perform the devalued work of butchering.

With these arguments behind me, I arrived at D Acres as an ovo-lacto vegetarian who was very dependent on nut butters, dairy, and egg products to meet my protein nutrition needs. It was our intention to provide a sustainable diet from the land in Dorchester based on these same constituent food groups. We purchased two goats, Bella and Clover, to begin our dairy herd; several older laying hens; and some feeder pigs and planted black walnut trees in the forest.

The results were not the rich Eden we had imagined. The black walnuts suffered from lack of sunlight; the older, free-ranging birds laid few eggs and resisted our attempts to indoctrinate them to farm work; the feeder pigs were profitable neither economically nor to the environment and spirituality of the land; and the goat-breeding process and milking commitment required further planning, ethical introspection, and skill building to become successful. While these actions were intended to provide

for an ethical, practical, sustainable food system, we were naive. We did not understand the realities of plant and animal husbandry. Our streamlined approach to food did not address the fundamental limitations nor the advantages of the farm system's skills, infrastructure, geography, and climate.

The diet of our region must be based on viable annual food production and preservation techniques. I envision a worldwide food system that maintains the vibrancy of the culture of individual regions. The traditional diet of New England colonists was a dairy-rich diet that worked in conjunction with the region's climate and geography. The hay field's capacity to be harvested and stored as forage for the dairy ruminants over the long winters allowed the regional dairy culture to manifest.

Although the food system at D Acres continues to evolve, we currently do consume animal protein from meat raised on-site. As an opportunist I will eat local game harvested via roadkill or crop deterrence efforts. Depending on the year, I also eat pork or chicken at least two to three times per week. We preserve pork based on our pig population and the quantity of food that is available for both the swine and the people. Chickens are culled as their production decreases with age, or if they develop a tendency to cannibalize their own eggs. Importantly, all of the animals on-site are active contributors to the health of our farm system, beyond their production of protein.

Currently I am satisfied with the evolution of D Acres' food system. I believe we are making systematic improvements based on reasonable human and animal potential. By methodologically developing our soil fertility and plant species diversity and abundance we are building long-term food production capacity. Animal manure and forest biomass are utilized to produce fertility, which in turn feeds the next generation of protein-producing plants. Anticipating the future, I envision less animal protein consumption at D Acres. Although deer and turkey will likely be harvested as necessary to prevent them from destroying the gardens,

1. Spellman, Frank R., *The Science of Water: Concepts and Applications* (Boca Raton, FL: CRC Press, 2008), 165.

2. United States Department of Agriculture Economic Research Service, "Agricultural Fact Book: Profiling Food Consumption in America," www.usda.gov /factbook/chapter2.htm.

the presence of domesticated animals will decrease as our land use stabilizes and our push to convert forest to field ebbs.

Presently we are engaged in a mutualistic relationship with the on-site animals as we expand and transition the farm system toward a perennial landscaped garden of abundance, using plant species and a soil-building design to proliferate a regenerative and rejuvenating system of agriculture. Our soil-building design promotes soil microorganisms and fungal mycelium, as well as the amplification of plant and animal diversity. We are providing our sustenance by building soil vitality and amplifying biologic abundance. This system of evolution involves a transition from the Northern Forest into a forest garden model that creates designed, ecological abundance.

Our food system also encourages insects and wildlife. Natural systems depend on healthy vigorous populations of birds along with beneficial predatory insects that compete to control pest problems in the farm ecology. While extreme pests such as Colorado potato beetles are dealt with expeditiously to ensure crop viability, we are working to further integrate the crop into a system in which their impact is diminished through designed ecological progression, by which pest problems are minimized. As a result, our food system is constantly in transition.

During the growing season successive crops from salad greens and peas to summer vegetables and the harvest of fall, the annual cycle has many phases. Likewise, some phases of the food system require years to transpire. The conversion of the native perennial forest to vegetable field and subsequent fruit and nut orchard is a twenty- to twenty-five-year process that will continue to mature well beyond our lifetimes. As fruit trees mature in this scenario there will be a succession of species from annuals to a reintroduced perennial emphasis. It is this circular cycle that must be understood so that maximum benefits can be realized over the duration of the evolving design.

When I started at D Acres, Edith often worked with a neighbor in her garden. The neighbor was in her sixties, had been raised in a rural existence, and was the 4-H youth agricultural club leader in our small town. But while she had been raised in the woods of New Hampshire, she was also part of the twentieth century food production model. She bought her groceries in the store and kept animals for pleasure. When I arrived I shoveled out years of barn manure that had been piled in her back woods. This neighbor was so disconnected that she would kill the nonpoisonous snakes that roamed the garden with ruthless intensity, hunting them down and cutting them in half with a shovel. I am not sure if it was fear, dislike, or lack of understanding that compelled her to murder these beneficial animals. To me such behavior is reflective of the dominion we have assumed as the central figures in the narrative of the garden of the earth. Instead of relishing our role as an enabler of the ecological bounty, we have exploited and destroyed simply for our individual benefit as a species. Instead of pursuing the conventional model of success through commodity yield production, we must endeavor to create a bountiful ecological system that yields an assortment of perpetuating benefits. We must come to accept our capacity for positive and proactive engagement as participants in the ecological system from which we are nourished.

We endeavor to produce quantities of food from the land through a slow, incremental method of plant speciation that introduces food, fodder, medicine, and apiary crops to the landscape, encouraging a regenerative symbiosis with the land. We accomplish this through recognition and mimicry of the existing dominant ecology. Spaces for the cultivation of nonnative agricultural crops can be difficult to maintain from the competition of the native forest plants. Developing a system in which introduced beneficial species can thrive and perpetuate allows a framework for development of an evolved farmed ecology.

To accomplish this partnership between humanity and plants requires recognition and an inventorying of the existing botany, which is extensive. Plants serve as the interpreters and expression of the efforts of our agricultural system. To know and act in conjunction as partners with the plant kingdom is the key to unlocking humanity's quest for sustainability.

We are also developing practices of gardening techniques that are applicable to all people regardless of physical ability. We would like to achieve a food production system that encourages outdoor exercise without overexertion. Through thoughtful division of tasks, all the garden work can be accomplished by young and old. We need people at all stages of the life cycle to be participants in the food production cycle, harnessing both the enthusiasm of youth and the patience and wisdom of the elders. Elements of the food system such as seed saving or food preservation and preparation can be reserved for experienced and less mobile human cooperators, while compost turning and potato hilling can be nimbly attended by vigorous youth.

To construct a reliable food system that can perpetuate requires a conscious shift in our approach and methodology. In the conventional food system, success is equated in terms of annual profit without regard for the exploitation of the soil base. The holistic approach is one in which capital is reinvested in the growth of ecological abundance with subsequent perpetual gleaning. We are assisting the design of a system intended for ecological maximization.

Labor of Love

Farming requires mental work, knowledge, and manual labor to accomplish the tasks at hand. The capacity to produce food from seed to plate or birth to butcher requires time, energy, and skill. Through designed effort, the energy invested in permaculture food systems should yield increasing production with reduced inputs over time. Likewise the investment in people to power the food system is a critical component of lasting farm success.

There are many ways to selectively utilize people based on their interests and aptitudes. Systems can be designed to specifically address difficulties imposed by physical limitations or accessibility issues. Other characteristics that define a contributor to the food system are diligence in observation, intuition, and creative problem solving. The strenuous nature of the work also demands a motivating factor; agricultural labor needs positive energy, enthusiasm, and love. Positive attitude not only affects the day-to-day atmosphere of the farm system but is contagious and can become institutionalized and encouraged as a lifestyle. An optimistic work environment is essential for organizational success.

It is important to fully discuss farm practices so that rationales for decisions and processes are apparent. Without common understanding of why certain practices are necessary the difficulty of accomplishing these tasks can make the attainment of goals impossible. Unless there is discussion of *why* we would like to prune the apple and *how* we need to prune the apple, the apple will not likely be pruned with a shared vision or enthusiasm.

To succeed in the physical and mental challenge of farming, it is advisable to be prepared. Physically, we need to address injuries promptly and appropriately. Making sure to drink enough water is a crucial ingredient for physical health. Regular, healthy, sit-down meals with shared conversation are part of the reward for this lifestyle and necessary for optimal personnel performance.

The level of difficulty a farm chooses to undertake should be preplanned and determined. We have chosen to limit the mechanization in our farming methodology, preferring an in-the-sun-and-dirt model for our farm system with a hands-on evolving, nonindustrial approach. While we enjoy the physical labor, we are not working simply for the sake of doing. It is important to focus and prioritize the tasks that can be completed. Physically, farm work is as strenuous and challenging as the design dictates. All choices are continually made as the system evolves to maximize the potential for enduring operations.

Within a culture that realizes the value of food, the people who grow and raise it will also be recognized. Unfortunately in our modern society the means of production are devalued along with the product. Quick and easy food is valued highly, as is the latest fad fruit or diet. This valuation process, which has created an era fueled by cheap calories, undermines a stable, prosperous, ecologic farm system. The forces that pull people from the land base and push them into the urban factory model of humanity have demeaned and exploited the profession of food production.

To work the land for subsistence and ecological abundance is a daunting task. To devalue this necessary, fundamental, and rewarding work discriminates against those who dedicate themselves to the collective benefit. Agricultural laborers deserve respect as civic servants. Butchers and field hands should be treated as well as soldiers and attorneys by our society.

Food Choices

One of our overall goals is to provide a localized food system that is unique, savory, and nutritionally balanced. In achieving this goal we hope to not only meet the community's minimum caloric intake but offer delicious food based on our environment.

We are not food self-sufficient, however. Although D Acres produces more of some vegetables than its residents can eat, other foods are not produced on-site. While our intention is to participate in a sustainable food system, we may never produce all the food we need at the farm. Our focus is to build a strong resilient system for the region, and we realize that we will probably not specialize in all the food production possibilities. Our goal is to produce what is reasonable on-site and trade with or purchase from others to acquire additional products that we consider essential.

It is helpful to consider our evolution as a continuum that begins with our initial enthusiastic yet naive steps and moves toward developing a strong and vibrant system based upon experience and infrastructure. When we arrived at the property in 1997 the only food crop that we planted that fall was garlic. For the first winter we bought 100 percent of our food at the local grocery. In the spring we began harvesting native plants and planting perennial and annual food crops. Later that season we built a greenhouse and began experimenting with season extension. We raised pigs and chickens for meat and purchased goats for dairy. By the end of that season we had accumulated some storage crops from a healthy summer garden.

We have come to accept the reality of following the prescriptive limits of a localized, seasonal diet. A local, seasonal diet for New England depends on crops that can be grown in this region with the land that is available and that have the capability to be stored for a long winter. Winter and summer squash, for example, are aptly named based on their storability and time of consumption. In the sections that follow, our methods of preserving the harvest demonstrate the variety achieved by our seasonal diet.

It is helpful when individuals in the group can agree on dietary necessities versus wants. While there have been a range of diets from veganism to omnivory practiced at D Acres, the overall ethos is a preference for localized options as a guide for our purchases. We seldom purchase fruits or vegetables that have been imported from beyond the region. Our goal is to purchase food products that are grown close to home. We will also buy food products that we currently do not grow but that can be grown commercially in the region.

Food items such as coffee, cane sugar, chocolate, olive oil, and even lemons are viewed as luxury products. While these are very easy to purchase, the New England climate does not easily support growing these crops. If we do decide to purchase or use these products, we strive to be conscientious spenders, supporting sustainable economies and workers. For example, coffee for events at the farm is generously donated by local roasters who import beans from their relative's family farm in Peru. In turn we promote their coffee shop to the public. In general we do not buy coffee, ice cream, or sugar as a group, as it is the intention to use group funds for collective purposes to which we can ethically agree. The exploitation of coffee growers and sugar cane farmers, as well as the health concerns from consumption, restrict our capacity to indulge economically as an organization. While we love the oil produced from olives the energy consumed from transporting this commodity inhibits excessive use of this good.

Preserving the Harvest

Preservation provides a mechanism to extend the food availability beyond fresh picked. A goal of food preservation is to embody the least amount of energy in the process while maximizing the flavor and nutritive qualities. The process allows the wealth accumulated in fresh food to be stored until the next harvest. There are multiple options, recipes, and methodologies for preserving the harvest that have been utilized by humanity throughout history.

Food products must be preserved with care. Improper storage can result in spoilage and the subsequent loss of time and energy invested. Undetected spoilage can also result in food contamination, which can be a serious health hazard. In particular, meat and low-acid foods such as tomatoes, green beans, and cucumbers should be processed and stored with care to avoid contamination issues.

Freezing

Freezing is a relatively new option to preserve food. The post-World War II industrial era brought freezer storage capacity to the general population of the United States. Freezers saved the time and effort of canning and other means of preservation. Frozen food was relatively easy to process and prepare and was popularized for our generation by TV dinners

Kale and collards are blanched, drained, and squeezed of excess liquid, then packed into freezer bags for winter eating.

and microwaved conveniences. Aisles in groceries lined with freezers catered to these profitably packaged products.

Freezers consume large amounts of energy to maintain this luxury. Every household and store in America relies on coal and nuclear energy to maintain these frozen spaces. Freezer trucks are necessary to transport the subzero cargo across the United States. This fragile, insecure network is dependent on continuous inputs of energy to deliver caloric value to the people.

For all the energetic consumption, the benefits obtained from this option in preservation are also dubious. While the initial processing may be quick and simple, the quality of the food is always diminished by the process. Colors fade and textures become limp and soggy as a result of freezing. No food product is as nutritious or flavorful as it once was after being frozen.

Freezer failures can also result in catastrophic food losses. By choosing the mechanization of this food preservation method, our food security depends on our capacity to repair and maintain this technology. In addition, freezer operations rely on a supply of electrical energy that can be interrupted by grid failures.

But for our purposes we use the freezer extensively to preserve meat and greens such as kale and collards, as well as fruits and berries. Meats are cut, processed, wrapped, labeled, and frozen until they are dethawed for use. After the hard frost in the fall, remaining greens are blanched and bagged to fulfill our yearning for chlorophyll over the winter. Fruits, such as blueberry, raspberry, kiwi, and rhubarb, are frozen on metal trays, then bagged for use in pies and other midwinter culinary delights.

Dehydration

Drying is an ancient form of storing the harvest. This preservation option can be used for many meat and vegetable food products as well as medicinal, tea, and culinary herb plants. The drying process can be accomplished through various methods, depending on the moisture content of the plant tissue to be dehydrated. Generally, look for space that allows for ventilation and that will reach temperatures between 90 and 110 degrees Fahrenheit but not greater than 120 degrees.

Dehydration can occur in any location that meets the criteria of heat and ventilation. This decreases the moisture content before the plant tissue begins decomposition, thus retaining nutrients.

The dehydration process is useful for storing the bounty produced from vegetative material. The design of infrastructure must correspond to seasonal variations in ambient temperature and solar incidence as well as the particular crops that will be processed. At our northern latitude, raspberry leaf drying in the hot month of July requires a different process from that of attempting to preserve tomatoes in September, when the sun begins to pass lower in the sky and rainy, cloudy days are the weather pattern.

As a result shade is also an important component of dehydration. The sun will actually deteriorate the color and nutrient quality of the plant material during sun-dried processing. Ideally, therefore, drying is done in an environment that is designed to limit the penetration of sunlight. Exceptions are juicy fruits such as grapes and tomatoes, in which the sun assists in adequately drying their succulent sugary tissue before it can attract mold during those damp early fall days.

To dry leafy material that is still attached to the stem, such as nettle and raspberries, we cut the plants at the base, then hang the stems in bunches in the upstairs of the Red Barn. During summer days the shaded heat of the barn space allows for adequate ventilation and ideal drying conditions. It is important to hang the bunches to allow air passage and to not group more than five to ten stems in a bunch. My experience with this methodology is based on my childhood in North Carolina. Tobacco barns were designed for dehydrating large volumes of succulent leaves. These structures were a ubiquitous part of the landscape, and their rudimentary form provided reliable results.

In addition to the barn we have a solar dehydrator on-site. This unit is designed to heat air with the sun, then draft the hot air via convection into a drying chamber. The drying chamber has shelves to hold frames of plastic screening to encourage ventilation. The entire structure is constructed of conductive and durable metal painted black to absorb the sun's energy. There is also a wood fire chamber located underneath the drying chamber for use when the fall weather inhibits solar drying.

The solar dehydrator functions reliably and effectively. During hot and sunny weather in the summer the solar collector requires a shade cover so the unit does not overheat above 110 degrees. The metal is weatherproof, critter-proof, and fireproof, meeting our objectives of sanitation and safety. The

In our solar dehydrator design, perforated metal baking sheets can easily rest atop the homemade screen frames. Here, black currants and raspberries are being dehydrated.

process is also primarily passively energized with minimal supervision.

Building materials for the drying trays are based on availability, design, and the product to be dried. The trays can be created by framing plastic mosquito screening. It is important not to use metal screening that releases toxic residues from processes such as galvanization into the dried product. Plant tissue such as wet, juicy tomatoes will leave a residue that must be scraped from the screening. For wet materials another tray option is standard restaurant oven baking trays with perforations. Ultimately the trays must provide ventilation but also be durable to endure cleaning and multiple uses.

Drying can be accomplished in any location that provides the proper conditions. Attics are an excellent opportunity for summer drying in many locations. Cars can also be employed as a zone for drying. Clothes dryers and boilers offer opportunities to capture hot air produced for other purposes. Anywhere the temperatures climb above what's comfortable and where shade can be provided can function for dehydration, although it is advisable to avoid using combustion exhaust gases that contain pollutants.

Once the material has been dried it is typically stored in glass jars to be used within a year's time. The glass prevents the rehydration of the plant material. Brown paper bags can also be used to store dehydrated herbs.

Drying materials requires evaluation during the process to realize success. The degree of dehydration is a process that can take place over many days. With changing weather and different plant materials we must pay close attention to the process as it transpires.

To dry moist fruit requires additional measures at times, and we use several designs. We sometimes use the cob oven after a day of baking when its internal temperature is falling. Although the cob material retains sufficient heat to dehydrate without cooking the plant material, this process requires practice. Depending on the quantity and type of food being dried, a small fire may need to

be rekindled over a few days. For example, we had five trays of sliced apples drying in the cob oven over one full week. The weather was damp, and a small fire had to be lit a couple of times each day to ensure the ambient temperature did not dip below 100 degrees Fahrenheit and that the moisture from the fruit continued to evaporate. It is important to pay close attention to the temperatures to avoid roasting the product. We also set up cold frames and mini–hoop houses elevated from ground contact. These structures intensify the sun's energy while providing protection from precipitation. By elevating the structures ventilation is increased and the vegetation is further removed from ants and mice.

To evaluate whether the plant material is dried requires practice. Leaf materials should not be dried

A mini–hoop house sun-drying structure can be set up using recycled materials.

so severely that they crumble to dust yet should still be colorful and crunchy to the touch and sufficiently dried to resist mold and mildew. Fruits such as grapes and tomatoes should retain a supple, leathery, stretchable texture versus brittle and bone-dry. The art of retaining flavor and nutrient quality for storable duration is different for each plant, and seasonal variability complicates the undertaking.

Mushrooms, apples, elderberries, and rose hips can also be readily preserved in the dehydrator. Cutting this food into slices reduces the time necessary to dry these materials. With small fruit such as blueberries and grapes, skins should be pricked to allow moisture to be released from the fruit.

Dried beans are an easily stored source of protein. Depending on the variety, for each bean planted the yield can average between eight and twenty-four beans per plant. Beans are sown in the early spring and left until early September to dry on the bush or

Storing the finished product in clean glass jars, out of sunlight, will help to preserve the integrity of the plant.

Gooseberries dried in the solar dehydrator

A variety of peppers hung to dry in the kitchen

vine. We generally harvest by cutting the plant at the base, avoiding wet weather, and well before a significant frost. The pods are then stripped of their stems and left in a covered, well-ventilated space to dry completely. Removing the beans from their pods is a productive winter activity for a group. The beans can then be stored in clean glass jars.

Root Cellar and Dry Storage

The root cellar is a food preservation space that mimics the ancestral habit of mammals to stockpile food in caches for the winter. Squirrels and chipmunks in this climate seek cold storage for the acorns and seeds that they harvest during the growing season. They burrow holes and diligently fill these banks with nutrition to preserve the summer's bounty. Similarly, humans have historically sought to create a location for the harvest with ideal conditions for preservation. The basement of Edith's Red House served as the initial root cellar at D Acres, since the house was constructed in the early 1800s. The damp, poorly drained cellar was dug to provide adequate protection from freezing temperatures. As the house was modernized and a central boiler was installed, the ambient heat was moderated by the lack of insulation, therefore maintaining an inefficient though usable space for storing root crops. The basement did not provide adequate rodent protection or accessibility, however, and the unsanitary space was muddy and cobwebbed.

The root cellar has ample storage for winter crops and overflow of canned goods. We use the space to store canning jars and other preservation-related equipment.

We constructed the root cellar in the Community Building based on research primarily derived from Mike and Nancy Bubel's book entitled *Root Cellaring: Natural Cold Storage of Fruits & Vegetables*, a comprehensive guide to the storage principles and process, including construction plans. Our root cellar is designed to meet the storage needs of the resident community and to support our value-added, farm-food cottage industries with produce throughout the winter months. The floor space is approximately 14 × 20 feet, with shelving and bins along three walls and a center table.

The root cellar is located in the full basement of the Community Building, with an access door in close proximity to both the first-floor stairs and the basement entry. The access door is insulated and sealed to help prevent the ambient heat of the basement mechanicals from offsetting the cold storage capacity. The walk-in basement allows for wheeled accessibility in loading the produce in the fall.

Each vegetable has conditions it prefers for longevity in storage. The root cellar provides rodent-free space, with high humidity and cool temperatures. Fall, winter, and spring temperatures hover just above freezing, while late spring and summer temperatures are steady around 55 degrees. These conditions are ideal for root crops such as carrots and turnips and for tubers such as potatoes. Cole crops such as cabbage prefer storage in these colder temperatures. In addition, tree fruits such as apples and pears do well in root cellar storage.

The process of loading the root cellar begins in earnest after the first fall frost. These freezing temperatures indicate that soon the soil will be frozen solid and unworkable. While some crops can be maintained over winter in heavily mulched beds, the window of opportunity for root cellar utilization closes when the winter snowfall begins.

Crops are brought into the cold storage area with packing materials. Sand, straw, leaves, soil, and sawdust or planer shavings can all be used as a packing medium for storage. These mediums are intended to offer a degree of sterility and to aid in moisture retention. The choice of mediums is based on several factors, including cost and availability. Sand is heavy but sanitary and has the ability to retain moisture. We tend to sift and reuse sand yearly. The sand that spills as we dig through the root cellar buckets is incorporated into the sand floor. Leaves, straw, and sawdust can be messy, though they are preferable for finger fruits such as apples and pears. We attempt to minimize the spillage of soil that is utilized to "replant" cabbage for storage.

Potatoes are stored in bins constructed from quarter-inch wire mesh designed to resist rodents and provide ventilation. The bins are elevated so air can filter upward through the piles of potatoes and prevent rodents from chewing the spuds from below the bin. Cabbage can be stored in a root cellar via a number of methods. In addition to providing the cold storage for sauerkraut and kimchi, unprocessed heads of cabbage can be uprooted before the ground is frozen and replanted into a soil medium in the cellar. Alternatively, these whole plants can also be hung upside down along the cellar walls. Both these processes stretch the plant's vitality into the winter months, holding at bay the impending decomposition process. Heads, trimmed of roots and loose outer leaves, can also be wrapped in newspaper and stored in boxes with adequate ventilation. Because the heads are a short-term storage crop that is particularly vulnerable to rot,

Root crops such as carrots, beets, turnips, and rutabaga are brought into the root cellar with the greens cut off; they are packed into planer shavings, leaves, or sand.

we prefer the open-air method of storage to make daily inspection easier.

When storing root cellar items it is important to avoid damaged and blemished produce that is particularly perishable. This fruit should be culled and preserved by other means. Likewise it is important to regularly check and inventory the root cellar so that produce is consumed in a timely manner: One rotten apple can indeed spoil the batch.

Our root cellar also has the capacity to provide refrigeration. Cellar temperatures hover between

Our largest potato bin can hold a few hundred pounds of potatoes. The quarter-inch mesh allows for ventilation and provides protection against mice.

38 and 42 degrees from November to March, consistent with the standard operating temperatures of industrial refrigeration. In November, therefore, we move the contents of our refrigerator into foam coolers designated as "leftovers," "dairy," "fresh produce," and "condiments," then unplug our refrigerator until March or even April. The quality of root cellar storability is superior to the refrigerator because of the introduction of storage mediums. By creating an adequate and appropriate design for humidity, temperature, and ventilation, functionality is provided without the carbon footprint. For suitable crops the root cellar is the optimal choice for a simple preservation technique with low annual energy inputs. The long-term yield from this durable infrastructure requires minimal

Cabbage can be uprooted and hung for storage. It can also be "planted" by the roots into compost or trimmed of its roots and outer leaves and wrapped in newspaper. When pulling the cabbage out of the ground for hanging, retain some of the soil so the roots don't dry out too quickly. Cabbage harvested in November and stored this way can keep through December.

maintenance and preserves the food whole for fresh consumption.

Root cellar ventilation should be controlled to maximize cold storage potential. To induce cooler temperatures in the spring and fall, the space can be ventilated with cold air at night. The ventilation may be as simple as opening a window or as sophisticated as thermostatically controlled fans. The temperature of the root cellar can drop below freezing during the winter if it is not monitored, especially in conditions of severe cold with minimal snowfall. To avoid this circumstance we open the main access door to the ambient warmth of the mechanicals in the basement as necessary.

Some crops prefer warmer and drier conditions for storage. Garlic is harvested, then hung in bunches until the skin dries to a papery consistency. The stems and roots are then trimmed, and the garlic is stored for winter consumption

Garlic hanging in the barn to dry before we cut the stems and store them in a cool, dry space

Our potato harvest curing in the warm, dry basement

The winter squash harvest curing in the G-Animal, protected from the elements and wild animals

in mesh bags in dry conditions and an ambient temperature of between 50 and 60 degrees. We generally store the garlic in upstairs closets or an attic crawl space where the temperatures do not fall below freezing. Rodents are not normally a problem, and the garlic will keep until spring temperatures encourage the bulbs to sprout anew. Other alliums such as onions are also dried until the outer skin becomes brittle and stored at a cool room temperature. After harvest, winter squash such as pumpkins and butternuts are cured with sunlight and cool temperatures (often laid out for two to three weeks in our cob/glass greenhouse) so their skins harden, then are stored in upstairs closets to avoid freezing temperatures.

Storing Meat

Meat is a valuable, trophically derived food commodity. While in some cases the entire animal carcass is consumed directly following slaughter, in most instances we need to preserve the meat for the future. In the United States we typically rely on frozen or refrigerated meat to enable our consumption, but there are many methodologies to preserve meat that were developed before the advent of the freezer and refrigeration. Meat products preserved for room-temperature storage are more readily transported than frozen food and can often be consumed without additional cooking. Freezing also is only effective for a limited duration before the quality of the meat starts to deteriorate.

Alternative techniques for meat storage include making jerky and cured and smoked products. Smoking is a low-temperature process that preserves the meat for flavor and storable longevity. When I lived in Spain the local bars would hang cured and smoked hams from the ceiling. Throughout the long summers of intense heat the mold-covered hams would hang, occasionally dripping grease. These hams were consumed regularly by cutting through the thin, outside layer of distasteful meat to access the delicious flesh of the interior.

Our experimentation with smoking and the production of jerky has been minimal. The smoking process requires a salty brine to keep the meat moist and flavorful and a firing process that allows combustion smoke to penetrate without overheating the meat. This process can be performed with a rudimentary structure and a wood fire or industrially with natural gas and nitrate-based brine, depending on the scale and scenario. We have recently constructed a smokehouse in the backyard and are very pleased to enjoy the flavors it provides.

Small-scale smoke houses offer an extraordinary artisanal product. The process requires time and practice. The slow, smoldering fire is designed to produce more smoke than hot flames and must allow for temperatures to remain below 100 degrees Fahrenheit—flames will burn the outside of the material, making it difficult for the cure or brine to penetrate and the meat to smoke evenly. Producing a smoked ham can take several days of slow-burning fires that must be meticulously tended. Various types of aromatic wood can be selected to further enhance the flavor. Regulations can also complicate the process. All products smoked for resale must be smoked at a licensed USDA processing facility, as salmonella, botulism, trichina, and *E. coli* are all real concerns.

To preserve any meat, by dry curing, brine pickling, or cold smoking, salt (sodium chloride) is a necessary ingredient and should never be overlooked. Sodium nitrite and sodium nitrate are natural chemicals derived from salt that are often used in the curing process. They are, however, not necessary ingredients and are toxic in high concentrations. Sea salt is a safe and reliable resource for all curing. Following recipes provided by reputable sources ensures a safe and successful preservation process.

Fermentation

Fermentation is an art that has been practiced for food preservation and harvest celebration throughout recorded history. As the cofounder of permaculture, Bill Mollison, noted in his book *The Permaculture Book of Ferment and Human Nutrition*, animals have always enjoyed fruit fermentation. Fermentation is often used to store the harvest in the form of a beverage such as wine or beer, so the common perception of fermented products is that they are alcohol based. However, fermentation is a process that creates a range of nutritional food and beverage products. For example, sauerkraut, kimchi, kombucha, yogurt, sourdough bread, and tempeh are all the delicious results of the fermentation process, yielding nutritious probiotic organisms.

The fermentation process builds proteins and carbohydrates that are then embodied in the fermented mixture. The nutritive value of fermented food is a result of sugars combining with a catalyst. Fermented foods and beverages are also intensely flavorful, can be stored for long periods of time, and are easier for many people's bodies to process and digest.

Cabbage is a principal example of a food crop that produces its own brine, a compound that not only assists with storage longevity but also builds nutritional value over time. Sauerkraut processed in October will stay edible through April in proper root cellar storage. For products like sauerkraut and kimchi, salt is once again a necessary ingredient for preservation. Other ferments, such as sourdough bread, kombucha, and wine, thrive on natural sugars in the fruit and flour and on introduced sugars such as honey or cane sugar. Without a doubt fermentation is a preservation technique

A large crock is a special tool for fermentation. This crock is holding a large batch of sauerkraut. Regina is pressing down the chopped cabbage so it will be submerged below its brine.

Regina is using a method of placing clean whole leaves atop the chopped cabbage to add a barrier layer between the cabbage mixture and the air. After this a plate or wooden lid is placed on the leaves and weighted down to keep the brine above the cabbage.

Sauerkraut can last a long time in cold storage. This batch was processed in October 2012 and taken out from the root cellar in August 2013. It was still crunchy and full of life.

and can encompass anything from apples, beets, and corn to cabbage, meat, and rice. It is advisable to use tested recipes. While improvisation is a part of the music in the kitchen, proven recipes tend to produce edible results.

Canning

Canning is a process of preservation that resulted from technological advances in food storage containers. Glass and steel cans have allowed heated food products to be sealed in anaerobic conditions, which prevent the introduction and growth of spoiling microorganisms.

Food safety can be maintained by adherence to proper recipes and procedures. Improperly canned food can sometimes be difficult to detect, yet there are obvious signs—a broken seal, evidence of mold, odd smells, or fizzing inside the container. The most feared contaminant is *Clostridium botulinum* bacteria. The best way to avoid the growth of botulism spores is to (1) adhere to recipes, and only vary ingredients for taste; typically ingredients such as vinegar and lemon juice are measured proportionally to the vegetable or fruit to maintain proper pH (botulism thrives in low-acid environments); (2) fully process the product in a boiling-water bath or pressure canner to ensure that the temperature stays high enough and for long enough to kill the botulism bacteria.

At D Acres we utilize reusable glass containers, though we do purchase new rubber caps annually to ensure a proper seal and avoid spoilage. We will

Regina works with two volunteers, Ken and Maria, to preserve garlic scapes in the outdoor kitchen. We used the raw-pack method, pouring a hot vinegar-water brine over the scapes and spices, then sealing the jars through a hot water bath.

When canning at the cob oven, organizing the necessary tools in advance is crucial.

Sydney Bennett works in the big kitchen to assist in making applesauce and relish. We often combine several fruits such as raspberries, currants, and blackberries into a mixed-fruit preserve. We shy away from utilizing sugars and extra pectin, simply making a canned compote that can be sweetened as desired after opening.

Not only should you have your tools ready in advance, but your ingredients as well. Here Theresa Rinaldo is helping to can pickled garlic scapes; we have dill, garlic, hot pepper, and other spices ready for filling the jars before packing the scapes.

Garlic scapes ready for the hot brine, clean lids, then processing in the hot water bath.

use most heavy-duty mason-style glass jars that have a continuous screw top. While the negative pressure of the canning process against the rubber cap is what maintains the seal, the continuous thread of the proper mason jar provides a solidity to bear down on while the canning jars are air-cooled to a self-seal. We solicit old canning jars from the community for reuse. There are also new companies that make and sell reusable canning lids.

Canning is labor and energy intensive. Jars must be cleaned and sterilized while the canning recipe is cooked. Then care and attention must be paid to filling the jars and adding the lids. The next step is to heat the jars so the lids will seal as the product cools to room temperature. The energy involved requires time spent in a hot kitchen preparing large volumes of food as the harvest intensifies through the summer and into the fall.

Canning also requires infrastructure. Fuel for heating the water and potable water for cleanup

Once items have been processed, they must cool completely before being stored. Label each jar with the contents and the date, and organize in whatever manner makes most sense. Our canned goods are stored in an easily accessible shelving area of the basement.

Spicy cucumber relish, cooling after being processed

Plums canned in a simple honey syrup

Food preservation can sometimes feel like a daunting task, but with a small group of enthusiastic people, the chopping goes faster, and the time passes with more smiles.

is necessary. Old cookstoves and commercial-scale ranges provide working spaces for the heavy, wide, cumbersome pots. Additional countertop space for food prep and canning accoutrements is ideal. We have chosen to develop outdoor, wood-fired equipment options for this type of hot summer work. The shaded open-air environment of the Summer Kitchen is ideal for this process.

Rocket stoves can be used for canning operations. These units can efficiently produce the large amounts of BTUs necessary to boil large volumes of liquid. They require attention to feeding the high-efficiency, small-capacity firebox and must be constructed to support the heavy, wide pots necessary for several jars.

Pickling

Pickling depends on a salt-and-water brine to add flavor and storage capacity. Both vegetables and meat (think corned beef) can be pickled, though meat must be further processed by either smoking or cooking. Some pickling is considered a short ferment in which the food can be consumed within a week, whereas longer-term ferments such as sauerkraut require three to four weeks to attain optimal flavor. To pickle food we immerse it in a properly prepared brine, which anaerobically prevents microbes from spoiling the food. Again salt is required. Pickling is extremely effective at preserving the crisp texture of fresh vegetables. Pickling brines can also be augmented with seasonal herbs such as dill or basil.

Pickling provides a storage solution for a smaller range of food products than canning. In particular pickling is effective for snap beans, garlic scapes, carrots, cauliflower, and cucumbers. It is a simple process and does not require boiling water to seal jarred food. The only infrastructure necessary is the raw ingredients, appropriate storage location, and glass, ceramic, or food-grade plastic containers. Pickling is a simple process that embodies low energy in a nonheated process while supplying reliable superior flavor and nutrition. In contrast to perishable raw brine pickles that can be kept in an unsealed crock and last a couple of months, store-bought pickles have been sealed via a canning method such as a boiling-water bath or pressure cooker.

Processing Dairy

Milk, like meat, is a trophically gleaned animal product that is perishable in the short term. Humans have developed various ways to modify and enhance its nutritional capacity and longevity. Yogurt, cheese, and butter are all examples of methodologies that are commonly utilized to preserve milk. All three can be performed at the home scale to preserve and enhance the nourishing quality of milk.

Yogurt and cheese are classified as ferments. Making yogurt is a relatively simple process that increases the shelf life of milk while also providing beneficial bacteria to the human dietary palate. Yogurt culture bacteria are introduced into heated milk, which is then cooled to allow the culture of bacteria to multiply. The only infrastructure necessary is a stovetop, a thermometer, and a starter yogurt to inoculate the liquid. We make yogurt to supplement our diets from raw milk that we purchase in the area. Yogurt increases the longevity of milk from seven to a maximum of twenty-one days. I recommend making it to anyone who wishes to consume dairy products on a regular basis.

Cheese and butter are also tasty, transportable versions of mammal milk. Butter is made from the thick cream that rises to the top of fresh milk. Our experience in the production of cheese and butter has been limited by the lack of surplus in our food system. Without on-site dairy animals, we have not had to contend with excesses of milk to preserve.

CHAPTER FOURTEEN

GARDEN DEVELOPMENT AND SOIL STRATEGY

Soil is the building block of terrestrial biology. The concept that all flesh is grass is rooted in soil science. A living soil system is one that is vibrant and capable of producing healthy crops sustainably by refining existing minerals and nutrients, making them palatable and accessible to plant roots. Soil is then biologically renewed as it cyclically decomposes the detritus of terrestrial life.

To produce the abundance of calories and nutrients that humans need for survival, plants utilize many resources. In addition to the sunlight, air, and water provided by the earth's energy, nutrients are drawn from the soil to create a healthy crop.

Nutrients are not drawn, however, straight from the rocks of the earth's crust. As a fellow gardener once told me, "Rocks don't make tomatoes." The web of life that provides the nutrition for plants is based on a naturally occurring process that has evolved over eons. In the highly active ecological zone that is the top stratum of the earth's crust, the edge of the two distinct biomes of land and sky, there is an interface that produces unimaginable microscopic diversity. The first two to five inches of soil, known as topsoil, is where the water, air, warmth, minerals, and organic material all mingle to foster the abundance of life. It is estimated that the number of microorganisms in a tablespoon of

healthy soil is greater than the population of New York City.

Soil can be built by nature from the top down. Plants create soil by mining nutrients from the earth's crust. A typical plant's roots are volumetrically equivalent to its aerial parts and tend to penetrate to at least the depth equal to their aboveground height. Plants grow as profusely below ground as they do above. The biologic quest for nutrients that propels plants to search the depths of the earth for food creates abundance on the surface. The vascular system of each plant reaches deep for stability, water, and minerals, creating channels that allow for vertical energy exchange. After the vascular system sucks these nutrients into the aerial parts of the plant these nutrients are eventually deposited on the surface when the plant decomposes. Leaves, branches, and fruit also fall and are composted by the biology at the surface of the earth. The roots create an invaluable network for vertical supply of nutrients through the soil horizons.

When developing farmland, we aim to be judicious, conscientious stewards of the soil resource that provides our sustenance. We have learned from the errors of rapid growth and unmaintainable expansion made by the industrial system. We endeavor to expand our agricultural operations in

Potato harvest is a prime example of incorporating the mulch layers into the living soil. Over the course of the summer, several layers of hay or straw have been placed around the potato plants, covering any bare soil. During harvest we are mixing the many layers into a fine blend of biological material. After harvest we may seed a cover crop, plant garlic, or add another fine layer of hay or straw atop the reshaped bed.

a methodical, incremental fashion. Our capacity to expand depends on our ability to maintain areas of our disturbance. The poor stewardship of bare soil and subsequent erosion is an indicator of improper soil management. Crop soils must be protected and regenerated by a living mulch or one composed of decaying biomass. This carpet shields the soil from the forces of erosion and builds the soil for the future.

Erosion can have drastic effects on efforts to build soil. Soil structure is destroyed by powerful rain and running water. A thin layer of mulch can dramatically improve soil resilience to rain, while surface water must be allowed to channel through

alternative routes or be methodically retained. Furthermore, mulch offers an energetic basis for future crops. As the mulch begins to weather on the surface of the soil, the hungry ecology of a healthy soil system will begin to incorporate it.

Soil can be examined in several ways. Vertical soil profile assessment, microscopic analyses, and soil lab tests are all methods to evaluate soil. Looking at the soil structure in profile is a typical way to assess its capacity, by examining the layers below the surface. Definitive strata are formed that vary in water retention and nutritive qualities. We can also assess the soil through visible surface analysis. All loam is a combination of sand, clay, and silt or organic material. This combination of ingredients can vary tremendously on a single site or be consistent for miles dependent on the region.

Loam with higher concentrations of sand drains quickly and does not retain water as well as clay-rich soils. Clay-rich soils are more prone to compaction and can retain too much moisture, making field space difficult to cultivate during wet periods. The premium soil is a mixture that combines the benefits of water retention and drainage, giving the soil structure without making it prone to compaction.

No-Till Agriculture

Tillage-free agriculture is a refinement that ensures continued production of food sufficient to meet humanity's needs while also reducing cultivation practices destructive to soil health. Eliminating the practice of tilling reduces soil erosion, increases fertility, and builds topsoil.

Beth is harvesting spinach from a cold frame after a snowstorm in October 2010. This cold frame has harnessed the sun's energy and protected the hardy greens from severe cold and frost, extending our harvest season.

It is true that the process of tilling and turning the soil can create a friable, loose topsoil layer, combating existing vegetation and allowing a new crop to become established. Tillage can also be used to incorporate compost and green manures into the soil profile, to produce a loose, enriched soil layer that mimics the soil created for container gardens.

Tillage, however, ultimately destroys soil health. When we penetrate and disturb the top layers of the soil we unleash a flourish of aerobic bacteria activity. As the profiles of soil are disturbed and the organic biologic material from the top layers is oxygenated and blended with deeper layers of soil, the activity creates an explosion of microscopic activity. This high-energy production technique results in substantial yields in the short term, while inevitably depleting the resource in the longer term.

The tilling process is reminiscent of a blender or a washing machine, shredding the webs of mycelium and mycorrhizal hyphae that connect the horizontal and vertical flow of essential plant nutrients and raw materials. Earthworms do not regrow if cut into many pieces. In addition their networks of tunnels, which supply aerobic and water channels, are destroyed by the equipment. Tillage techniques decimate existing ecosystems by creating a catastrophic, cataclysmic event, then exploiting the imbalance that ensues. This creates a cyclic imbalance that demands steady inputs of organic material to meet the needs of the persistently hungry, aerobically inspired bacteria.

Furthermore, tillage creates a sheer glaze on the soil profile. The repeated force of a tiller's blades on the same soil depth creates a hardpan that effectively forms a shell that resists root and water penetration. This is evidenced on farms where standing water floods the fields in the spring because the hardpan holds the water like a kiddie pool. Unable to penetrate the top cap of the soil horizon, crops become nutrient deficient and susceptible to drought. Because rainwater is unable to penetrate into the soil, it is not retained and rapidly evaporates back into the atmosphere. This process of annual compaction and glazing of the top layer of the growing medium creates a lifeless, resource-intensive system similar to container gardening. The mechanized process limits the ability of the living, natural system to self-regulate, regenerate, and provide nutritionally rich food.

In addition, tillage exposes the soil to erosion. Organic roots and plant residues that bind the soil, as well as the existing webs of mycelium and mycorrhizae hyphae, are mechanically severed, their natural networks destroyed. As these bonds are cut, soil particles disassociate and begin migrating because of the influence of wind and water. The endemic problem of desertification witnessed globally is caused by this abuse of the soil and subsequent degradation.

D Acres' process of cultivation does not involve mechanization. Instead we utilize hand tools to minimize compaction. We also practice the technique of mulching to build and protect the soil. In addition, we amend the soil with compost and green manures to add life and fertility. Through practicing, we are developing alternatives to the destruction of soil.

Developing Arable Land

Our practices for developing arable land and healthy, productive soil are alternatives offered against a historical backdrop of different techniques. Looking deep into history, neither the pre-Columbian nor postcolonial methods offer sustainable models that can address the global food needs of the modern world. Indigenous farmers, for example, used slash-and-burn techniques to develop agriculture zones. The trees and vegetation were cut, then burned, releasing carbon dioxide, heat, and particulate matter into the atmosphere. Applied today, the technique would not only pollute the atmosphere but also incinerate the biological potential encapsulated in the vegetation. Instead of regenerating soil fertility, the energetic value of the mass is burned with minimal benefit. Our current carbon crisis implores us to sequester organic material in the soil.

The colonial method of agriculture was an experiment inspired by the migrations of the era.

Europeans fled to the New World seeking freedom and opportunity from a crowded continent where the resources had been plundered, where native stands of forest were limited and water had been spoiled by human activity. Centuries of activity had diminished Europe's natural resources, fueling wars and feudalistic forms of government to manage the land.

During the late 1700s the mountains of New Hampshire were settled by enterprising colonists who had inherited the agricultural traditions of Europe and were attempting to mimic them in a new environment. The forest was harvested extensively to produce charcoal and clear the fields for sheep. The fieldstones were used to build stone walls, and sawmills were built to process the wood products. Wool was shipped to the fiber mills of northern Massachusetts, where it was processed.

This era of agriculture in New Hampshire did not endure for multiple reasons. Principal among them is that the destructive system of agriculture destroyed the soil resource. By adopting the techniques of slash and burn, as well as overgrazing, the temperate forest of New Hampshire was in danger of desertification. European settlers attempted to recreate an agricultural system similar to that of the Old World but with a vastly different climate and vegetation. If New Hampshire's population had not dispersed in the second half of the 1800s, these agricultural techniques might have altered the landscape even more severely. This was an era of exploration and exploitation. The earth's reserves of energy existed to be consumed for the progress of man. Seduced by the idea of infinite manifest destiny, the settlers exploited the landscape. Their style of agriculture depended on the unrestrained opportunity for migration in pursuit of new resources to exploit.

As the settlers moved west, the advent of the plow signaled a new era in agriculture. The plow increased the capacity and the economic advantage for food producers who possessed deep, rich soil. This exploitation of the soil led to the catastrophic Dust Bowl era of the 1930s. Despite attempts to mitigate the effects of industrial agricultures, the war on soil through these cultivation techniques continues to cause massive soil loss.

With an intellectual understanding of this history, we arrived at D Acres seeking alternatives to destructive conventional farming techniques. We did, however, possess a Farmall Cub Cadet from the 1940s era that was produced by International Harvester. The tractor was an investment in time, fuel, knowledge, and pollution that could offer either immediate gratification or mechanical nightmares. During the first spring we felt it was worthwhile to attempt using the machine as a plow to expand our garden space. We felt the equipment would save time and energy in our attempts to immediately develop arable ground in a grassy field. We nervously tested this powerful machine on several occasions, though we were dissatisfied with the quality of the work we achieved. The plow dug a ditch four to six inches deep and a foot wide into the earth and flipped the sod, leaving chunks of grass and roots. We were left with exposed subsoil that was lacking in vegetative competition but also biological and organic material essential for optimal plant health. This was the antithesis of our goal.

Our experiments in transforming land for arable uses took two directions, sheet mulching and partnerships with animals. Sheet mulching is a significant, one-time investment of energy that yields results with time. The process is advantageous because it can be timed to meet your schedule and the availability of materials. The potential downside of sheet mulching is the time horizon necessary to suppress native vegetation (see subsection in this chapter titled "Cardboard Sheet Mulching").

Our second technique was the use of animals. Animals require time, resources, and effort, but they can yield exponential benefits in fertility, food, and companionship and as a farm attraction. The process of utilizing animals for land transformation requires containment. Though the animals are not forced to work, they are confined in an area until their natural tendencies persuade them to engage in foraging endeavors for our mutual benefit.

This process is commonly referred to as using "animal tractors." These human-animal designed partnerships limit the use of fossil fueled mechanization while providing dividends for all participants. We are seeking maximum advantage based on this specific, evolving farm food system.

In pursuing creative ways to transform the land, we must also address our region's lack of organic material. The hillsides at D Acres have been stripped by glacial and settler erosion. The topsoil that exists is a brown, sandy soil that contains limited organic material. We are attempting to raise the organic content in the soil for overall health and stability. Increased organic material will energize the web of life. Once that happens, nutrients will become available, and water retention will increase.

Over the years we have attempted various means to introduce organic material to the soil. We have experimented with a wide range of materials, including recycled paper products, barn manures, wood products, wood ash, seaweed, hay, straw, leaves, cover crops, green manures, biodynamic accumulators, and soil amendments. Each ingredient has its costs and benefits, which we have assessed as follows.

Paper Products

When utilizing recycled paper products as a soil conditioner it is important to assess the risk. We avoid colored inks, glossy papers, office papers, and junk mail. Generally we prefer natural brown paper and cardboard as well as black-and-white newspaper. Even with these minimally dyed products, however, we recognize the potential for contamination of the soil. This is a risk we are willing to take in the short term, however, as we glean these free resources for immediate efforts to remedy our lack of organic material on-site.

Paper can be introduced to the soil system through sheet mulching, composting, or worm bins. Paper is a woven wood product composed primarily of carbon.

As it degrades, the carbon becomes available to the soil medium. We source paper from area businesses that would otherwise have to pay for disposal and also ask the public to drop off newspapers.

Barn Manures

Barn manures can be a great resource for soil fertility and conditioning and are as distinctive as the many types of farm operations that exist. The quality of the material is based on several factors. Manures can be evaluated based on their age and storage situation, as well as by the volume and composition of bedding and animal. Cows, goats, and rabbits produce the premium barn manure, which is free of weed seeds because of the digestive processes of the animals. The manure is rich and well balanced in nutrition. Pig and bird manure is more chemically caustic and rich in phosphorous. Horse and pig manures contain undigested seeds that can include problematic weeds. My preference is to obtain manure from many different sources to compost and create a nutritional medley for the garden. The availability of barn manures depends on the type of agriculture practiced in the region. Most farm-scale operations have a manure management plan and either compost for sale or reintroduce the manure into their own fields. Smaller, hobby-animal farmers and horse farmers are more likely to part willingly with their own fertility. Animal lovers are prone to collect critters without considering the fecal ramifications of this compulsion. We promote our interest in manure with announcements at the local feed store and through advertising outlets. We also scrounge our region for animal farms, where we solicit the owners for their excess or unwanted energetic potential.

Along the continuum of manures from fresh to finished are many possible stages and circumstances. Manure contains different amounts of carbon and nitrogen necessary to provide finished compost. Proper mixing and processing of the barn manure can result in optimal compost: stable, soft, black soil conditioner that is free of weed seeds and ready for garden application. Unfinished barn manures are also valued for specific applications, such as energizing the decomposition of existing piles. Each farm offers different opportunities to obtain manure. At some locales the manure is aged to perfection and readily accessible. At others poor-quality manure may be located where retrieval is more difficult and energy intensive. It is essential to evaluate the potential costs and advantages of retrieving manure, depending on its quality and site conditions.

Once the manure has been located, the work of moving the material begins. We have negotiated many arrangements to obtain barn manure. Generally, we seek manure within a thirty-minute drive that can be safely accessed and loaded. The typical problems that can ensue include vehicles getting stuck in mud or from overloading and difficulty accessing the material. The challenge of manually loading a vehicle is accentuated by poor accessibility, although some small farmers have tractors they are willing to use to mechanically move manure. On one occasion we found a source of barn manure that had been piled for many years. This large deposit warranted hiring a neighborhood machine operator to mine the area and truck it to D Acres.

Wood Products

Wood products are readily available in the Northern Forest for use as soil amendments and have a range of value and usability. Wood is not purely nutritive but is used rather as a soil conditioner that induces stability of nutrients in the soil profile. This is due to the high proportions of carbon in relation to nitrogen contained in wood products.

When we first moved onto the land at D Acres the by-products of the wood industry were easy to obtain. The electric and telephone companies eagerly eyed options for dumping chips generated by their daily work. Local mills happily loaded vehicles with mountains of their by-product, and we happily received these gifts gleaned from the waste of the system. After the year 2000, however, the availability of wood by-products was drastically reduced. The demand for materials to be processed into wood pellets or fed into cogenerating plants eliminated this source for our agricultural process. The perpetual economic shifts forced by global resource depletion directly affected the availability of resources within the local economy.

This loss of gleaned biomass material also created a problem for our animal husbandry practices. Cheap wood biomass had been the status quo for animal bedding and for conditioning volatile manure. We did not have the facilities to adopt a sand-based bedding system, nor did we see the environmental, economic, or philosophical merits of doing so. As we assessed the situation we became convinced that we must partner with the Northern Forest without the large-scale, commercial forestry industry as a middleman.

The methodology of forestry that we first adopted at D Acres was the standard best management practices of the era. Brush and slash was cut and left in the woods to regenerate. In areas to be cleared of a forested understory, brush was burned or piled in parallel stacks and dangerously chainsawed into smaller pieces. These compacted piles served as habitat for forest critters in addition to longer-term fertility for the forest. To facilitate logging with the oxen it was often necessary to laboriously stack the brush just for their accessibility. The problems of unrealized yields from our forestry program and the change in availability of industrially produced biomass led us to seek alternatives.

In 2003 we purchased a mechanical chipper, allowing us to speed the decomposition process for wood and transform it into a product that can be utilized directly in our food system. Brush is transformed into usable bedding and mulch from the limbs of our forestry program. Limbs less than three inches in diameter are referred to as detritus. The highly active metabolism and plant biology in this wood offers more immediate energetic potential as

a soil amendment, while larger limbs impart value through the years as a soil conditioner.

Wood biomass can assume a variety of distinct forms depending on the species and how it is processed. Each form has qualities and traits that can be used for specific purposes. Wood can be incorporated as whole trees into the soil system. Large swales and terraces can be constructed using whole tree parts that are incorporated over time into the system. This process of nutrient and carbon recycling provides a resource for decades. To utilize wood biomass within a short period of time, we process the wood into chips, planer shavings, and sawdust. These are by-products of the forest industry and wood manufacturing process and can be used as follows.

Chips that include leaves and bark compost more rapidly, while durable hunks of white chipped sapwood can be utilized to build paths. Planer shavings are produced from shaving the sides of milled lumber to a smooth finish. Planer shavings from hardwood provide premium habitat for fungal activity. Sawdust is residue produced from milling wood into dimensional lumber. The rough sawing produces small, dusty particles. These particles will compact and begin a slow process of anaerobic decomposition if they are not mixed with other materials.

While wood products are valuable they can also detract from the overall soil health under certain conditions, retarding the soil biology by unbalancing the soil chemistry. With the exception of detritus wood, freshly cut wood products should not be incorporated directly into the soil. The entropic condition in such fresh wood attracts and absorbs available nitrogen to be employed in the wood's decomposition. This reduction of available nitrogen limits plant growth in the soil system. This is particularly apparent with freshly milled sawdust or wood chips.

Wood Ash

Wood ash is a commonly used soil amendment. Wood ash is typically well-combusted material that is gray in color, fine, and dusty, distinct from bio-char or other partially combusted material that is black and chunky. Wood ash has many adherents among gardeners, though I am cautious, since many suppliers offer ash from trash incinerators that contains high quantities of heavy metals. I am also hesitant because of the chemistry of the ingredient. Wood ash has an alkaline pH of 10; it will balance acidic soil but only for a short period of time. I am reluctant to apply large amounts of wood ash because it can be caustic to the biology of the living soil system. For this reason we perform minimal applications of ash to pastures and field space to dispose of ash produced on-site.

Seaweed

Seaweed and other aquatic vegetation can be a premier fertilization tool, although we use it infrequently because of our geographical distance from available sources. The plant tissues that are formed in the suspended liquid have absorbed tremendous quantities of nutrients. Seaweed is more elastic than woody tissue and decomposes into a beautiful soil conditioner. Although additional sea salt on the seacoast may be problematic, the salts can be beneficial where they are lacking farther inland. If you are receiving lake milfoils or other species from an invasive species eradication project, it's important to ascertain if herbicides were used on the aquatic vegetation.

Hay and Straw

Making hay is one of the principal methods for feeding livestock during the winter in New England. Hay fields can be cut up to three times every summer. The intention is to cut the fields before seed heads form on the vegetation. The dehydration process retains nitrogen and the greenish color in the vegetation. Hay comes in square bales that weigh thirty-five to seventy-five pounds or round bales weighing several hundred pounds that roll out. Hay is packaged for transportability and a typical square bale costs from four to six dollars.

Straw is a by-product of food production. We typically receive straw that is the residue from an oat harvest. This golden-brown, reedlike material comes in bales that can weigh up to ninety pounds. Bales are available at our feed store, though the price has climbed from less than five dollars to ten dollars during the last ten years.

Straw is advantageous for mulching because the weed content is minimal. The seeds that exist are oats that are easily weeded or utilized as a cover crop and tea ingredient. Hay is more likely to contain an assortment of persistent weeds. Depending on the timing and species mix of the hay field, the preponderance of seeds can limit the hay's utility as a garden mulch. Typically hay is used first as animal bedding or feed, then composted. The long fibers and energetic potential of the green hay make it a crucial ingredient to blend into the compost.

Leaves

Leaves are another soil-building resource. We choose to harvest leaves from locations where they are unwanted, such as roadsides and suburban lawns. We typically avoid harvesting leaves that if left alone will benefit their native habitat as a source of fertility. Consequently we rake the roadsides where the leaves accumulate. We dialogue with neighbors to rescue leaves that would otherwise be burned, bagged, or pushed over an embankment. We also attempt to assist municipalities and institutions in the area that want to dispose of this valuable resource. The leaves are packed in outdoor bins and introduced as a mulch and compost amendment.

Cover Crops and Green Manures

These two genres of vegetative plantings are designated as crops grown for the benefit of the soil, not for food production. Cover crops are typically a rotation tool used to prevent bare soil. Seeds are heavily sown to prevent an encroachment of opportunistic weeds and to protect the soil from erosive forces. Green manures are focused efforts at

This clover pathway is ready to be cut with a hand sickle. A vigorous nitrogen fixer, it also provides erosion control and a soft footpath. The cuttings can either be fed to animals or laid immediately on the garden bed as mulch.

After potatoes were harvested from this field, we seeded a fall cover crop of oats and buckwheat. After several hard frosts, the biomaterial will lie flat on the bed, adding to the diverse soil structure.

building organic matter and nutritive capacity in the soil. Legumes that fix atmospheric nitrogen through symbiotic root zone activity are a typical choice for a green manure.

To accomplish the goals of this vegetative process requires timing, soil preparation, and seeding density. Adequate germination depends on the temperature, soil moisture, and preparation of the soil to receive direct seeding. Soil should ideally be relatively leveled, free of weed competition, and bare to connect the seed roots to the earth.

As choices for cover crops and green manures there are many options. On paths we typically use white Dutch clover, which responds well to being pushed slightly into the earth via a roller or light foot traffic. The seed is expensive, though the small seeds cover a large area per pound. These perennials are low growing and leguminous. When the growth becomes too plentiful it is cut and applied to the beds or fed to the animals. We are also considering rabbit path tractors, long movable cages that enable us to rotate foraging rabbit through the clover path system.

In our garden beds we typically use buckwheat, oats, and annual legumes as our vegetative soil builders. These crops are annuals and do not persist over the cold winters of New Hampshire. We use perennial vegetative crops such as alfalfa and red clover where we are not planning to cultivate in the near future. For reliable germination and quick coverage of the soil, rye is an excellent cold-tolerant choice.

After seeding onto bare soil, a light mulch of straw or hay is important. This coverage helps ameliorate the destructive forces of wind and water erosion in the interim until the vegetative roots are established. Even a thin addition of mulch provides the blanket necessary to reduce erosive forces.

Cutting Greens

Grass clippings and other vegetative cuttings provide essential boosts of energy to the farm system. As opposed to straw and hay these cuttings contain the fresh moisture and nitrogen content of living plant tissues. Piles of fresh grass clippings rapidly heat up and become putrid. This blow-off of energetic potential is best utilized as animal feed or a compost amendment.

By cycling the energetic potential of fresh-cut greens through compost and animal digestion we are capturing the full vegetative potential derived from these solar collectors. We cut greens by hand or use the oxen to mow the lawn by grazing. While finely chopped lawnmower cuttings are less appealing to cattle, they are appropriate for smaller animals and as a compost amendment.

Biodynamic Accumulators

All plants are accumulators of the sun's energy; however, particular plants have been identified for their greater capacity to amass nutrients and energetic potential. Comfrey is a preeminent example of such a soil-building plant. The vigorous vascular system of the plant digs deep with its roots to bring quantities of nutrients to the surface. These nutrients are incorporated in nonwoody tissue that is shed in the fall to decompose into the soil food web. Accumulators such as horseradish, comfrey, rhubarb, parsnips, chard, and dandelion are identified by their powerful probing root systems and lush vitamin-rich foliage. The roots of horseradish penetrate nearly fifteen feet seeking nutrients, which are then brought to the surface. By bringing up nutrients that were restricted in subterranean locations the plants allow their vitality and energetic potential to be utilized by other plants. Accumulators such as comfrey leaves can be used as mulch, animal fodder, and compost accelerator, while the poisonous leaves of rhubarb are excellent mulch and compost ingredients.

Soil Amendments

We have purchased bagged soil amendments for many years. Originally we mixed ingredients à la Eliot Coleman, making a dusty concoction. We purchased and mixed proportions of bagged amendments such as greensand; Azomite; black rock phosphate; and bone, blood, and plant meals such as cottonseed and alfalfa. This exotic collection of measured ingredients resulted in a collection of partially used bags that were difficult to dry store without creating a disorganized mess. The process of mixing was a dusty pulmonary hazard. Consequently, for the last several years we have purchased an all-purpose, seed-starting amendment produced by Paul Sachs at North Country Organics. This preblended mix has many of the same ingredients as Coleman's recipe at a comparable cost with less noxious mixing and storage concerns.

The decision to utilize conventional agriculture by-products and industrially mined amendments is an ethical compromise. We are choosing to enhance our food production capacity through fossil fuel derivation, animal abuse, and land exploitation. The Azomite mined from the prehistoric remains of a lake in Utah provides a boost for our soil food web, as do the cottonseed and alfalfa meals that are industrially produced on large-scale agricultural operations. While the blood and bone meals are considered by-products of the meat processing industry, we are complicit by buying these products sourced by the abuse perpetrated by the industrial food system. As such, through the years we have purchased less and less of these amendments, instead focusing on building up our compost, worm casting, and compost-tea capacity, which is essential fertility particularly for our annual seedlings.

Cardboard Sheet Mulching

Sheet mulching is a method to transform an area of highly competitive vegetation into workable garden space. We use cardboard sheet mulch extensively as a transformative, progressive element in our farm food system.

Currently, cardboard is abundant in the United States and Canada, and the price per volume as a recycled product is marginal. The paper resource and means of production for virgin forest products is still plentiful, and as cardboard heads down the cycle its transportation and collection costs make it a marginal recyclable in comparison to metals. This current economic valuation of cardboard provides a steady stream that is available for collection.

Generally, local businesses pay the disposal costs for waste management specialists to collect their rubbish and recycling. We have found that the businesses would prefer to collect the cardboard for us rather than pay a disposal service. So when we drive to deliver farm products and conduct our errands, we also collect municipal cardboard for soil building and weed suppression.

But there are concerns regarding the sustainability of this system. One concern is based on availability and the other on possible contamination. Cardboard may not always be a plentiful, free by-product of the industrial machine. And in fact we are supportive of a system that would reduce such wasteful externalities. In our current system these external wastes of production are disregarded and devalued. That

may not always be the case, however. When I was working at Gaia Ecovillage in Argentina, for example, they had opted to use black plastic and carpets to sheet mulch the voracious grasses of the pampas. The lack of availability of cardboard and newspaper waste in a resource-competitive economy had made it impossible to source sufficient quantities to protect their pecan trees.

We are also concerned about soil contamination from the inks and glues used in the production of cardboard. We are optimistic that the era of dioxins in the newsprint industry has passed and the inks are vegetable and soy based. We avoid glossy newspaper and waxed cardboard, seeking primarily brown cardboard with minimal if any black printing. Boxes are also often taped together. The packing tape can be removed as the cardboard is applied or can be plucked from the mulch as the cardboard decomposes. The glue that holds the boxes together is also a potential contaminant to the soil. There has been discussion that the glues contain formaldehyde, which has prompted the USDA Organic policy makers to prohibit the practice of using cardboard mulch.

Although I have been cautious regarding the potential contamination of the soil food web, my perception is that the biomass of cardboard and newspaper is a beneficial soil conditioner. This blanket of mulch enhances soil quality in several ways. It provides total darkness that allows soil microorganisms that are intolerant to sunlight to flourish up to the surface. The organic material also allows vertical transfer of air and water. Earthworms thrive under the brown layer as it decomposes into flaky organic tilth. In comparison, plastic traps water, heat, and air, creating a biological suppression of not only weeds but also the health of the soil.

My personal philosophy is to use cardboard when available and if necessary to deter weed competition where animal tractor or handwork is impractical. Cardboard should be utilized in areas of expansion to extend the perimeter of the garden and to act as a barrier to the progression of weeds along the edges.

I do not encourage repeated applications of cardboard to maintain garden areas except in the case of fruit trees and shrubs that should be protected for several years until they become established. Garden areas should be sheet mulched thoroughly; then weeds can be maintained. Sheet mulching is a process for species suppression, nurturing transplants, and building soil and is a tool for specific applications rather than a blanket measure for all garden situations. This transformational technique requires energy and resources to initiate but provides enduring yields.

The possibilities for utilization of cardboard are immense. For long-term weed suppression it can be used around perennials and along expanding garden borders. As a transformative element there is a time continuum for vegetation suppression that must be considered, which depends on the existing species' resilience and the season it was applied. Some plants are particularly vigorous and difficult to eliminate. The rhizome roots of brambles and quackgrass can require repeated weeding and reapplication to suppress. Comfrey and apple sprouts require reapplication and vigilance over a course of years to suppress with sheet mulching.

Sheet mulching around shrubs and trees allows them to develop with less vulnerability to vegetative competition. The process of establishing a mulch ring provides nourishment and protection for the infant plants. Mulch rings, however, can harbor voles and other rodents, so it is important to safeguard juvenile tree bark during the winter months. We also use sheet mulch as a barrier along garden edges. We weed the edge, then overlap the cardboard by about a foot so the edge is protected from the extending vegetative competition on the garden perimeter. By applying the sheet mulch along the edge of the garden we prevent weed encroachment from the perimeter and create a perpetual zone of expansion for the garden.

Sheet mulching can also be used to retake a garden area that has become overgrown. This method is a logical response to weed invasions that

These blueberries are planted between an open grassy pasture and the forest edge—they need consistent care to help suppress encroaching grass and forest regrowth. Heavy cardboard mulch has been the solution, as well as planting an herbaceous layer of iris, daffodil, hollyhock, and lupine.

can occur through garden neglect. In place of the time-, labor-, and energy-intensive process of weeding or tilling, sheet mulching requires patience, planning, and materials.

To appropriately apply sheet mulch it is necessary to first evaluate the existing ground cover. Vegetation should be cut to the ground. This biomass and additional organic nutrients can then be laid on the ground, provided that the cardboard can be laid flat.

Planting new root stock in the spring is a yearly task. We use a thick layer of cardboard once the tree has been planted and watered to create the first layer of the mulch ring.

Depending on available materials, we will add a second layer of organic material, such as wood chips, manure, or animal bedding (a mixture of hay and manure). Regina is layering fine wood chips atop the cardboard mulch ring.

In April 2012 we began converting a pig pasture into perennial and annual food production space. Every fifty to seventy-five feet we planted a hedgerow of fruit trees that followed the contour of the slope. A mixture of apples, pears, and cherries make up the large orchard space.

In the sheet mulching or "lasagna method" the first step is to identify the area needing remediation.

For this particular bed the group laid down the cut grass and included some comfrey cuttings as the base layer.

Next the group added newspaper, wetting it to mat it down.

While some are still laying down newspaper, others in the group can begin layering the organic material: in this case, leaf mulch.

Once the surface is prepared, the cardboard or newspaper is placed on the ground in overlapping layers. It is necessary to be generous with the quantity of sheet mulch applied. Four to six layers of cardboard and up to a dozen sections of newsprint are necessary to deter weeds. These layers must be overlapped to prevent the competitive vegetative roots and aerial components from seeking escape. Depending on the vigor of the existing vegetation, if you sheet mulch a field of vegetation in spring, by the following spring the vegetation may be decomposed sufficiently to pull back the remnants of mulch and directly seed vegetables. If you sheet mulch in midsummer the field may be insufficiently decomposed to accept seed cultivation, though you will likely be able to transplant seedlings directly into the space.

The final sheet-mulched area with cardboard and wood chip pathways. This bed could now be seeded with lettuce or bok choy or even transplanted into with a variety of annual crops.

Sheet mulch is difficult to apply on a windy day. We use water cans to douse the newspaper as it is laid so the papers are bonded in place, à la papier-mâché. Newspaper is more flexible and can be utilized in small spaces to provide maximum protection to smaller plants. Cardboard is more durable and appropriate for larger areas and paths. It is also important to cover the sheet mulch to prevent the wind from moving the mulch and to build soil by improving the aesthetic. Wood chips, mulch hay, or other abundant, weighty, wind-resistant biomass can be used to hold down the sheet mulch.

In place of a tillage, which depends on dry weather to cultivate soil, sheet mulching is flexible because it can be applied at any time of the year. The volume of cardboard necessary to guarantee

Maintaining the perimeter edge and pathways is a yearly task. In this forest garden, the North Orchard, the perimeter is layered with cardboard, wood chips, and a decomposing log, while the interior has been weeded and lightly layered with wood chips.

adequate coverage is immense, so ample storage area is necessary. Limitations in space for storage often dictate a necessity to utilize cardboard. Cardboard can be stored outdoors, though wind will spread the material haphazardly if it is not confined.

Compost

Compost is what we use to fortify the soil. This ingredient is crucial for our potting soil and overall soil enrichment. Compost is a living resource of organic, biologically active decomposition. The process of decomposition and re-creation affects all biology on this planet. Plants depend on raw elements and symbiosis with bacteria and fungi to exist. Compost concentrates both the biologic and mineral enrichment that food production plants need to thrive.

Humans have been practicing compost techniques for thousands of years. Thomas Jefferson explored the use of manures to regenerate croplands. There are many books written exclusively on the topic of composting. In the modern era Sir Albert Howard's *An Agricultural Testament* (1943) was a companion to the writings of J. I. Rodale and set the standard regarding the creation and usage of compost.

Through the solar cycle of decomposition, rejuvenation, and recomposition, energy can be gleaned for human purposes. Compost is the tool with which we can harness the solar cycle's daily biologic trophic composition. Ultimately the use of compost as a resource to invigorate maximization of ecological abundance and food production is without limits.

Our food production system depends on continued enrichment of our soil system. As stewards we must compensate for the energetic value that we continue to derive from the soil food web by inputting sufficient energy to rejuvenate the system. Compost serves as a biologic and vitamin source as well as a conditioner for the soil medium. The organic material of compost serves a function similar to silt by helping to balance the sand and clay ingredients that form the basic loam beneath our feet. Sand lacks the capacity to retain moisture, and the cohesive nature of clay locks out attempts to cultivate. By combining clay, sand, and the organic material of decomposing biology, nature creates the ultimate medium for renewed regeneration: *living soil*.

Quality compost depends on the carbon-to-nitrogen ratio. Finished, stable, biologically active compost retains a ratio of twenty-five to thirty parts carbon to one part nitrogen. To achieve this balance, many ingredients with varied ratios must be blended. Woody brown materials such as wood chips and brown leaves are higher in carbon, while manures, green leaves, grass, and kitchen wastes will be higher in nitrogen. The ratio of sawdust can be as high as four hundred parts carbon to one part nitrogen, while manures can contain as few as eight parts carbon for every one part nitrogen. Balancing these elemental ratios creates an environment conducive to the biologic activity necessary for the compost process.

These ratios are derived from molecule ratios and are not translated in terms of volume or weight. Thus there is no simple measurement that will define the proper ratios. Rather, it is a subjective judgment responsive to the continued evaluation of the degradation and subsequent rejuvenation of the materials. It is also important to consider the structure, size, and shape of the pile and its constituent parts. The hollow structure of straw,

for example, encourages aerobic conductivity that assists the overall decomposition process. Stones are difficult to compost, but rock dusts can be introduced directly into the compost creation process. A successful aerobic compost process allows the contents to achieve a hot core temperature and "cook" to completion. The finished compost that we are seeking is a dark rich color and has a spongy, soft texture, a crumbly material with particulate sizes of less than a quarter inch. The material should no longer be recognizable as its original ingredients, and the smell should be pleasantly rich and earthy, like a forest floor. The art is to produce a masterpiece from the ingredients at hand.

The biology of compost and living soil is difficult to conceptualize in human-size scales. The amount of relational activity of bacteria, fungi, protozoa, nematodes, arthropods, and earthworms that transpire within living soil is immense. This intense biologic activity is concentrated in the composting process. Composting creates an abundance of available fertility that serves as the web of life.

When engaging in the compost process, it is important to consider the purpose of your compost. For what plants are you creating the meal? For what do you intend to use the compost? Is it for potting soil and annual field applications or for perennial mulch rings? Would the compost need to be available in a short amount of time, and how much effort should be expended to obtain the yield?

Thermophilic composting, for example, is a hot aerobic decomposition process that is rich in bacteria and nitrogen. Also, horses, people, and pigs do not digest some weed seeds in their diet; thermophilic hot composting destroys these seeds along with disease pathogens by attaining temperatures in excess of 140 degrees. Mesophilic composting, by comparison, is a slower, cooler fungal process that is rich in carbon. Thermophilic compost is faster acting and dissipates more rapidly, while the mesophilic decomposition results in compost that is more stable and enduring. The bacteria-rich thermophilic process is used primarily for annuals

This entire compost pile is the accumulated bedding from the oxen. With a mixture of hay, wood chips, and manure, as well as coffee grounds hauled from the weekly town run, it is a biological chemistry experiment. With the catalyst of a pitchfork and a willing volunteer, aeration takes the decomposition process further.

and is composed of nitrogen-rich animal and green manure ingredients; the fungal-dominated mesophilic systems are intended primarily for perennial systems with carbon-rich wood-derived ingredients. The choice of whether to pursue a mesophilic versus a thermophilic decomposition depends on your goals and resources. Thermophilic production requires energy inputs, while the mesophilic process requires more patience.

To create compost we blend primary organic and mineral ingredients based on their availability and our needs. This system has evolved through the course of our history as opportunities have arisen. The composting process intends to convert and enhance biomass into usable soil fertility for plant nutrition and soil conditioning. The inclusion of kitchen and garden debris into the compost can create several problems. Kitchen waste attracts rodents and vermin, and garden debris can harbor disease pathogens and persistent weed seeds and roots poised to regenerate. At D Acres we incorporate animals to help process kitchen and garden waste. This trophically designed energy layer refines the material, making it more suitable for ingestion into the soil food web.

To construct a thermophilic pile it is important to understand the principles of gravity, convection thermodynamics, and aeration. The concept is to create a vertically supported environment with abundant surface area that allows oxygen to be introduced, thus speeding up the bacteria process. If compost piles are built to be tall, heat generated in the base of the pile rises through the higher aspects of the pile and adds thermal potential. Air is also integrated through horizontal surface area and rises through the pile, nourishing the aerobic bacteria activity. Creating the conditions that enable biologic invigoration requires sufficient critical mass. The size and shape of the thermophilic pile requires both surface area for aerobic introduction and sufficient scale for recombinant heat generation. Typically 120 cubic feet is deemed sufficient to generate ample mass for the compost process.

The typical progression of building a thermophilic pile is to introduce layers of available organic material, imagining the finished mixture will have a carbon-to-nitrogen ratio of 30 to 1. Ingredients can include most any organic biodegradable material. Carbon-rich, woody materials should be shredded or chipped and utilized with caution and in moderation proportionate to the nitrogen. It is important to develop a pile-building process that can thoroughly blend the material as it is turned or decomposes. Compost piles can stall and need resuscitation to achieve complete decomposition, particularly if the carbon levels exceed the nitrogen capacity. To encourage further decomposition high-nitrogen-content materials such as fresh manures, grass clippings, and comfrey leaves can be incorporated into the pile. To integrate the fresh material it can be introduced as layers when the pile is being turned.

Structures have been developed to assist the construction of successful piles. The three-bin system has become a favorite for smaller-scale operations. The three bins serve as markers for the progression of the compost process. Essentially, raw materials are layered into the first bin until they are turned into the active pile in bin two and eventually deposited into the third bin as finished product. When three separate enclosures are constructed for the first, second, and third stages, the piles are clearly delineated and can be piled high against the walls on the enclosure. Removable walls can assist with turning the material. This ideal system epitomizes the process, though it requires a thoughtful, interactive progression to maintain.

Another structure for creating compost is the wire bin method. Typically we source 1- × 4-inch rectangular grid wire cut into sections to create a circle that is a minimum four feet in diameter and three feet in height. Wire fencing such as chicken wire must be supported by posts. If larger-diameter wire mesh is used, more material will escape the vertical integration. When the compost is ready to be turned, the "bin" can be relocated adjacent so that the material can be turned directly into the next location. Because of the portability of the structure, the fertile residues, and the suppression of plants areas where the pile has been located can be systematically transformed into garden space.

Elevated plastic compost tumblers are also an option. These systems are quite functional in urban environments and as worm farms because they are rodent resistant. Introducing worm cooperators is advisable because the capacity and usage are geared more for the concentrated nitrogen of household kitchen waste versus the designed dilution and dispersal and reintroduction of large volumes of animal manures.

Generally we build thermophilic piles that are exposed to the rain. The piles require massive inputs of water to maintain the biologic activity. Steam generated by the heat of the process exhausts the reservoirs of water in the compost piles. At times water can be added to accelerate the process. On the other end of the spectrum, compost should not be inundated with so much water that it becomes soggy and anaerobic in its decomposition. As the material decomposes, its innate capacity to retain moisture and evenly translocate humidity levels becomes apparent.

Evaluating the pile's stage of decomposition requires the senses of smell, touch, and sight. Smell allows us to assess if the pile has become anaerobic. If there is high moisture content and the smell of ammonia, steps should be taken to reduce

A three-bin compost system

the moisture levels. Compost should not contain puddles of standing water or the organic material will begin anaerobic decomposition. By touching the compost material we can evaluate the moisture content, water retention capacity, and structure, as well as the texture, to assess how it will serve as a soil conditioner. Our visual observations interpret the appearance of the material degradation. Visual inspection also yields information regarding biologic compost activity such as mycelium and earthworm density. The temperature of the pile can be visually determined in the form of steam rising and referenced both by thermometer and in relation to body temperature.

The composition of materials and humidity levels are important to diagnosis the stage of decomposition and possible remedial measures necessary to induce or accelerate further decomposition. Balancing the carbon:nitrogen ratio is necessary throughout the process. Since our farm is blessed with abundant carbon, we incorporate as much as possible into the carbon-to-nitrogen compost ratio cycle to maximize the quantity of compost we can produce.

As experience with composting continues to evolve we discover the variability of the process based on the ingredients, weather, and construction affecting the compost. By recognizing the stages of decomposition and adjusting the process, we can maximize our composting system. Enhancing the composting process usually requires adjusting the carbon-to-nitrogen ratio by adding ingredients, introducing air, and eliminating anaerobic conditions but ensuring sufficient moisture levels.

The NRCS-funded compost facility will allow the farm to consolidate large amounts of material for proper decomposition without water contamination. The large bay is designed to hold the first and second stages of material, while the smaller bay will contain the finished material, ready for distribution to sites around the farm.

The purchase of a skid steer makes for easier compost turning and movement of material throughout the farm. Even in the early months of winter we are able to turn the hot manure material.

Random piles of barn manure do not rapidly generate superior compost. Developing an effective system of composting that is scalable for each operation depends on certain factors, including time and the availability of resources and equipment. For many years we complied with the national organic standards, which require careful record keeping on each compost pile. The records consist of temperature and visual analyses that monitor the safety of the compost. The standards dictate that if temperature extremes are attained, the compost material will be safe from pathogens and diseases. The organic standards are set for the safety of the public at large in light of commercial operations where dead animals are routinely composted in the piles. However, for our small-scale composting program, such standards and regulations became burdensome. It is our experience that compost production is an art defined by practice and situational response to the land's fertility needs and the resources at hand. Ultimately our credo for crop safety shifted from compost monitoring to an emphasis on not applying any compost that was unfinished onto crop soil within ninety days of the coming harvest.

Turning and moving compost requires energy. Whether the compost ingredients are obtained through mechanical means or human, the process must be orchestrated for efficiency. It is important to be able to easily access compost piles so that additional material can be added and the finished material can be retrieved. Whether this is accomplished with wheelbarrows or a dump truck depends on the situation. We like to build compost piles in well-drained locations that are vehicle accessible. It is helpful if the compost is located close to the ingredients and point of usage. Ideally, piles are built uphill of their destination, allowing for easy distribution with the assistance of gravity.

To turn compost it is important to consider the parameters of effective pile construction, even if machines are being utilized. The materials should be thoroughly blended so that no individual ingredient is concentrated. The piles should be tall and loosely stacked to enhance aeration and convection of heat. There are commercial machines designed to turn compost that straddle windrows of material that are ten feet tall and twenty feet wide. Backhoes and loaders can be used for compost turning, though it can be important to access the material from all sides or construct a heavy retaining wall for the material to be pushed or piled against. Machines will quickly destroy compost structures constructed of pallets or other plank wood.

In the past we have used a hayfork to turn compost on the farm. The labor is arduous, and when there are over twenty piles on-site, time-intensive turning has often been neglected. At times we have had volumes of material in stages of partial decomposition and none of the high-valued finished product. To ameliorate the situation we chose to seek assistance from the Natural Resources Conservation Service to refine our compost operations.

The new compost facility completed in July 2013 offers several improvements to enable the efficiency of our soil fertility program. The design is a concrete pad with retaining walls on three sides. The open-sided access to the concrete surface allows for materials to be easily unloaded in the area. The flat impermeable surface and durable walls are ideal for mechanized turning of the pile, while a floor drain allows for the effluent liquids to be collected and reincorporated.

Animals as Agricultural Partners

Using the various techniques and resources previously discussed, we have experimented with varying scenarios and scales, developing a system of soil building that blends the concepts of no-till gardening and animal tractors. By weaving these two established principles we have evolved a framework to build soil and expand the arable land base.

At this juncture in our land-building process, we are heavily partnered with our animals. The pig, as discussed in chapter nine, is nature's rototiller. Pigs use their bodies as implements for soil cultivation.

Their natural tendency is to forage the soil-seeking insects and vegetative roots. Chickens are also specialists at cultivating the earth. Their powerful legs and talons work to rip apart the soil as they forage the ground.

Chickens

By encouraging chickens' natural instincts in areas of confinement, we were optimistic they would perform the tasks of tractor-powered cultivation. In the first years at the farm we experimented with small chicken tractor enclosures that were placed throughout the gardens. Depending on the existing conditions and goals these tractors were marginally effective. At times the tractors were utilized in field or lawn space as precursors to sheet mulch transition. We would tractor the animals about two birds per 4- × 8-foot footprint and allow them to forage the vegetation. Completely eliminating the existing native vegetation in an area could require two to three weeks. We also used the chickens in established gardens, particularly to eliminate bolted salad greens or other crop residue. These enclosures provided a feast for the chickens; however, the nature of their digging destroyed soil structure and bed shape.

We also developed a larger-scale chicken tractor. This tractor was a house on wheels that we were able to move with vehicles and oxen. This movable house provided a shelter to up to twelve birds, and a fence was attached to allow access to forage and tractor-type activities. This tractor was helpful in larger and more remote areas where we would leave the chickens to denude the vegetation and supply fertility.

Chickens dig holes, seeking dirt baths as relief from the itching mites that bother their skin. Where they expose the earth to rain, compaction ensues. Thus, by the time the vegetation is removed, the soil that remains is an undulating hard landscape that must be improved if it is to be directly seeded or utilized for transplants. The soil should be vertically forked to aerate it and physically reduce compaction.

This small chicken tractor is located in the vicinity of the permanent chicken coop for easy access to the chickens. The tractor can be moved in succession and followed by a process of sheet mulching to create a new garden bed.

The larger chicken tractor can be moved by the oxen or a truck to various locations. This area is planted with mulberries and was once completely grass covered. The chickens were grazing between July and November.

Following the tractor and the chickens is a heavy layer of cardboard and organic material. The following spring we transplanted lupine, mullein, and comfrey into the understory.

Following the forking, the soil should be raked and shoveled into position to provide a level soil surface.

Pigs

Pigs are very aggressive in their desire to obtain fresh forage. They can be severely destructive to stone walls and fencing when they are employed as foragers. Even in barn situations with adequately supplied feed the pig will actively seek an escape route. There are various movable cages constructed of wood and steel that are adequate for small hogs, though as pigs grow, durable, mobile structures are more difficult to implement. The advent of the electric fence allows pig tractors to be much larger enclosures than those limited by the durability of movable structures.

In these larger enclosures the pigs find their preferential locations. By moving the housing and feeding locations the pigs can be induced to concentrate their activities in areas they had neglected to cultivate. Pigs will gather in wet spots and seek shade or sun depending on the ambient temperatures. Famously, pigs dig holes for mud baths. This connection to the earth provides soothing, cool relief from insects and sunburn.

The amount of time that pigs require to remove vegetation depends on the breed of pig, its size and foraging interest, and the density and resilience of the plant species. Grassy fields may be cultivated by systematically grazing the animals in grid patterns. Pigs can denude large sections quickly, though they may have to be encouraged to forage evenly. Resprouting tree roots in the Northern Forest such as red maple require more than two growing seasons to completely eradicate. These roots will regenerate without continued pig presence to forage the sprouting wood and curtail the natural rejuvenation.

After pigs have adequately eliminated existing vegetation in an area, it is time to respeciate the zone or build the soil through sheet mulch. As with that of chickens, the pigs' activity will result in undulating soil and severe compaction. We can address the compaction by vertically forking the soil to physically aerate the medium. The undulated soil should be addressed by leveling the medium with a hard garden rake and shovel. After the compaction

Morgan Casella works with volunteers from a local high school to prepare the G-Animal space after pigs did the initial earth work. Aerating with a broadfork, raking the soil smooth, and shoveling in large holes are part of the labor.

In an area we call the Pig Pasture, the pigs have done the initial work of eliminating weed pressure and fertilizing. Beth (background) uses a rock bar to pop loose, small stumps, while Regina uses a hard rake to smooth the compacted surface and rake loose debris into contoured swales.

The excess debris from the Pig Pasture is lined into small aboveground swales that follow the contour of the slope and are in line with fruit tree plantings. Over time this organic material will break down, adding billions of microorganisms to the site, slowing erosion, and retaining water along the hillside.

from pigs and exposure to wind and rain it is important to cover the bare soil.

The soil should be covered with light mulch and seeded with preferable species. Cover crops can be utilized at this point to build biomass, suppress weeds, and fix nitrogen from the atmosphere. We have had particular success growing potatoes in the postpig areas.

Our methodology for planting potatoes is fairly simple. We cut the eyes from the potatoes and place them directly on the ground twelve to eighteen inches apart. These potatoes are spaced in three rows about twelve inches apart so that four-foot beds are approximated along the contours of any perceptible slope. If water erosion capacity exists

because of slope and volume it may be necessary to also weave in larger swales and erosion control and absorption structures. The rows are separated by one-foot-wide paths seeded with white clover.

We seed these potatoes directly onto the ground around the first day of May, when the serviceberries are blooming, and cover them with a mixture of loam, compost, and soil amendments. This mixture provides water retention and root surfaces for the potatoes to begin growth. We use about one gallon of soil for each eye. Then we add one to three inches of hay to provide further biomass and erosion control.

Throughout the ensuing summer we add more biomass to hill up around the potatoes. On the first of June, July, and August we attempt to cover the

This Forest to Field Pasture was clipped, shoveled, and raked over. The next phase was a cover crop of oats, white clover, forage beets, and sorghum, a diverse covering that would grow through the summer. The pigs were then put back into this field to forage the grasses.

plants with biomass. This biomass accumulation not only provides the cultivating technique for potato production, it builds the garden beds for the future. At the end of the growing season we harvest the potatoes, reshape the beds, and plant cover crops or garlic for the following season. This process depends on the reasonable acquisition of large quantities of biomass for the hilling and bed building. Adequate planning and preparation must be made to source this quantity of material. There are no given rules for bedding composition. Yields and potential disease problems vary, depending on the weather and adequate bedding composition and quality. The yields from this system are generally slightly less than the standard for potatoes. We typically harvest

at least 7.5 pounds for every pound of seed that was planted, though gardeners in cultivated soil can yield up to 10 pounds for every pound seeded. This reduction in yield is offset by the transformative, expansive process that will yield productivity for many years. This investment in an annual crop yields dividends long after the first crop is consumed.

Within these potato fields are also the seedlings of the perennial crops of the future. Fruit and nut trees as well as shrubs are quickly introduced after the pigs vacate the landscape. After the potatoes are harvested the contoured beds can be maintained for annuals or transitioned into perennial herbs and flowers that will flourish in the developing orchard for years to come.

The Upper Field was the first large pasture space to be converted from pigs to field crops. Here residents are creating paths of cardboard and wood chips while simultaneously planting potatoes. Straw is being used to cover the compost-covered seed potatoes.

The Upper Field completely planted in potatoes

Regina observes the progress of the Upper Field potatoes at midsummer.

Potato seed is placed directly on the bare ground after the beds and paths have been designated.

Seed potatoes are covered with approximately one gallon of soil per seed.

Planting potatoes in this manner is labor intensive, but it creates a fairly quick conversion of unworkable soil into immediate food production. In designing these spaces we are careful to follow the contours of the area. Much of our land mass is sloped, and by working with this shape we can retain nutrients and moisture and minimize erosion and soil loss.

After harvesting, the many layers of compost and hay are intermingled into the soil to add biological diversity and an opportunity for complex microactivity.

The garden beds are then cover cropped or planted with garlic and heavily mulched.

CHAPTER FIFTEEN

THE FARM ECOLOGY

We are still in the infancy of our understanding of how to create a self-perpetuating, sustainable food system in this climate. Creating such a system will require a transformation of the landscape and ecology as well as an evolution of knowledge, skill, and lifestyle for the human inhabitants. This chapter outlines a vision of progression for the farm system at D Acres.

The success of our permaculture system is rooted in our care for the soil and our commitment to the land's fertility. This is ongoing work, a rhythmic part of our seasonal cycle. Tandem to our development of the soil is the continuing evolution of our farm system itself. The designed succession and planned evolution of species amplifies biodiversity as we develop the fertile ground from which it grows. In the section that follows, this biologic design will be detailed, along with the annual and perennial cycles that seasonally unfold across the farm landscape.

The task is to design and implement strategies of succession that maximize benefits for the farm organism without denigrating the resources or regional ecology. This ecological, successive process is dependent on the circumstances of the landscape. The existing ecology and the resources of the environment provide a starting point to begin examination of the potential for system implementation.

To maintain the grassy pastures and meadows of New England, human intervention is generally required. Otherwise the forest moves into field spaces through a process of succession. This ecological progression is dictated by the availability of resources both energetic and biologic. The energetic resources include solar and nutritive inputs, while the biologic transition is dependent on species density and diversity.

Through this succession the relative concentrations of carbon and nitrogen change dramatically. As the field turns to woody forest the carbon that is embodied by the organic material increases exponentially. This process is also correlated to the transition from bacterial to fungal dominance occurring within the ecological system. In the shade of the forest, fungus must serve the role of distributing the resources that the nitrogen-powered bacteria had assumed in the field.

As we examine the phases of transition that compose the succession process, assessing the horizontal profile of the ecologic system is beneficial. The vegetative profiles can be loosely defined as seven layers: the root/soil zone, the herb layer, the shrub layer, the bramble layer, the vine layer, the midsuccession tree layer, and the overstory. Each of these layers intersects spatially, functionally, and potentially competitively over a time continuum, as part of an evolution.

The vegetative layers of this terrestrial system are categorized to simplify the intertwined dynamics of the plant universe. The root zone consists of the soil and plants' subterranean elements. This

The diverse herbaceous layer nestled between two apple trees and a pear tree with lupine, aster, bee balm, dandelion, and goldenrod

layer is the foundation of the living system. The soil provides the biological and structural basis for all the vegetative terrestrial activity. The soil serves to distribute water and nutrition to the plants' roots while the crops act as soil builders and conditioners for the future. Root crop food can be gleaned from the abundance of this system while the living soil decomposes and recycles the detritus of biomass being shed from above. No plant exists as just a root so the distinction between root and herb zones is indefinite. Bulbs and food crops such as potatoes, garlic, and parsnips are standards of this category.

The herb zone consists of a diverse array of plants. These are low-growing plants that provide weed suppression and a high proportion of potential yields as the larger plants in the system are being established. Plants such as lupine, sage, valerian, bee balm, mints, and lemon balm, as well as most annual vegetables, are examples of the herb layer.

The vine layer consists of plants that have adapted by adopting strategies to procure sunlight through rapid, vertical growth. Because their vegetative structures are unable to attain freestanding height, these plants utilize existing structures to seek elevation. Vines such as hops, grapes, kiwis, and beans can overwhelm existing vegetation and trellises; it is therefore important to consider the timing of their introduction. As an example, with the three sisters of corn, beans, and squash it is important to start the beans once the corn is well established—perhaps even knee high—otherwise the beans will

overwhelm the corn. A late planting of beans will yield fewer beans, though the cumulative benefits of the system provide the justification. As for the perennials, vigorous vines such as akebia or hardy kiwi will strangle trees, so it is important to consider how to design a beneficial vine habitat that is not counterproductive.

The shrub layer consists of multistemmed plants that grow less than twelve feet tall. These plants are typified by the blueberries and *Ribes* species such as currants, gooseberry, and jostaberry. The currants can be particularly appropriate over the time continuum because of their shade-tolerant fruit production capacity.

Brambles are colonizers of nature's field-to-forest succession. These opportunistic plants travel horizontally with running root systems. New England brambles are typically *Rubus* species such as blackberry and raspberry.

The south side of the house is an ideal place for vining growth, creating a shaded pathway. Climbing the trellis are several species: grapes, schisandra, hops, and wild yam. On both sides of the trellis an understory of walking onion, lemon balm, and a perennialized leek grow vigorously.

The layering here in the Lower Garden is complex. A native apple tree stands tall with a shrub layer of blueberries to the left and a patch of raspberries lower. Below the shrubs is a layer of herbs and flowers such as chervil, bachelor's button, and kale; in the background are potatoes and sunchokes.

The forest continues to surround all open growing and living spaces at the farm. This is a 2010 view of the North Orchard, twelve years after the first plantings were made. In the foreground an older oak continues to tower above a butternut, a crab apple, and filazels, while farther up the slope are a range of apple and pear trees, cherries, and quince; below that are currants and gooseberries, and below that layer are many herbs and flowers, such as echinacea, mullein, peony, lemon balm, daylily, crocus, hyacinth, horsetail, black cohosh, and mint.

The fruiting trees will grow into the twelve- to thirty-foot height range. These trees are typified by apples, pears, cherries, and peaches. Nuts are generally categorized as shrubs, fruit trees, or over-story species, depending on their size. The overstory consists of the largest trees. These trees can tower up to 150 feet if they mature. The maximization of yield in this layer depends on use, which could include firewood and food crops.

While the vine, herb, and root layers are often mixtures of annual and perennial plants, the vegetative layers of the shrubs, brambles, and trees are reserved for perennial species. Whether annual or perennial, plants are selected based on their characteristics. Some plants are chosen because they provide ground cover and resist erosive forces, while others possess soil-building characteristics such as biodynamic accumulation and nitrogen fixation. Human food and medicine potential, as well as fodder for animals, are valued in this system. Plants are also chosen to provide habitat and food sources for beneficial insects. Ideally, specific plants can provide multiple benefits to merit their inclusion in the system.

We still see moose on the property, but this open field has made a drastic transformation over twelve years. Through the works of sheet mulching, this roadside field has become the Roadside Orchard.

Plants compete in this system for resources. Depending on their aerial and subterranean growth habits, plants can be matched to complement their resource utilization capacity. For example, hardwoods whose leaves are slow to emerge, such as black walnut, make an excellent overstory for daffodils that flower early in the spring. Likewise, the taproot of the black walnut can be an excellent complement to the sprawling root system of *Rubus* species such as raspberries.

The intention is to combine the capacities of all seven layers of vegetation over time to maximize their productivity with minimal resources. This combination demands that resources are partitioned, shared, and symbiotically provided. It also

demands comprehension of the time continuum that will evolve the actualization of the system. As the system transforms with time it is important to react and reassess regarding the developments of the biologic succession.

While specific combinations of plant species are almost limitless, we have had particular success with the following interplantings, especially combining the fibrous root system of the blueberries with the upright aerial growth and penetrating taproot of the heartnut. In one twenty- by twenty-foot space we planted this combination in 1998. The six blueberries were planted in a semicircle to the south side of the heartnut. The blueberries were about twelve inches tall, and the heartnut stem was smaller than

Roadside Orchard, spring 2007

my pinkie. We had ordered the heartnuts as a pair from a glossy magazine that proved disreputable, and the mate died the first winter. Since that time we have planted other heartnuts in the vicinity. We also planted lingonberry and a spicebush to the east of the heartnut, Siberian peashrub to the west, and rhubarb along the chickens' fence. A few seedlings of elecampane and some daffodil bulbs that have naturalized in this zone were also planted nearby.

Since the bed was established we have weeded and mulched thickly with wood chips once a year to keep the area free of competition and build the soil. The spring succession begins with the reliable daffodils, which are followed by the flowers of the blueberry and the Siberian peashrub. The rhubarb and elecampane powerfully emerge and begin their solar collecting activities. Eventually the heartnut wakes from winter slumber to produce leaves that cast minimal shade from the sun rising at its summer declination. During the summer the elecampane and rhubarb grow large, abundant foliage that prevents weed intrusion while also cooling the area for water retention. After the summer of harvest the leaves fall to the earth, creating a thick, rich mulch that is rapidly recycled into the soil food web. The spicebush is thriving, though perhaps due to shade, the lingonberry has not established itself with vigor.

We are optimistic that production in this zone will continue to grow for the next seventy-five to two hundred years, until the heartnut begins to decline from age. This zone provides very productive blueberry plants with the least labor of weeding

Roadside Orchard, summer 2011

The heartnut stands tallest with elecampane, blueberry, and spicebush in the understory.

or mulching. While we have not harvested a heartnut, we are hopeful that adequate pollination from other heartnuts maturing in the area will eventually result in nut production.

There have been futile attempts to empirically quantify and compare this system of successive agriculture to the industrial model of food production. The reductionist approach to analysis contradicts the complexities of the holistic approach to ecological abundance. The maximization of the design is dependent on a cost-benefit analysis of the endeavor that requires considering a complexity of factors and externalities. While a basic assessment of inputs in time, energy, and materials is relatively simple to quantify, it is difficult to gauge the value of this successive system implementation as a replacement for the predominantly fossil fueled system. A simplified yield analysis does not derive and quantify the benefits of habitat, water retention, erosion control, species diversity, soil building, and aesthetics, nor the security of a stable perennial food system.

The Evolving Design of Nature

The design and manipulation of this natural process involves humanity's interpretation of our role. Each of us makes choices regarding that role: At what level do we wish to intercede as active participants in the processes of the natural world? At D Acres our concept is to develop a system that maximizes the trophic potential of each biologic niche. There is

no particular climax to this succession but rather an evaluation of the cost benefits over the continuum. Each phase in the succession offers opportunities to harvest from the productive ecologic systems. Each layer offers possibilities for collaboration simultaneous to the creation of resiliency and fertility.

Definitions of ecological succession provide a reference for examining the tendencies and productivities along the continuum that typically defines nature's course. The farm organism is a landscape undergoing perpetual succession and evolution influenced in part by the articulated planning of the human organism. Ours is a cognitive choice to interact with our environment as a conscientious element in the perpetuation of the living systems of

The North Orchard in 1999 with the butternut and small plantings of herbaceous annuals and perennials

Earth. It is a conscious decision to reap what we sow in the ecological matrix.

By accepting our role as participants in the process of ecological succession we can begin to examine a vision of that continuum. By establishing an initial interpretation of the possible successive pattern we have a baseline for continual assessment of the evolving design. This approach broadens our thought process and allows an evolution not stipulated by the parameters of the initial design. Our limited understanding of nature's complexities creates the necessity to react and reassess along the time continuum.

The maximization of plant diversity and biomass at each level provides multiple benefits to the farm organism. Plant species provide benefits as food and medicine, as well as heating fuel and fodder for animals. Plants also provide food and habitat for beneficial insects and a visual and aromatic beauty to the landscape. Plants build soil, prevent erosion, and provide opportunity for diverse ecological interactions. The habits of roots and aerial growth provide opportunities for plants to navigate toward biologic potential by fulfilling various niches in the vegetative layer.

The energetic potential of the farm can also be maximized through the storage and transportation of the sun's energy. To achieve sustainable energetic consumption levels we must limit our utilization to that which is supplied daily by the sun. We choose

The North Orchard in full bloom during the summer of 2011, with the butternut standing tall in the middle of the image, but not as tall as the oak tree behind it.

to develop a system that harnesses the solar energy through existing natural systems such as photosynthesis and the planetary water cycle.

Our daily dose of solar energy is stored by photosynthesis in plant tissue. The plant tissue can then be fed to animals, enlivening their metabolism and producing manure. Manure is then used to make compost that fertilizes the soil that we use to grow crops, powered, once again, by the sun. The cyclic process relocates and stores the solar energy allotment that we are supplied each day. The capacity for storage and mobility of energetic potential creates opportunities to maximize yields of vegetation throughout the farm system. This vegetative potential is what feeds the farm system. Our role as farm stewards is to glean from that perpetuating abundance.

Potting Soil

Annual plants require an early start to bear fruit in the short growing season of New England. The planting medium, daylight length, and temperature conditions are crucial factors for a healthy nursery of seedlings. In an attempt to localize our annual production we start seedlings on-site. This production requires specific timing and appropriate conditions for each species of annual.

Our indoor seedling nursery is a compromise between our attempts to provide a food system that is expansive and one that is acceptable for the conventional consumer's palate. We grow many vegetables, such as eggplants and tomatoes, that are not native to this climate and require a longer season to bear fruit. We also invest in indoor operations to extend the growing season and maximize the production of our operations. This investment is substantial in terms of time, energy, and effort.

There are many choices regarding how to grow nursery starts. We have resorted to the standard plastic trays that are common in the nursery industry. The trays provide individual cells so seedlings can be easily transferred for transplanting without

A kale transplant, ready for the ground, has been nurtured through the entire process of seeding in the basement, potting up, and hardening off in the Big Cold Frame. The fine root hairs have expanded throughout the soil.

root damage. These containers provide a shape into which the soil and root fibers can mold and resist erosion from watering. As an alternative to plastic we have also made newspaper pots by rolling the paper into an origami-shape container. These pots are laborious to construct, though fully functional for seedling production. We also have used large quantities of recycled containers that were originally purposed for yogurt or soy milk. These containers can be functional, though they require substantial drainage holes to be drilled to allow aeration and bottom watering capacity, which encourages healthy plant growth. Round pots can also increase the complexity of the spatial relations in the nursery whereas square pots can stack together in higher volumes. We have also considered purchasing gadgetry to make soil blocks, which would eliminate our need for plastic trays and containers; however, we have resisted this course of action because peat moss or coconut fibers are typically used to retain moisture

Resident workers start seeds in the Big Cold Frame. We use the cold frame when spring temperatures are warm and we can raise starts in full sun.

and the block's structure in this system. We have been reluctant to add another layer of dependence in sourcing this ingredient material as a component in the process of food system. That said, we will continue to experiment in choosing alternatives incentivized by circumstance and ethical inquiry.

To build our soil medium we start by considering the needs of the plant roots. The medium must have the capacity to retain moisture while also being porous and well drained. The medium should have structure to allow the roots to grasp the material and prevent damage during the transplanting phase. This medium also provides a nutrient and energetic boost for the initial growth of the plant.

To create this medium we have developed an all-purpose potting soil recipe. The recipe we are currently using consists of 5 parts loam, 5 parts compost, 1.5 parts vermiculite, 1.5 parts perlite, one cup humates, and a quart of Paul Sachs's Pro-Start

The compost sifter is a homemade tool. The A-frame can be moved to a finished compost pile, and with a wheelbarrow placed below the hanging frame, the sifted material can be wheeled easily to the Big Cold Frame or basement seeding stations.

2-3-3. This mixture allows us to build up the existing soil medium by incorporating portions of the limited supplies of compost and imported soil amendments.

Loam is sourced on-site. The local loam is sandy and well drained, providing a basic building block for our soil fabrication. This material, however, lacks sufficient nutrients, structure, and water retention capacity to maximize the health of seedlings. Consequently, we mix in additional components to improve the medium.

The compost that we reserve for our potting soil is typically the highest quality that we can attain. It is important to source compost that is nutrient rich and free of weed seeds. The structure of superior compost provides particulate material that helps retain moisture and provide soil structure. We generally sift the compost through a quarter-inch screen so the large indigestible particles are eliminated from the mix that is to be returned to the compost heap.

Sifting compost can be laborious. We use framed screening to sift it through. To augment the process, screens can be suspended by a structure: As the compost swings in the screen, the motion encourages sifting. Other mechanized tumbler structures can provide this filtration process, producing a fine, soft, black, crumbly compost.

Vermiculite and perlite are soil amendments that we currently source from the agri-industrial complex. Vermiculite is organically certified and is created from heating the mineral mica to absurd temperatures. It is relatively inexpensive and compensates for high proportions of loam by providing superior moisture retention and lightening soils as a conditioner. Perlite comes from volcanic glass and reduces compaction by offering high permeability and low water retention.

We also incorporate various soil amendments in our potting soil medium. These amendments are intended to provide energetic components and nutrients that are readily available to plant roots as well as long-term nourishment for the soil food web. The amendments are generally rated by their proportions of nitrogen, phosphorous, and potassium (NPK). While the NPK proportions provide a measure of the amendments' capacity, this oversimplification of the seedling's necessities does no service to the reality that the plants require over fifty elemental nutrients for health.

Cycle for Annuals

Future growing seasons begin with seed saving and seed acquired during the short days of winter. Threshing and sorting of beans and other seeds is accomplished after the hectic harvest time has subsided.

The difficulty of seed saving depends on the plant and the conditions. Seed saving varies from orchestrated artificial pollinating intended to preserve open-pollinated heirlooms to rudimentary collection of bolted kale seeds and typically involves planning and processing over the whole time continuum. New England's rainy weather can deter effective saving of seed crops such as lettuce that are susceptible to mold and mildew. Timing and fortune is crucial to produce viable seeds before the wet weather of fall, as are structures such as greenhouses to provide artificial protection from the weather.

Because of the infancy of our seed-saving program, we also order seeds to meet our annual plant needs. We generally prioritize whom we order from based on a preference for seeds grown locally by smaller-scale operations. Our belief is that the seeds will be more adaptable to this region and will support the localized food production model. The Fedco cooperative based in Maine is our ideal choice for sourcing seeds. Their catalog provides the opportunity to decide between options of locally or industrially produced seeds. A second option is the worker-owned distributor Johnny's Selected Seeds, which offers an extensive collection of seeds packaged in quantities suitable for northeastern commercial growers.

Historically, our national seed-saving capacity has been fostered by land grant state universities that

Growing veggie and flower starts in the basement

maintained and enhanced seed libraries and researched and developed specific food crops. However, the era of state agriculture schools providing the security of seed selection and distribution has been replaced by corporate models that seek to eliminate competition and increase stockholder dividends. As a result, the seed industry is increasingly monopolized and profit driven. Often seeds are produced internationally in countries such as Iraq and Afghanistan, where the dry climate is ideal for seed saving.

With seeds in hand, whether bought or saved, our growing season officially starts in February, when we begin our indoor planting regimen. The outdoor greenhouses are generally still frozen solid when we begin seeding the first annual crops into

flats. We start long-season, slow-growing crops such as onions that benefit from the season extension to reach maturity and hardy brassicas such as kale and spinach that will be able to endure the cold spring weather. These plants are started in cells numbering forty-eight or seventy-two per tray. Once a seedling has germinated and produced its first true leaves, the plant can be transplanted into the field or potted up into a larger container. The seed calendar is an example of the chronological succession.

Indoor seeding takes place in two distinct microclimates. We start hot-weather crops such as pepper, basil, tomato, and eggplant in our boiler room, where the temperatures remain above 70 degrees. The temperature can be accentuated by

Seed Calendar

Jan. 30: Fedco seed order

Feb. 6: Preliminary cleaning of seeding shelves
  Seed leeks, onions
  Seed greens for greenhouse transplanting (kale, chard, Asian greens, arugula,
  spinach)

Feb. 13: Seed greens

Feb. 20: Seed greens and lettuces
  Watch allium germination, reseed alliums
  Basement cleanup, boiler setup, lights/
  timers tested and operational

Feb. 27: Seed brassicas (kale, broccoli, cabbage,
  brussels sprouts, kohlrabi, collards)
  Seed choys
  Seed peppers, basil

Mar. 5: Reseed leeks, onions, hot and
  sweet peppers
  Chard, spinach, New Zealand
  spinach, celeriac
  Cabbage, broccoli, brussels sprouts
  Stevia, sage, rosemary, lemon grass, basil

Mar. 12: Continue brassicas (cabbage, broccoli,
  kohlrabi), celeriac

Mar. 19: Continue brassicas (broccoli)
  Seed kale, collards, chard
  Seed parsley, basil

Mar. 26: Continue brassicas
  Seed peas in flats
  Seed cleome, zinnia, strawflower, calendula

Apr. 2: Seed tomatoes, eggplant
  Seed lettuces
  Seed peas in flats
  Seed rice, ginger, sage
  Seed daisy, cosmos, marigold,
  statice, zinnia

Apr. 9: Continue

Apr. 16: Seed lettuces, flowers, herbs
  Seed nasturtium

Apr. 23: Continue
  Outdoors: seed peas

Apr. 30: Continue lettuce
  Seed cucurbits (cucumber, winter and
  summer squash, sunflowers; melons)
  Outdoors: potatoes; lettuces in cold frames
  Nursery: Echinacea

May 7: Seed curcurbits
  Continue lettuce
  Outdoors: salad; seed carrots, beets,
  turnips, radish through July
  Seed flowers outdoors (calendula,
  echinacea, etc.)
  Reseed peas as necessary

May 14: Outdoors: seed dill, cilantro
  Seed beans
  Block plantings (salad)
  Seed carrots, beets, turnips, radish
  through July
  Indoors: lettuce

May 21: Block plantings
  Seed beans
  Seed carrots, beets, turnips, radish
  through July
  Indoors: lettuce

May 28: Outdoors: salad; seed parsnips
  through July
  Block plantings
  Direct seed flowers outdoors
  Seed carrots, beets, turnips, radish
  through July
  Indoors: lettuce

June 4: Block plantings
  Seed carrots, beets, turnips, radish
  through July
  Seed parsnips through July
  Indoors: lettuce

June 11: Block plantings
  Seed carrots, beets, turnips, radish
  through July
  Seed parsnips through July
  Indoors: lettuce

June 18: Block plantings
  Seed carrots, beets, turnips, radish through July
  Seed parsnips through July
  Indoors: lettuce
  Seed flats for fall: broccoli, funjen cabbage
  Seed peas for fall
June 25: Block plantings
  Seed carrots, beets, turnips, radish through July
  Seed parsnips through July
  Indoors: lettuce
July 2: Block plantings
  Seed carrots, beets, turnips, radish through July
  Indoors: lettuce
July 9: Block plantings
  Seed carrots, beets, turnips, radish through July
  Seed parsnips through July
  Indoors: lettuce
July 16: Block plantings
  Seed carrots, beets, turnips, radish through July
  Seed parsnips through July
  Direct seed dill, cilantro
  Seed peas for fall
  Indoors: lettuce
July 23: Block plantings
  Seed broccoli raab

  Indoors: lettuce
July 30: Block plantings
  Indoors: lettuce
Aug. 6: Block plantings
  Indoors: lettuce
Aug. 13: Block plantings
  Indoors: lettuce
Aug. 20: Block plantings
  Indoors: lettuce
Aug. 27: Block plantings in greenhouses
  Indoors: lettuce
Sept. 3: Block plantings in greenhouses
  Indoors: lettuce
Sept. 10: Block plantings in greenhouses
Sept. 17: Block plantings in greenhouses
Sept. 24: Block plantings in greenhouses
Oct. 1: Plant garlic
Oct. 8: Plant garlic
  Seed spinach/greens to overwinter

Notes taken during 2012 for improvements in 2013:
- Seed nasturtium as noted here, two to three weeks later than in seeding log
- Seed basil two to three weeks later
- Seed peppers two to three weeks later
- Seed eggplant earlier
- Seed more greens for transplanting into hoop house
- Seed cukes two to three weeks earlier
- Seed tomatoes one week later

the use of heating mats to stimulate germination. Cold crops such as cabbage and other brassicas are started in the cool ambient temperature of the basement, which ranges between 50 and 60 degrees. In both of these nursery environments it is important to provide adequate ventilation, essential to prevent the common condition of damping-off. A light breeze also conditions the plants, preparing their tissue by mimicking outdoor conditions. Electric oscillating fans can provide the ventilation necessary without unduly damaging the plants with excessive airflow.

We use four-foot fluorescent bulbs to start seedlings. We don't buy special fluorescent grow lights or utilize other bulbs such as high-pressure sodium. We hang the lights above shelves so they can be raised as the plants grow, maintaining a distance of about one inch from the plant leaves. The lights are a major downside of this system. The bulbs and fixtures are disposable and contain toxic elements, while the energy to produce the light is derived primarily from fossil fuels. Our goal is to reduce our consumption of this technology so that our food system evolves toward a localized system. With the advent of the solarium addition—a space that has already proved to maintain temperatures above 40 degrees Fahrenheit even without an introduced heat source—we are confident we will actualize this goal.

Cycle for Perennials

The distinction between perennial and annual crops comes from the hazily defined classification system that we use to describe the behavior of plant species within the shifting of seasons. Annuals propagate by setting seeds in abundance so the plants can regenerate from reproduction when they are signaled with an appropriate climatic stimulus. The energetic potential of the plants is directed toward an annual cycle of reproduction. Perennials return yearly after withdrawing into dormancy during the shorter days of winter, to reemerge in the spring to thrive and grow from existing plant material. There are also biannuals that set seed and reproduce during their second summer of existence, as well as annuals that will overwinter and persist without reproducing via seed.

Annual production is generally accomplished by starting seeds, while perennial production and amplification can be accomplished through several different methodologies. Perennials can be started from seed, divided, transplanted, or grafted. The goal is to multiply the diversity throughout the landscape using the techniques most effective in any given circumstance. The goal of the permaculturalist is to be able to create an ecological bounty through the cultivation of plant species.

Perennial plant genetics are the valuable basis of this propagation. Once we have obtained a species we can begin the process of reproducing and proliferating that species. The goal is for a plant inventory to become available through the localized production and proliferation of these valuable species.

Division and Naturalization

Naturalization occurs when an introduced plant begins the process of self-promotion and propagation within the ecology, occupying available niches. By self-selecting to propagate in an area, the plants save human cultivation time and effort.

Division is a method humans use to assist a plant's introduction to the landscape. It is not appropriate for all plants, though when performed in the right circumstance and timing, this process is a very reliable means of propagation and proliferation. Herbs such as comfrey or bulbs (daylily, iris, and daffodils) can be dug and divided to produce more plants. To proliferate these perennials we generally seek out locations where there is thick, robust growth and denote these locations during the summer months. We return to divide the plants in the spring as they emerge or in the fall when they have retracted for a dormant winter of rest. To ensure plant health and vitality this process is best accomplished during spring and fall versus when the plants are in full flower and summer growth mode.

Division can be a necessary exercise in plant maintenance. Some plants, such as those with bulbs, will crowd themselves out, choking access to space and nutrients if they are left undivided. It is also necessary at times to dig plants so that weeds embedded in the root-ball can be removed, providing an opportunity for division.

Ideal scenarios for transplanting and division are cool, cloudy mornings with moist conditions. Root disturbance damages the vascular system that powers the plant, so care should be taken to minimize the effects of bare-root conditions. Roots should be maintained in cool, shady locations, and plants should be returned to the soil in their ultimate location as soon as possible. Full sun, wind, and dry heat stress plants that have already been disturbed. Upon returning the plants to the soil in their new location, it is important to firmly press the soil around the roots so the soil level is the same as where the plant was previously embedded. Ample watering is important to mitigate plant stress and ensure consistent soil-to-root contact and to reduce bubbles of air trapped below the surface. Shade and water can assuage the trauma of summer root disturbance. Though some perennials such as comfrey are rugged and will survive rough treatment during division, care will increase the probability of propagation success.

Directly from Seed

We buy and save seeds to propagate perennial herb plants. Seeds can be purchased from a variety of sources, including Fedco. Starting seeds is a great way to introduce new plants to the existing ecology for a low investment. We start seeds either in flats or by direct seeding. By utilizing the flats we are able to more closely control growing conditions as the plant is first established. This limits competition and enhances the care that can be directed toward the juveniles in a nursery situation. Direct seeding offers less climatic control of the physical seedling environment, though the process can yield results with less work. Direct-seeded perennials do not have to endure transplant shock and are able immediately to begin penetrating the soil in search of nourishment. By seeding perennials in this manner, we hope that they will regenerate and establish themselves. Developing a nursery in this way affords increased inventory and diversity without laborious human intervention.

Ribes and Rubus

Rubus species plants are classified as brambles. These hardy plants spread by rhizomes that are easily transplanted. The *Ribes* species that constitute the currants, jostaberries, and gooseberries can be propagated through the process of layering. If the stems are laid on the ground and a section is covered with soil, the plants are prone to set roots into the soil. After a season of growth the stem can be clipped from the mother plant and transported to a new expansive location.

Woody Plant Proliferation

The propagation of select fruit cultivars requires access to the parent's genetics. Varieties such as McIntosh or Red Delicious can be traced to the woody material of a single tree. Apple fruit contains seeds that are genetically distinct from their parent tree. The seeds of a Red Delicious could produce a tree that bears small green apples, so to preserve the character of the fruit that has been selected, perennial fruit trees such as apples are derived by grafting the wood tissue of the selected variety. In addition, new fruit varieties can be added to existing trees; for instance, a Red Delicious can be grafted onto an old Granny Smith.

These fruit cultivars are then attached to standard rootstocks, which provide the support and vascular network that allows the aerial tissue genetics to achieve fruit production. This methodology is common in the agricultural industry with both the orchard and nursery segment. Tomatoes are also another plant that is commonly grafted so aerial production can be matched and maximized through selective root grafting.

Evaluations must be made to judge the importance of the plant with regard to the time spent necessary to effectively propagate. Is it more effective to start the plants from seeds, seek divisions from the neighborhood, or order bare rootstock directly from a nursery?

We buy bare rootstock to augment our species diversity on the farm; every winter we preorder trees for delivery in the spring. There is a huge variance in the quality of nursery stock that is available through mail-order catalogs. When picking a source for your perennials consider the location of the nursery to see if the stock is grown in a climate similar to where they will be planted. We order bare rootstock from Fedco, St. Lawrence Nurseries in Upstate New York, or Elmore Roots in Vermont. On occasion we will also order from Oikos in Michigan.

Bare-root purchases are a comparatively expensive cash outlay. Typical apple trees cost in excess of twenty dollars. The product will also vary tremendously, depending on the nursery. It is important to seek a reputable nursery that offers a strong, healthy stock for the expensive plant material. Pay close attention to the size of the plant described by the catalog company. A one- to three-inch plant is barely sprouting, versus a one- to three-foot plant, which has an established root system. Glossy catalogs also deceive customers by offering ubiquitous plants that are labeled with obscure names.

When trees and bushes arrive via mail in the spring, we try to plant them as soon as possible. The first step is to unpack them and store them in a cool, dark space such as the root cellar with their roots covered by wet, shredded newspapers. The nurseries generally send the plants out in a dormant state, before they have leafed out for the season. Although we source stock from nurseries with a similar climate, because of the cold of Dorchester we will at times receive stock that is further along in seasonal development than the plants on-site. When trees arrive already showing leaves, we have little choice than to plant them directly outside. If the plants cannot be planted within a couple of days, the best option is to lay them on their sides in a shady, sandy, well-hydrated outdoor nursery until they can be directly planted in a timely fashion, within a week.

When preparing the location for planting it is important to consider proper spacing and access to sunlight. Nut and fruit trees depend on adequate sunlight to produce fruit, as do shrubs, with the exception of *Ribes*. All plants can be restricted depending on space availability. Before planting it is essential to consider the size of the woody perennial plants, as they continue to grow upward and outward toward maturity.

We dig the holes for the plants according to the size of the rootstock that is provided, using primarily native soil to refill the hole and plant the tree roots. We do not believe in digging an excessively sized hole and filling it with compost and soil amendments. We are of the opinion that a large, disproportionately fertile planting hole will discourage the roots from exploring the surrounding subsoil seeking nutrients and establishing a resilient, foundational root structure.

After digging the hole we insert the tree so that the original depth of planting is about two inches above the grade. We then backfill the tree roots so they are well covered. The height attained by the planting depth is sloped down to grade at the diameter of the planting hole so that the root crown is slightly elevated to avoid submersion in water and subsequent crown rot.

Once the tree is in the ground we provide a perimeter mulch in a circle around the base of the tree. This mulch provides nutrition and protection from competition for the tree seedling. By layering biomass such as barn manure and compost on the surface, then covering the material with cardboard and wood chips, we give the tree a mulch that retains water and suppresses weeds. The trees are

Covering young trees with fine wire mesh in the late fall protects the bark from voles.

then marked with labels and wooden stakes so they are easily identifiable.

Pests can inhibit woody plants even though fruit production is still substantial. Producing beautiful commercial-grade organic apples is a difficult proposition given the humidity of the East Coast. Apples are severely affected by pests and disease. The apple borer is a particularly dangerous pest that will kill trees. Voles can be equally harmful, girdling trees during the winters; protective screening wrapped around the base of the trees will keep them at bay. The foliage of grapes and raspberries is severely affected by Japanese beetles. Tent caterpillars can also wreak havoc on foliage, though they can be removed and burned if attended to in a timely manner.

It is important to map, label, and mark the location of perennial plantings. The time pressures of the farm mandate that adequate records be kept so that trees can be maintained into the future. Proper identification and locating ability is essential to compensate for fading memory as well as seasonal and personnel changes.

Mushrooms

Mushrooms are an essential part of the forest ecology in New Hampshire. These fungal delicacies recycle and recirculate the mineral and energetic potential that was created by the collaboration of the land, sun, water, and plant biology. This process serves to exemplify humanity's role as gleaners

A crew work to inoculate logs with shiitake mushroom spawn.

of the trophically derived natural energy cycles. By harvesting the fruit as caretakers of the fungal process we serve a natural role in the ecology.

Shiitakes are our foremost cultivated species. We purchase sawdust that has been inoculated by a local mycologist. For shiitakes we generally cut down the wood before the sap flows and allow it to season for four to six weeks. Oak and sugar maple, species that we otherwise avoid cutting, are considered optimal wood for mushroom inoculation. Ash and red maple have also provided acceptable results. Once wood has been seasoned and situated, we then drill $7/16$-inch holes four to six inches apart and pack in the sawdust inoculated with mycelium. Wax can be utilized to seal the spawn so rodents are not attracted to the sweetened sawdust.

We are proponents of using the ground to maintain adequate moisture in the logs, though many growers recommend regular soaking of the logs to encourage a predictable flush of mushrooms. Mushrooms can mature in less than a week, so it is important to monitor the logs regularly during the growing season. We position our logs along paths in the woods that are regularly traveled; the traffic prevents the rapid development of the mushroom from going unnoticed on the busy farm. We have found that moving the logs, especially as they decompose, is difficult. Ground contact, therefore, allows the logs to retain moisture and thus reduces the maintenance of moving and soaking the logs. In this system we only inoculate the top half of the logs that we use to line our woodland paths.

Predatory Insects and Pollinators

Diversity and abundance are welcomed into the energetic cycle of the food system. Nature has developed a complex system of collaborators that feed from the energetic cycles of the earth. These creatures offer beneficial characteristics toward the proliferation of the food system. As occupiers of diverse niches in nature, they have evolved to subsist and depend on their partnerships in the

Plant diversity attracts many different pollinators at different times throughout the season.

planetary garden. We aim to support the pollinators, predators, and even parasites of the natural world.

Among the collaborators in our food system are legions of predatory insects. Most pests have natural predators that can be encouraged by avoiding the use of insecticides and herbicides. An abundance of pests signals an imbalance and an opportunity for the natural world's pest predators. An overshot population will self-correct into a system that is balanced by the grace and justice of nature. Ant, beetle, and wasp populations all depend on the available food supply. Parasites such as the Tachinid fly, *Istocheta aldrichi*, can be beneficial, laying their eggs directly into nonbeneficial Japanese beetles, where they hatch and burrow.

Often people conceive of pollinators strictly as honeybees. However, the New England region is host to over fifty species of pollinators. Our goal as stewards and food farmers is to provide pollinators with exceptional opportunities for forage, meaning flowers in bloom, between the spring thaw and the

The Lower Garden is home to a diverse spectrum of plants, all producing a flower at different points in the season. A rotation of annual crops complements the perennial species of kiwi, rose, walking onion, rhubarb, blueberry, chive, lavender, acacia, lemon balm, daisy, and sunchoke.

We utilize every possible space for growing food, flowers, and medicine.

autumnal freeze. The window of apiary opportunity is designed to provide compelling attraction for pollinators to occupy the garden habitat.

We have experimented with raising the European honeybee at D Acres, though our attempts have met with mixed success. As novice beekeepers we consistently lost bees that swarmed away. To maintain the bees required white sugar fortification. There were also problems with mites and colony collapse disorder. In recent years this epidemic of uncertain origin has decimated hives globally; instead of focusing on maintaining colonies of European honeybees we have opted to encourage the native populations of pollinators. While almond growers in California that monocrop almonds on thousands of acres of bare soil depend on imported honeybees to pollinate the flowers of their trees, we have opted to provide habitat instead of a rescue remedy.

Tools

The Hori-Hori is a Japanese gardening tool that I find essential to accomplish transplanting. The long, narrow knife blade serves both to dig the hole and extricate plants from pots. The blade has a serrated cutting edge ideal for smaller-diameter roots, and the tool also works to divide root-balls of daylilies and comfrey.

The garden fork is the most commonly used tool in the garden. This multipurpose tool is our primary assistant in the task of weeding and soil preparation. The four rugged steel tongs of the fork are used to aerate and loosen the soil so weeds' roots can be extracted to prevent regeneration. This is particularly important with rapidly reproducing rhizome weeds. Garden forks can also be employed similarly to a rake to flatten and shape the soil in preparation for planting. Garden forks should not be utilized to dig large rocks or the tines will bend and the handles break, reducing the overall effectiveness of the tool.

A good solid fork will cost in excess of eighty dollars new, so take care of the equipment. We still have the original forks from when we moved to the farm.

Bypass pruners are also essential to maintain the farmscape. While they are primarily used in the spring, they will have applications throughout all four seasons. Pruners serve for small-diameter woody materials (unlike the scythe, which is effective for grasses and nonwoody vegetation).

Hand scythes are crescent moon–shaped tools designed for cutting tall grasses. They allow for precise edging of garden areas and collection of the forage for animal feed. By developing techniques to collect the grass as it is being cut, we can efficiently collect feed while maintaining the excess vegetation. A leather glove is essential to avoid cutting the hand that grasps the grass to be cut.

To maintain the tools we regularly sharpen and clean them. They are stored daily under cover from the rain and raised so they are not submerged in contact with the earth. By respecting the tools and maintaining their cleanliness we honor the value these instruments provide as essential aspects of our daily life, preserving and maintaining them through proper usage and care.

CHAPTER SIXTEEN

FORESTRY

Forestry is a principal component of the natural resource–based economy in the state of New Hampshire. The forest is the predominant ecology of the region. As the economics and the ecology are intrinsically woven together, they provide a basis for the regional identity and nature-based culture.

Perpetuating a forested landscape that serves our needs requires a conscientious assessment of our interactions with this specific geography over a time continuum. Forest-based agriculture involves not only a full-year cycle of activities but also the ability to plan years into the future.

The forest serves a multiplicity of functions within this region. The Northern Forest provides a niche platform for an orchestra of life forms, including bacterial, fungal, floral, and faunal species. This diversity creates a web of relationships strengthened by connections that bind the forest system together, enabling it to respond as a whole to environmental changes. The multifunctionality of the forest system, with its myriad of components and connections, offers an example of a regenerative, resilient natural system. The forest cleans air and water, prevents erosion, and builds soil. The forest provides shelter from the drought conditions of the summer. The competitive foliage of the forest provides shade, cooling the surface of the earth and preventing excessive evaporation. The trees also act to shelter the landscape from the intensity of wind.

The forest has developed a system of rejuvenation and regeneration that utilizes genetic diversity, solar energy, atmospheric gases, and water, as well as the upper crust of the earth's surface. Our relationship as humans to this regenerative system should be carefully considered. Rather than exploit this resource for short-term gains we should seek designed solutions that enhance rather than degrade the resource, creating abundance and perpetuation.

In an economy designed to perpetuate, proper management enhances the utility of the resource. A permaculture approach to the Northern Forest region recognizes the immense value of the unique abundance provided by this ecological entity.

Philosophy of Holistic Forestry

An evolving design recognizes existing and predictable circumstances, then capitalizes on this knowledge to coordinate and maximize synergistic, mutually beneficial relationships with the forest. This includes refining the possibilities of whole-tree harvesting and utilization, addressing the potential of detritus wood and working toward species recomposition. In our quest to optimize a holistic system, the Northern Forest is full of seemingly countless connections that can be strengthened through conscious design. By defining the goals of our relationship with the forest, we can then design an implementation strategy.

Auguste and Henri yoked and ready to walk and work

The perpetual management of a natural system requires a dynamic-systems thinking. Our best management strategies are designed to recognize the evolving nature of the forest and the personnel and tools available. These strategies also recognize seasonal variations and an evolving prioritization of goals. Our system marries traditional techniques with specific modern technologies. Instead of extraction for profit our motivation is to upgrade and upcycle the abundance of the resource.

Holistic forestry practices require envisioning the forest's evolution through two-hundred- to four-hundred-year cycles. We must be cognizant of our efforts within that continuum. While our lifetimes are a small part of that cycle we must anticipate the issues future generations will face. By enhancing its food

capacity, access, and species composition we hope to prepare the forest for use by future inhabitants.

As a component of our forestry practices, we make use of domesticated work animals. D Acres' oxen team represents our recognition of natural existing strength and abundance and our positive redirection of that energy. Using the oxen as an integral component of the forestry system epitomizes permaculture principles. This traditional practice has a long history in New England, where the vernacular and knowledge of teamsters still resides, and where instructional information and equipment is still accessible to novices. By preserving these traditional skills we not only develop a direct relationship with the animals but also a positive connection to the landscape and our historical predecessors. Oxen are a keystone of the forestry system because they serve a multiplicity of critical functions. Their strength allows for mobility and transportation of resources across the landscape over difficult terrain to sites where they can be more readily accumulated and processed. Moreover, the oxen accomplish this task with minimal disturbance of the soil, limiting erosion that would be inflicted by mechanization. These bovines also provide fertility to the forest without the air and noise pollution of the typical diesel-powered machines. Since we are able to source their food and tack from the region readily, all aspects of the system are local.

In addition, there is a spiritual and philosophical connection that we are attempting to manifest in our relations to nature. The oxen are patient members of a team, working to clear the landscape while enhancing the nutritional capacity of the farm system. Their inclusion provides a connection beyond the human realm, a synergetic union crucial to our permaculture design. The employment of working animals also offers an alternative to the traditional spring cycles of New England dairy operations, in which newborn cattle are confined for veal production. Instead of a cruel dominion over these animals, we are attempting to foster long-term partnership and mutualistic approach in our relations.

Forest Composition

The Northern Forest's diverse combination of deciduous and coniferous trees is due to the overlap of two forest types. In our region the boreal forest typical of more northern latitudes mixes with the deciduous forest that is more common in Appalachia, resulting in a remarkable array of tree species. There are twelve common deciduous trees at our location, including red, striped, and sugar maple; beech; white ash; pin and black cherry; aspen; hop hornbeam; paper and yellow birch; and red oak. A notable smaller tree is the serviceberry, and there is a high occurrence of the imported *Malus* species (apples), particularly along field edges and homesites. We have also imported many nut trees as well as black locust to the property. The coniferous natives on the property include hemlock, white pine, balsam fir, and red spruce, and we have planted cedar and tamarack. There is also an abundance of low-bush blueberries and cane fruit available from patches of *Rubus* species.

The maples all display variations of the iconic maple leaf, but they are remarkably different in their roles within the Northern Forest. While the striped maple is highly valued by moose for its delectable bark, it can be a nuisance for modern forestry. Its ability to grow rapidly in low-light conditions; its large leaves, which shade out other species; and its competitive nature have led local loggers to consider the tree a weed. When cut, striped maple resprouts from the stump with vigor, sending multiple stems arcing up toward the sunlight. The tree is highly susceptible to rot and has a low Btu per weight but is relatively short-lived and rarely grows larger than eight inches in diameter. The red maple is a grander version of its striped cousin. Its tolerance for wet soils has earned it the nickname "swamp maple." The tree rarely is utilized as a saw log but often is used for firewood. The sugar maple is one of the gems of the forest: Highly prized sweet sap, while also available from its redheaded cousin, is more concentrated in the sugar maple. The sugar maple is also aptly known as the rock maple because of the hardness of its wood, which can be used for furniture and offers a high Btu-per-volume ratio as a fuel source.

Quaking aspen is a fast-growing tree that is underappreciated in the Northern Forest. This poplar is a pioneer species and rarely lives more than a hundred years. The tree, while appreciated for the aspen foliage in the Rockies, is most commonly recognized in the Northern Forest by its summer leaves, which dance uniquely in the slightest breeze. Bigtooth aspen is an exceptional hardwood for use as beams, as it resists sagging, a trait that is common in most deciduous species. The paper birch is another pioneer tree that initiates the process of Northern Forest succession. These trees are easily recognizable for their white bark, which peels in sheets. These trees also have a limited life span and are highly susceptible to rot. The yellow birch is significantly harder than its white cousin and is valued for woodworking because of its character and strength. Hop hornbeam, also in the birch family, grows sporadically in patches and rarely attains more than an eight-inch diameter. Its wood is extremely hard and resistant to cutting.

Black and pin cherries occupy substantially different niches within the Northern Forest. Pin cherries populate readily though they are short-lived. The black cherries are less common, and their reddish wood is highly prized for woodworking. White ash is a common tree that grows majestically in the Northern Forest. While it is slow to attain girth the resilient wood extends rapidly, high into the canopy. This lightweight yet strong wood makes an ideal baseball bat.

Beech is another undervalued species in the Northern Forest. The trees are durable and grow robustly even in shady conditions. While the wood is fairly hard, it is generally structurally inconsistent and susceptible to fungal activity, thus rarely utilized for milled lumber. Beechnuts provide a source of protein for bears, deer, and smaller mammals.

Red oak is valued for woodworking, acorns, and firewood. While oak is slightly more rot resistant than sugar maple, the value of both for their combination of uses for construction, food, and fuelwood stands out in the Northern Forest. Along with yellow birch, red oak and sugar maple are well adapted to the boulder fields and ledges common in the uplands of our region below two thousand feet. Above two thousand feet the composition shifts to a more conifer-dominant forest.

White pine is common in areas where there is a high percentage of sand in the soil strata and is the tallest species of the Northern Forest. These fast-growing trees are lightweight and strong and provide a degree of rot resistance. The hemlock is shade tolerant and tends to grow in wetter areas, along with cedar and tamarack. There is also abundant spruce and fir. While fir grows quickly, it is susceptible to rot, and trees must be harvested in a timely manner or they will rot from their centers outward. Spruce grows straight and is a dependable framing material.

The species in this region are frugal in their capacity to produce excess calories in the form of seeds or fruits. The short growing season requires an intense focus on biologic activity to maximize the limited solar energy entering the ecosystem. Of the hundreds of tree seedlings that attempt to populate a small amount of acreage, only a few will compete to become a climax specimen tree. Large specimen trees cycle over a two- to four-hundred-year period and are fairly rare in our landscape. During the two- to four-hundred-year span, species are in an ongoing process of succession, and a multiplicity of forces act to dictate which small percentage of the trees will survive to old age. Trees have adapted to endure natural catastrophes such as fires and windstorms with alternative regeneration strategies.

Reforestation is a consideration when timber is removed. In most cases the forest of our region rapidly regenerates via stump or seed. In areas where we would like to encourage alternative species we may plant some seedlings, though we have learned that transplanting is rarely effective without adequate sun because of the competition of the forested landscape. We have chosen to plant quick-growing black locust as an additional source of firewood and fence posts. The locust's flowers are also aesthetically pleasing and a benefit to the bees, while its roots help fix nitrogen in the soil. We plant cedar to utilize for fence posts and trellis material and for the wildlife habitat its thick foliage provides.

Areas of forest are often left to follow a natural succession. These zones are established so the natural processes will be maintained without human-designed interference. There are also areas in which forestry might only be accomplished on a limited or strictly seasonal basis. Steep rocky slopes may be better utilized as natural habitat rather than risk attempting to access this dangerous terrain.

The Northern Forest has the capacity to regenerate its species composition. To augment that capacity we consciously preserve mast trees and encourage selected specimens to procreate. Mast trees are large specimens that produce great quantities of seed as a genetically viable example of the species. We avoid the practice of high grading, which refers to the harvesting of the largest, most profitable trees in the forest. This practice reduces the genetic capacity of the forest to rejuvenate itself. By consistently harvesting the strongest, most vibrant trees, we would disturb their natural reproductive cycles, and the strongest genetic candidates would be removed. This practice diminishes the volume and vigor of the genetic pool available for perpetuation and is thus in complete opposition to the holistic forestry philosophy.

It is important to recognize the potential resources within the variations of the forest. For example, in some areas of our forest there are thick stands of softwoods. Spruce, fir, and pine grow in patches of tightly spaced trees engaged in a competition for survival. These stands offer ideal opportunities to harvest straight pole wood while also thinning the forest. In our quest for building materials we remove a percentage of trees that would otherwise succumb to natural competition.

Slower-growing trees generally have higher Btu-per-volume ratios. Trees in a field generally grow more rapidly compared to shade-grown trees. Field-grown trees are also more likely to produce lateral branches, which remain to scar the trunk of the tree, while trees growing in the forest are more likely to shed lateral branches in their haste to attain a competitive height. To demonstrate why all of this matters, the original trees harvested for Edith's house were spruce poles, which are extremely dense and strong. This wood is extremely insect and rot resistant compared to the second-growth and field-grown wood. The wide growth rings of rapidly growing wood have a loose grain, which is weaker and contains more digestible cellulose. It is significant to note the difference in Btus per volume in trees that grow rapidly in the full sun of the field versus trees that grow more slowly in the shade.

Weather conditions can induce or prohibit forestry activities. Severe cold temperatures allow for work to be accomplished in wetlands and marshy areas, while warm and muddy conditions may make areas of the forest impassable. Skidding logs with the oxen can be easier in frozen conditions, although slippery ice can be extremely dangerous: Imagine an ox on roller skates.

Developing a Forestry Plan

Forestry requires observation, evaluation, and implementation of a planned course of action. The natural landscape and intended uses guide the planning process and direct our activities. Many geographical factors describe the forest and must be considered. What are the typical tree species and their average age? What are the condition of the forest and the access routes? What is the soil type? Weather patterns? What is the slope, and how will the processing of the forest products be staged throughout the landscape?

In addition, owners of forested property must define objectives they wish to meet. What are the intentions for the forest? Is it timber sales, or water and air quality? Is wildlife a component of the goals? Is there storm damage to address, or insect/disease issues that must be dealt with? What times of year are better for woods work based on the weather patterns, farm needs, equipment, and personnel? How are forestry operations to be arranged spatially so that erosion is minimized while efficiency is maximized?

If the forest is young and thick, most of the work may entail thinning small-diameter trees. If the forest is more mature, the work may involve more harvesting, as well as developing the seedlings of the future. Ultimately, the decision of what to cut in a selective forestry plan depends on the goals for the landscape and the resources available. Dead trees, for example, may be left on the forest floor to provide habitat for wildlife. Sometimes particular trees or a stand of mature trees are a priority to be cut before their growth cycle declines so the live wood ratio is maximized. Trees that are leaning are undervalued because they take up proportionally more space versus trees growing straight up. Tall, straight trees with a central leader are more valued for sawing into lumber. Trees with excessive knots and twisted multiple trunks are often not suitable for milling into boards. Generally, log lengths of a minimum of six feet in length are sought to produce straight, defect-free dimensional millings.

Environmental factors such as fire, wind, and ice storms can also influence forestry plans. In severe cases catastrophic destruction can reshuffle the typical succession process. There are remediation techniques for these occurrences. Usually damaged wood is harvested and the area replanted. Dead wood that remains aloft and does not fall to the ground can become a fire danger. Once the dead wood comes into contact with the moist earth, the process of decomposition begins.

Our ability to do selective cutting thus far has been limited. Instead our focus has been on access restoration and enhancement and clear cutting. The skid trails used in prior conventional logging efforts have been improved, and additional woods trails and bridges have been constructed. Our demands for harvested wood have been met by the total removal of sections of forest for species recomposition. This process is relatively simple to accomplish. Once we have identified a location, usually a former field delineated by rock walls built in the 1800s, we begin by cutting down the smaller trees. We work to remove what we can maneuver via human power into a pile, then use the oxen to pull the accumulated biomass to the chipper. Our intention is to create enough space to maneuver the oxen and increase the ease of felling the larger trees. The oxen can be endangered by branches and sticks protruding at eye level, and great care is taken to cut trees level with the earth so they do not become dangerous pointed stakes.

Logging within the forest creates multiple challenges. For example, the tops of trees can become hung up in the branches of other trees so they are unable to fall to earth. To address this issue we first remove smaller trees that may impede the fall of the tree we wish to cut. After the initial cutting has eliminated the small-diameter material we begin harvesting the larger trees. By working in this manner we systematically remove the more manageable wood first, then address the larger, more difficult trees with greater flexibility.

Practical Aspects of Working in the Woods

The amount of knowledge involved with forestry is immense. When I arrived at D Acres I had no concept of forest management or the economic forces that shape the use of this natural resource. I could not distinguish the trees by name, nor did I have much experience with a chain saw. For the first couple of years my knowledge grew as I began to recognize the differences between spruce and fir and attempted to provide firewood for the homestead. A significant mentor during this period was our neighbor, Jay Legg. Jay provided expertise in usage and equipment, as well as assessment of our woodlot. During our first summer Jay worked several weeks on-site clearing a proposed house lot for my parents. This project provided the firewood pile from which I would begin my matriculation as a woodcutter.

The number of available personnel and the skills they possess can vary over the duration of the forestry plan. At times there are multiple people who can perform similar tasks, while at other times a single person may be responsible for several different aspects of the process. As an example, we may have a couple of people who are trained and capable of crosscutting firewood, but there may be only one person who can cut down trees and manage the oxen simultaneously.

Developing proper habits is essential for safe working in the woods. While accidents can be impossible to predict, precautions and procedures minimize the dangers. The health and welfare of all the team members is the primary concern in the woods. In addition to the human team members, care must be taken so the oxen are not injured in their efforts. It is important that the chain does not become tangled in their legs. If a limp or other injury is noted, due caution must be observed before proceeding and perhaps further injuring the animal. It is our responsibility to reciprocate the trust these animals have granted us with good care. The stamina and endurance of the animals depends on their age, level of training, and regular activity. It is important not to overexert the animals in hot weather or without prior adequate workouts.

Tools

Fluency with, and preference for, various tools depends on the personnel. Given the identical task of clearing trails of overgrowth, some prefer working with hand loppers to using the chain saw, depending on personal philosophy, safety factors, or the mood of the day. While mechanization can add speed to the process, we often seek alternative low-technology approaches when feasible.

Log Hooks

Log hooks can be helpful when moving large-diameter or long lengths of firewood. Their hooks grip into the wood so it is easier to transport. The cant dog has a movable hooked arm that adjusts to clamp and spin logs of variable diameter. This tool is effective in rolling logs that are on the ground or spinning a tree off the stump if it is wedged in the overstory. The full-size cant is ineffective in biting onto logs smaller than six inches in diameter.

Chain Saws

The chain saw is an extremely dangerous tool. Americans seem to have a love-hate relationship with their portable, gasoline-powered yard work

Safety is crucial during chain saw use, with proper footwear, clothing, and eye, head, and ear protection.

devices. As do lawn mowers and weed whackers, chain saws cause thousands of emergency room visits and deaths every year. While efforts have been made to incorporate safety elements into the equipment, safety is ultimately the responsibility of the user. While homeowners relish the opportunity to utilize the equipment, proper training and understanding of the equipment is necessary for even the simplest of tasks. Large-tree felling and pruning is only advisable after years of experience with a traditional saw.

The chain saw's loud two-stroke polluting engine is an odoriferous and auditory intrusion, although one benefit of the air pollution is its ability to discourage flying insects such as blackflies and no-see-ums during use. Long-term use of the vibrating equipment also damages joint and nerve endings. As with all machines driven by a centrifuged force, there is the danger of kickback. In the event that the chain momentum is impeded (for instance, by an old nail embedded in the tree), the chain saw can recoil violently, potentially in the direction of the user. To minimize this danger contemporary chain saws include a chain brake safety feature, which stops the rotation of the chain in the event of severe kickback.

This small gas-powered equipment requires proper handling and maintenance to ensure success. The fuel is premixed with a specifically formulated two-stroke oil that is introduced at 40:1 to 50:1 ratios. In addition, at each fueling, fluid is added to lubricate the bar. This oil reduces wear on the bar and helps keep the cutting teeth sharp. The bar oil is designed to discharge proportionately to the fuel so that both reservoirs are filled at the same time. The bar oil is discharged into the wood and the immediate area of cutting, thus contaminating the zone. We have experimented by using filtered waste vegetable oil on the bar, though its viscosity clogs the pumps in cooler weather.

Chipper

The chipper is a powerful tool for our farm system, providing a mechanically induced method of speeding the decomposition and reuse of wood biomass for soil fertility. For us to navigate the forest requires brush management. Teamsters and oxen do not enjoy working with various limbs and brush protruding underfoot. Scattered brush can trip and endanger walking animals. To deal with this issue the brush must be stacked, burned, or chipped. While building brush piles is an option, the chipper allows for the concentration, storage, and transportation of this organic material. We prefer to sequester the carbon and fertility on-site rather than burn the material.

The conventional practice of mechanized logging is to leave the slash where the trees are cut up to three feet deep. The high density of limbs helps protect the forest floor from the wheel and tracked machines, and brush is often used as a short-term patch for traffic on muddy terrain. The slash is cut down to this height to help reduce the risk of a high-temperature treetop forest fire. While the logging industry has perpetually voiced its pride in using the formerly valueless brush to invigorate the future forest, this practice is diminishing as the value of combustible biomass increases, caused by societal energy demands. As a result, whole-tree harvesting operations in which branches are chipped to be burned in electric plants or processed into pellets for home heating is increasingly common.

Before we purchased the chipper we dealt with brush by heaping the material into specific piles. Once the brush was laid down parallel, we would saw down through the pile every eighteen inches. By repeatedly jumping on and sawing into the pile we were able to compact the branch material. These piles provide a valuable resource for the landscape, serving as a nutrient slow-release capsule and animal habitat. Although the chipper adds a fossil fueled accelerator to the decomposition, we are seeking alternatives.

We have begun to utilize wood branches and debris more effectively to create swales and influence the landscape pattern. When we stack limbs across a watercourse, they strain and accumulate

Josh uses the chipper for small-diameter trees and limbs.

the debris otherwise headed downhill. In other situations wood laid across the landscape aims to slow or divert a watercourse. We prefer these alternatives to burning brush, as this reductionist technique does not sequester the carbon from the atmosphere, while delivering a swift but not enduring plant fertilizer.

We generally chip wood that is between one and four inches in diameter. Smaller debris that is inconsequential to the oxen travel can be left to decompose on the ground. We must continually evaluate whether to chip or process the wood for fire. With smaller-diameter trees and branches we

The chipper allows us to pile the chips into specific locations, or into the back of a truck or trailer for easy hauling and unloading.

use discretion in distinguishing what wood serves us best to process in which manner. Twisted and bent trees might be better used for chips rather than creating difficulties in a stack. On any given day our processing choices are also dictated by the location limitations and the needs of the farm.

We invested in a chipper that will consume up to eight-inch-diameter wood. The Brush Bandit 250 is powered by a diesel 4B engine. We made the purchase in the fall of 2003 for several reasons: We needed bedding for our animals, and purchasing chips was increasingly expensive. We also had grown tired of wrestling the brush into concentrated piles and yearned for the opportunity to use the material directly. As our forestry operations had grown with the addition of the oxen, we needed to upgrade our capacity to utilize the abundance of brush provided. We purchased the chipper in the era when the local dairies began using sand instead of sawdust, local sawmills were closing, and the demand for combustible biomass began to intensify.

Chipping in the summer creates a product with a high proportion of leaf material. This nitrogen-rich material breaks down rapidly, providing an initial heat to the brush pile that also acts to dry the overall material. Wood that is chipped in the winter is less likely to clog the machine, though the moisture content freezes the chips together for the duration of the winter.

Chipping—as with chainsawing—is loud and dangerous work. It is important to maintain proper safety procedures, including discussion of what to do in the event of an accident. We also utilize agreed-upon hand signals so that communication can be maintained during these processes.

Working with Oxen

While working with oxen is a continual learning process, my initial experiences with them had a steep learning curve. As described in chapter nine, our team came to D Acres in 2002, just as our need for firewood and our capacity to carry out woodlot management merged. The ability of oxen to move through variable terrain throughout the year while pulling a load is crucial for woodlot management. When Henri and Auguste arrived we had gone through the easily accessible field edge cutting and also burned the firewood that Jay had laid log length from the house lot clearing. We needed a low-cost, reliable mechanism to pull timber from the forest. After several years of cutting firewood and felling trees, my skills were sufficient to consider more extensive logging.

When our neighbor Steve initially approached me with the opportunity to adopt the oxen for a mere three hundred dollars with equipment included, the situation appeared too good to be true. After I accepted ownership of the oxen from Steve, I was given a brief demonstration in the severely undulating terrain of his backyard. The oxen were rowdy and unruly, and the demo was less than textbook. Steve struggled to keep control of the team, and we opted to call it a day after a difficult pull with a single small log. While this situation was unsettling, we were determined to continue on with plans to incorporate the oxen. Fortunately, during my time in South America that winter I was able to witness traditional techniques performed in daily subsistence agriculture. By watching the farmers, including small children, interact with these large animals I began to recognize and gain confidence in my ability to work with our own oxen.

My initial solo outing with the animals was extremely frustrating. While I was able to yoke them, when I tried to walk them up the road they drifted into the ditch. From that position they alternated between standing stock-still and walking awkwardly, half in the ditch despite my exasperated pleas and exaltations. After some duress I was eventually able to return them to their stable, where I proceeded to reread Drew Conroy's *Oxen Handbook* in search of solutions.

After months of effort the oxen taught us how they needed to be encouraged to perform the tasks at hand. While they did not necessarily begrudge

Josh drives the team to pull a load of brush.

the work, they had a difficult time coordinating their strength with direction in a timely fashion. They would often sidestep down across a slope as they were pulling a load in an effort to ease their burden rather than follow the route we preferred along the contour. The juvenile oxen were also unpredictable and rambunctious and required disciplined reinforcement of their training. Teamsters who were shorter in stature had particular trouble managing Henri because of their limited reach with the goad stick. In addition to voice commands we depended on body language and signals so the team could function in close proximity to loud equipment such as chain saws, and we learned that they appear to respond with effort based on the intensity of the goad and the tone of the teamster.

As the oxen have grown older they have developed means to communicate their need for rest. They have learned that if they step over the towing chain so it is between their legs or off to one side it will delay the pulling. They are inclined to use this ploy to buy time to restore their energy.

To maximize the efficacy and productivity of the oxen it is necessary to coordinate their activities within the landscape. The beauty of the arrangement is that the two animals can act independently for maximum maneuverability or jointly to maximize power. To make a sharp turn one ox can move backward as the other moves forward. They also can be directed to move laterally by sidestepping.

Logs are generally dragged by the bottom of the log so the pulling action elevates the heavy end of the timber, thus reducing friction and diminishing the chance that a branch pocket will erode soil or catch on a rock or another branch, resulting in a jarring halt. In our attempt to ensure the wood is pulled from the forest with minimal impact, we pay attention before the tree has been cut to the selection and direction of fall.

Skid trails are arranged to minimize erosion and protect the oxen from hazards while allowing removal of the wood by the most direct and quickest means possible. By directing trails across the slope with the contours of the land, soil erosion will be minimized compared to trails that provide a direct path down the slope. A load pulled directly uphill or downhill perpendicular to the contour might be the shortest distance, but it will also be the most physically demanding load on the oxen. In the event of a direct uphill pull in which the weight is overbearing, the oxen will be forced to back down until they are at a safe location. Pulling weight directly downhill can endanger a team that accelerates and essentially runs away with a downhill load. Logs can also slide into the rear legs of the oxen during downhill pulls, particularly in wet or icy conditions. That being said, if a pulled load rolls down the slope as it is being pulled across the slope, the oxen could still be endangered by the shifting load.

The trails utilize skid trees positioned along the trails to act as a fulcrum directing the trailing timbers around corners. The bases of the skid trees are damaged by the repeated passage and are sacrificed to help maintain a minimal width of the trail.

Careful attention is paid toward hitching and unhitching loads with two people when working with the oxen. Having two people hitching logs allows one individual to be steady with the supervision of the oxen while the other person focuses on the chains and attaching the logs. By standing directly in front of the oxen, a teamster can protect the individual working between the animals from a runaway situation. The oxen may be eager to start the pull and edge forward in their haste. It is important to hitch the ring on the yoke before attaching the wood to be pulled. Otherwise an individual might be run over by the wood being pulled if the oxen are startled or choose to begin the pull while an individual is trapped. Safe working habits and procedures as well as ongoing communication between these individuals is crucial in this circumstance.

Still, oxen are predictable in their habits. They are prone to follow the same paths if they have utilized the same course in succession. They are often eager and testy at the beginning of the workday, though they will calm down with work. Hungry and thirsty oxen do not perform well, and it is difficult to train them to remain at attention if there is abundant, enticing food in the area. Bloodsuckers such as large horseflies spook the oxen, who are averse to their painful stings. Blackflies also annoy and inflict pain on the bovines, particularly on the sensitive exposed tissues such as the nose and eyes.

We have broken several yokes and bows during our time working with the oxen. The equipment has generally failed when the oxen were jarred by an abrupt halt when they pulled into an unyielding object.

We normally use $5/16$-inch linked chain to pull wood. To attach the chain to the yoke we use a cross-link hook that is easily unfastened should an emergency arise. The length of the chain allows ample room so the load being dragged does not disrupt the oxen's stride. The load is attached via another cross-link hook, which is easily attached to a choker-chained log. The choker chain hook allows the chain to tighten around the log when it is pulled, though it is easily released once the tension is relieved. The chain is heavy, and for long walks we often drape the chains over the yoke to be transported. While nylon strap is lighter and has strength to be utilized when new, it becomes dirty and wears quickly with the friction of pulling across the terrain.

It is easier for the oxen to pull weight downhill versus uphill. With this in mind, it is important to consider the spatial locations of woodlot operations across the geography. Wood is generally pulled

The chain must be linked correctly for safety and proper pulling.

Matt Palo hitches the chains to the yoke, while Beth watches, preparing to lead the team.

downhill where the chipper is located and various piles of timber are stacked lengthwise. Piles are separated based on their intended usage. Sawlogs and building materials of either softwood or hard-wood are usually predetermined before they have been cut down. The remainder of the hardwood more than four inches in diameter is stacked to be cut for firewood processing.

Working with the oxen in the woods requires the agility and athleticism to direct the team as it weaves through the landscape. Footwork is crucial to maintain a consistent close presence while avoiding being trampled. It is also important to recognize that the oxen respond to the tone of your voice, and while they might respond with vigor when startled by screaming, that urgency should not be your normal working tone.

Beth drives the oxen during the summer.

Once a tree has been cut down, the next stage is to cut all the limbs from the main trunk and cut the top where it tapers to less than four inches. The brush is then piled so the wide ends match. The choker chain is then tightened around the end of the pile; often the oxen will be aligned to pull perpendicular to the length of the wood load so the chain tightens and the initial friction of the load is reduced as it pivots to follow the oxen.

We build our piles for processing the firewood by stacking the logs parallel as high as possible. We pull the logs parallel to the pile, and two people use iron bars or sticks as levers to elevate the wood onto the pile. This allows multiple logs to be cut without dulling the blade through ground contact. The wood can then be split or tossed into the back of a pickup truck and taken to its storage location. The wood for the Community Building is driven into the garage, where it can be dumped down into the basement.

To transport tall, smaller-diameter trees with the oxen we will often separate the limbs and pile several four- to six-inch-diameter thirty-foot trees together so a single choker chain can be woven around them. At other times we will pull a whole tree to the chipper and limb it alongside the unit.

Building Trails and Access

Accessibility is crucial for forestry operations. Roadways and access points allow passage for

These two logging roads have recovered from the oxen treading—they appear to be only pathways.

oxen and equipment to extricate the harvest. It is important that these routes are maintained and not degraded by traffic.

By only skidding wood with the oxen in dry or frozen conditions, we have substantially protected our wood-lot from erosion and soil degradation. The damage that can occur from wheel vehicles in wet conditions is catastrophic. Once the woodlot has suffered from soil disturbance with ruts and mud, it is difficult to repair without substantial earthwork. In addition, these scars remain susceptible to future disturbance.

It is important to utilize alternative routes if the continuous passage of skidding logs creates a plowed rut through the landscape. If water begins to accumulate in these ruts it should be diverted and dispersed via constructed water bars.

Logs can be loaded into this skidder and transported by the oxen or a vehicle to another location for cutting and splitting.

Valued-Added Goods

In terms of the forestry program, "valued-added goods" refers to items that we are able to sell for income or that we would otherwise be forced to buy to meet our needs. These goods are specific commodities that can be marketed or processed so that the value received from sale, barter, or usage is greater than the wholesale commodity value. To assess the costs we compare the time and resources spent in production to the costs of buying such goods on the retail market.

Firewood

We produce firewood for on-site consumption. While at times we may help a neighbor in need, our primary goal is to provide the fuelwood sustenance for on-site operations. Wood should be crosscut about two inches less than the maximum allowable in the device to encourage air flow and smooth loading. Wood should be split to maximize drying and as necessary to fit in the door of the stove or oven. Bases of the trees and knotty forks are the most difficult sections of the tree to split.

Lumber

The cash value of lumber is dictated by the global commodity market. Softwood is cheaper to mill because of its low weight per volume and the ease with which it can be cut. By comparison hardwood strains the mills and dulls the blades more rapidly. When we first began logging in earnest we sent a dozen marginal hardwood logs to Jay to be sawn into lumber. The volume of wood this produced was beyond our expectations, and we are still working through this stockpile nearly ten years later. We

This pig house construction is a great example of the use of pole wood as rafters and slab wood for board and batten siding.

will still harvest softwood for milling when we are clearing a field if it is reasonably straight and above twelve inches in diameter. Based on our ability to use roundwood and salvaged material, most of our needs can be met with one-inch stock utility wood.

Two by-products of Jay's mills that we utilize are sawdust for storing vegetables in the root cellar and the slab wood and offcuts, which are helpful for fencing and board battens.

Pole Wood

Pole wood is a valuable component of our forestry program. We use this wood extensively to construct the structural elements of our buildings. After we harvest pole wood, the bark is removed with a drawknife. The wood is then incorporated as posts, beams, rafters, and joists. To facilitate construction it is helpful to harvest wood that is straight in at least one dimension. Pole wood saves energy compared to milled lumber while providing additional strength. Compared to the wood and energy wasted in milling, the roundwood is stronger and is available on-site. Even milling only one side represents a quarter of the energy spent on a board milled on all four sides. The slabs of wood that are cut off the log are usually chipped or processed for combustion. Dimensional lumber, which has been squared on all four sides, is useful in certain circumstances, but roundwood is structurally sufficient and readily available. Irregularities that protrude because of knots or the twisting trunk of the pole can be rectified by removing the wood with an axe, chain saw, or chisel. We commonly use spruce and fir stands for lengths of pole wood up to thirty feet, which taper from eight to ten inches down to two and a half. We use hemlock and pine for posts and beams, with the hemlock being more effective in ground contact and the pine better for UV exposure and lightweight aerial elements. We have also experimented with maple, ash, and locust as structural elements, which offer aesthetic displays when the bark is removed.

Stripping the bark is most conveniently done in the spring when sap is flowing in the cambium. We utilize drawknives and a chisel tool attached to a six-inch straight handle. After cutting through the bark, we peel under to separate the protective skin. In ideal spring conditions the bark peels back beautifully in long strips, revealing a glistening, sculpted trunk. While the process can be messy with sap and bark, it is quickly accomplished compared to scraping small unlubricated chunks in the winter.

To complete a floor or roof system with roundwood, poles are cut and sufficiently spaced one to three feet apart, depending on the strength necessary. Size and spacing are determined based on prior experience, building needs, and testing. A six-inch-diameter pole will easily suffice for a ten-foot span at a two-foot space for the roof and a one-foot layout for flooring. To begin the layout I generally arrange the poles based on their girth in diameter. By arranging the poles by size we can then lay them out in position and adjust the spacing to attain structural integrity as needed.

Hardwood saplings less than three inches in diameter can be utilized for multiple purposes. Six-foot lengths can be harvested to use as handles for tools such as rakes and shovels. Saplings that have unique character can be utilized for walking sticks. We also harvest wood to be used for agricultural structures such as fences, trellises, and other garden support structures. Cedar, locust, and tamarack are highly valued in this region for these purposes, though we often are forced to choose inferior alternatives because of the current lack of sufficient harvestable specimens.

Mushrooms

We have also utilized tree material to grow mushrooms. Both roundwood and chips can be inoculated. Trees are best cut in the spring before the sap is flowing, then left to cure for six to eight weeks until they can be inoculated in warmer weather. Trees that are four to eight inches in diameter are ideal because of their basal-diameter-to-sapwood ratios. Larger-diameter trees offer less efficiency as a fruiting medium and are more difficult to handle and thus are better used for lumber or firewood.

Roundwood is inoculated by drilling holes and inserting an inoculating agent such as plugs or loose sawdust spawn. Ideally the inoculating material is then covered with hot wax to seal and protect the material. Without the wax we have had problems with rodents or insects eating the inoculating materials before they spawned.

In our initial attempts at growing mushrooms we followed the typical strategy of crib stacking logs with the intention of regularly soaking them to induce flushes of mushrooms. We found the crib stacking of roundwood to be precarious and awkward, however, so we then opted for a rail system in which we would lean the logs against a horizontal rail. Often these arrangements were situated in shady locations that were difficult to access with adequate water for soaking. Our most successful soaking locations were alongside a stream where we could readily insert and retrieve the logs. Unfortunately the remote locations of the logs prevented us from adequate oversight, and we consistently missed the flushes.

Over time we have sought other modes to reduce maintenance and ensure a timely harvest. Instead of remote locations we have more recently chosen to inoculate logs alongside paths. The traffic along the paths provides opportunity for daily reconnaissance to evaluate the state of the logs. We have also adopted alternative means of ensuring moisture for the mycelium. We will either lay a log directly in the ground or bury it vertically like a fence post. By directly contacting the ground, the wood is able to absorb moisture and induce fruiting.

Holes for plugs are drilled with a jig or specialty tool, so the material can be hammered into the wood. Loose sawdust must be tightly packed into the drill holes, which are spaced in a triangulated, diamond pattern four to six inches apart. The mushrooms typically fruit from the holes that were drilled in the log so it is not necessary to drill holes and inoculate on the bottom or below ground portion of the log.

Slugs are the biggest detraction in this system. The ideal weather for the production of mushrooms naturally coincides with a population explosion of hungry snails without shells. In a sense it is unfortunate that the ideal conditions coexist; to cope we pick the bugs as we harvest the flush.

While oak and sugar maple are the preferred species for growing shiitakes, we have generally selected less valuable species for our mushroom endeavors, reserving those prized species for perpetuation and other usages. Consequently, we often have chosen red maple, ash, and beech for our mushroom projects. These species have performed adequately, though they decompose more rapidly, thus limiting the years of fruiting potential.

Maple Sugaring

Making maple syrup is a traditional seasonal activity of the Northern Forest. By tapping the maples and boiling the sap, northerners are able to create a

Beth pours maple sap into an evaporator pan to boil down for syrup.

sweet-tasting confection and food enhancer. While the process is labor intensive and requires significant reduction via evaporation, the rewards are valued throughout the world.

Food and Medicine Crops

Planting the understory of forested lands offers the remarkable potential to augment system resiliency by incorporating elements beyond tree crops. There are species of plants that have adapted to capitalize on the sunlight that falls on the forest floor before the trees leaf out in the spring. Plants such as fiddleheads and ramps or wild leeks can be propagated to regenerate annually. Other highly prized medicinal species such as ginseng and goldenseal can be grown for harvest. Foraging for mushrooms such as chaga can also be a productive endeavor. Developing a successful long-term enterprise requires investment and personal interest. Crops such as ginseng can be finicky to cultivate and require up to seven years to harvest. Developing stands of ramps capable of enduring perpetual harvest requires steady management.

Woodworking

Woodworking produces potential value-added goods that are discussed in depth in chapter eighteen. Suffice it to say that the potential for wood products produced on-site is virtually limitless based on the abundance of species available in the forest. Black cherry, yellow birch, sugar maple, oak, and pine are some of the most highly valued in the world for furniture, cabinetry, and general woodworking.

Recreation and Education

It is important not to overlook the recreational and educational benefits of the forestry program. The

The marked trails offer a wide range of recreational opportunities on the property throughout the year.

Utilizing grant funding, we built sturdy bridges over wet areas throughout our trails system to allow for better ski and bike access.

The oxen can be used to enhance the trail system by packing down a skiable path.

woods provide a sanctuary to those interested in convening through hiking, snowshoeing, skiing, mountain biking, or camping in the forest. By offering this opportunity to the general public we hope that people will become aware of the rich wealth of resources that can be perpetuated through a designed management practice.

While the trails are open to the public year-round, guided tours with youth groups are a special time to share the forest. Through opportunities such as getting to know plants or seeing a beaver dam in progress, bridges to nature are built with the youth in the area. By bringing school groups into the forest we hope to elevate the pride we feel as stewards of this bountiful resource. We also bring volunteers on-site for annual trail cleanup and construction projects, which allow the public to be active stewards of the farmscape.

CHAPTER SEVENTEEN

COMMUNITY OUTREACH

Our outreach is designed to support, invigorate, strengthen connections, localize economics, network, and educate within our community. At all times the organization strives to match the needs of the farm with the desires of the community, and vice versa.

To help ensure an educational experience for the public requires planned design and implementation. Providing a welcoming, receptive atmosphere that invites dialogue is essential. The goal is not to find a location to build a community or move to a location where there is already a community; rather the goal is to nurture the community within. A fundamental component of the community outreach is actually working for the community. Whether we are stacking wood for the neighbor or helping build a community garden, we make time to prioritize volunteer work.

At its essence our work is to create relationships between people, build community resilience, and connect people to place. Connections between people are fostered by direct face-to-face contact and the network that each individual brings to the organization. Resilience is fostered by the demonstration and education programs, which inspire sustainable alternatives in the region. By offering the grounds of the farm to visitors, we provide the opportunity to commune with nature in a setting with a dynamic, active ecology.

The concept of a community farm can be nebulous and difficult to grasp. The organization provides support and nurturing to the community similar to that found within a family. It transcends our typical definitions and perceptions of an organization. While D Acres supports the general community in ways similar to a church or chamber of commerce, our outreach is secular, and not limited to business incubation and development. The lack of dogma in the organization allows a clear nonpartisan focus on sustainability and socioeconomic progress. In a sense the community farm augments the traditional farm by seeking methods to enhance and upgrade the community vitality. The community farm fills gaps in the existing system by providing educational services, networking, and exemplars that build the resilience and strength of a broad constituency.

While we encourage visitors to follow their interests freely we also like to present them with several themes; for example, sustainability will only be possible through cooperation; the process is challenging and evolving, not a textbook resolution; nature provides our home and nourishment; hard work is necessary to live within our natural environment.

The specific programs we offer are dictated by a variety of factors. We often make decisions based on personnel interests and connections. At times we have also been fortunate to facilitate events for outside groups and individuals. Of course, education is not free, and hidden costs such as planning and organization must be acknowledged. While there are opportunities to garner revenue from an endowed private school, there are also many cases of financial need in the public education system. Community outreach entails becoming a collaborator within existing networks and seeking niches that are not currently being served but would mutually benefit both the organization and the broader community.

To offer events to the community requires commitment to all aspects, from promotion to reception, departure, and cleanup. It is necessary to sufficiently promote the event to ensure success. People generally want to participate in well-attended but not crowded events. When folks arrive for any event, it is important for them to be greeted and any specifics such as fees or waivers to be addressed early on. When folks are leaving it is important to solicit evaluation and feedback.

A fundamental question to ask before committing to community outreach is whether farm staff can fulfill these outreach responsibilities. It is crucial to consider the time available and the prioritization of farming tasks, as well as the professional expertise necessary to successfully complete these types of projects. Can the farm staff meet the professional standard when organizing a conference? Can farmers become servers at an on-site wedding, or is it necessary to import specialists such as caterers? Which staff can handle youth education requirements? Do people possess the intuition and experience to learn the tasks, or do they require specialized training?

The goal of community outreach is to maximize the potential of the organization without overextending ourselves, through coordinated efforts with public and private entities. By addressing various segments of the region's demographics we hope to reach the full spectrum of the population. As we integrate with and reach into different segments of the population, information is spread laterally through various age groups, interests, and socioeconomic strata.

One of the benefits of outreach programs is that they often create unanticipated and corollary effects. Often group programs, such as the bulk-buying club, become their own microcommunities in which members share and communicate as a result of this association. The permaculture classes have developed into a community in which alumni gather and share into the future. The open mic and community potluck crowd has regulars who rarely attend other events but enjoy the camaraderie of the Last Friday. In this manner our goal is to catalyze and let community happen.

It is essential to define our goals as an educational and service organization. In our role as demonstrators our standard method is implementation. We wish to offer a transparent organizational structure of a working farm at which the processes are organic and evolving. In this respect we do not have the desire or economic capacity to provide a classroom-style curriculum for residents, nor can we provide an in-depth tour to every visitor who wanders on-site for a respite from the highway. While we will coordinate a plan for on-site residents that addresses their individual learning objectives, this course is designed to complement the overall objectives of the organization.

This role may seem ambivalent, but the elastic and flexible nature of the system allows for resilience and self-determination. Our philosophy is based on the perplexing reality that there is no one way to meet our goals and carry out the mission. While the farm has overall goals, crucial factors such as the weather, personnel, and the economy can change rather quickly. The challenge is to predict the patterns of evolution and supply alternative routes when circumstances demand.

Our community outreach programs vary not only with the four seasons but also with the perpetual evolution of personnel and organizational priorities. Thus, while many of these programs run concurrently, it is important to note that not all of these programs are operating simultaneously. Throughout the progression of the organization, we have evolved based on the personnel and organizational commitment and interests. At certain times we have shifted away from projects that are a burden beyond what we can maintain as an organization, while still finding ways for these programs to perpetuate. In other situations we have pursued programs and projects based on particular circumstances; these decisions are often the most challenging organizationally as we struggle to agree on where to prioritize our collective energies.

Reaching agreeable terms for collaboration requires continual communication regarding roles, commitments, and responsibilities, as well as the fluidity to evolve over time. As an organization we continue to strive to immerse ourselves as a positive contributor and an influence throughout the layers and networks in the web of our community.

While this section details a design to interacting with the general community, words do not adequately convey the tangible human relations that this entails. The organization is blessed with many community partners, generous souls willing to commit time and resources to forward the mission of the organization.

We have also been fortunate to develop working relations with several experienced professional contractors and experts in their fields. These people may become involved in a professional capacity and choose to share further with the organization. Or they may come to D Acres for personal interest and share professional skills. Whatever the combination, there are numerous examples of skilled volunteers sharing their time and expertise for the betterment of the community. As examples, Taylor Mauck of Sunweaver has been a strong supporter of the organization as a board member and solar installer. In addition to putting in the pole-mounted photovoltaic system and hot water heating, he has been a teacher in solar-related educational programs held on-site. Ronda Kilanowski has served the organization as our accountant and developed systems for managing the challenges of bookkeeping and tax payments. Her guidance and stewardship as a board member and organizational supporter has been extremely helpful. Jay Legg has contributed his expertise, ranging from chain saw use and road building to developing cottage industries.

Foundation and excavation specialist Bob Guyotte has also been a key contributor to the success of the organization. From the initial foundation work of the Community Building to the latest silo, solarium, and compost facility projects, he has offered clear, concise, common sense advice on many issues. While he has shared his fifty years of experience on the nuances of earthwork, machine operations, and building, he has also offered an example of steady hard work and systematic management of project operations. His voice regarding how to implement and accomplish projects echoes in my head.

We are fortunate to have become friends through the years with Kevin Maass, a mechanic, and Stacey Lucas, a graphic designer. These relationships have benefited the individuals and the organizations and provided rewards to the community at large. Both of these individuals could earn substantial wages for the high level of services they provide for nominal fees or trades. Kevin has been extremely tireless and faithful as a friend of the organization, providing mechanical support for our motor vehicle pool and other construction projects. Stacey Lucas continues to astound us with her creative capacity and enthusiasm.

By working with people who know our voice and whose voices we know, over time a high level of trust, confidence, and faith is developed. With these individuals there is no balance sheet of transactions; rather, an assurance of continued commitment to mutually benefiting interactions and relations. Listing contributors in this manner acknowledges the time and consideration of their investment, and there are many others who have also contributed time and expertise. The many positive investments representing a diversity of ideals and expertise have led to the evolution, character, and resiliency of this project.

Similarly we have developed strong ties over time with the Bread and Puppet Theater organization. The extended Bread and Puppet family serves as a model for a mission-driven land-based project. We often spend time at the theater's Glover, Vermont–based farm, working on the agricultural system. By partnering with Bread and Puppet through the years we continue to share goods and services, as well as personnel and information. Together with partners who share compatible missions we can extend the reach of our organization's capacities. It is also a pleasure to develop working relationships that become lifelong friendships full of mutual support. It is through this process that we can create a well-connected network that will thrive through generations.

When we arrived on Streeter Road in the fall of 1997 we knew that we did not wish to create an island or oasis that did not interact with the local community. Community outreach is designed to strengthen the community by sharing, not creating a lifeboat zone for isolated enclaves to be protected from marauding hordes and the zombie apocalypse. We knew our mission was to include the existing community. One of our first attempts to connect with the larger community in Plymouth was the construction of the Whole Village Community Garden. In the spring of 1998 the garden was constructed in a community barn-raising fashion. While D Acres' efforts were limited to labor we were exposed to the greater gardening community of Plymouth through this event.

Since then we have donated surplus time and food toward existing community food projects. We continue to regularly donate vegetables to the Whole Village food pantry, Bridge House, and Senior Center. When staff are available and interested, we participate in food preparation and serving at the weekly Meals for Many. We also donate products and services to various charity auctions held as benefits for nonprofit organizations in the region.

Community Food Projects

We have invested organizationally in initiating several community food projects. The concept has been to incubate a mechanism for empowering localization that can then become independent of D Acres. In essence, we aim to initiate projects that can then add value to the community without draining organizational resources.

Community food programs are intended to strengthen regional food security with a focus on the direct connection between those who produce food and those who eat it. Programs strengthen the capacity of both parties and build connections between them. These programs allow farmers to market their products and services collectively while reducing overhead and delivery costs. The organization acts as a catalyst in the crucial stages of the process that effectuate economic localization. By addressing specific points and cruxes that stymie the growth of a localized economy, our plan is to fertilize the strength, resiliency, and security of a regional permi-economy (an enduring self-perpetuating economy.)

Local Foods Plymouth

In the fall of 2004 we initiated a program known as Local Foods Plymouth (LFP). The project was formulated from brainstorming Internet possibilities for connecting farmers directly with consumers. I was imagining an online listing of various area farmers' products and prices to help consumers and restaurants shop. At the time interactive web platforms were in developmental stages, and we pursued an online shopping cart as a suitable platform. The concept was thus transformed from simply listing the information to providing the capacity to order directly from the farmers. To further this beyond the conceptual stage, we knew that we needed a broad community base, so we pitched the idea at a community brainstorming session held by a local group known as the Plymouth Area Renewable Energy Initiative (PAREI). The Energy Initiative was seeking to develop programs that focused on the food system and were eager to collaborate on a project.

During that winter Sam Payton, the resident woodworker at the time, worked to produce a website where local consumers would shop at various local farms, then be able to combine their selections on one bill. Abby, the farm manager at the time, began organizing local farmers to participate in the program. The plan was to list the farmers' products on Mondays and allow consumers to shop the website through Wednesday. On Wednesday we would close out the website and inform the farmers of their orders. These orders would then be delivered to the local farmers' market, where they would be consolidated for pickup on Thursday.

During the pilot year the program received a large amount of media exposure and was featured on National Public Radio. The Plymouth community, largely drawn from PAREI's base of participants, responded with enthusiasm to the prototype. It is important to consider the effort involved in administration and actualization. Especially when launching a program or trying a new event, there is potential for unforeseen challenges and unrealistic expectations. It is important to recognize how budgets and enthusiasm can be drained by the burnout from coordinating efforts such as conferences or program launches.

Currently the program requires an administrator to adjust the website products and inventories. While there have been more recent programs that allow farmers to directly update their own inventories, this stage is critical so that inventory and availability is accurate. Personnel are also required to help sort and package the various orders so they can be picked up in an orderly manner. In addition farmers must be punctual at the drop-off, and consumers must remember the hours for pickup.

After our involvement during the initial year we made an organizational decision to remove ourselves from the day-to-day operations and participate as farmers/vendors in LFP for the summer of 2005. While we continue to participate on an advisory board for the organization, PAREI is the administrator of the program.

In developing programs such as LFP the intention is not to supplant or replace the existing connecting mechanisms between consumers and producers but rather to augment the networks in place. Local Foods Plymouth was not designed to replace the farmer's market, but instead to accentuate the

market by providing additional options and flexibility for consumers and producers.

The benefits of the Local Foods Plymouth program are immense in terms of efficiency. Instead of being forced to shop at every booth in the market, consumers could preorder, pick up the staples, then shop at their leisure. Through shopping online, consumers could also make the choices, then send a friend or teenager to perform the pickup. The Local Foods Plymouth program also includes many farms that do not attend the market regularly. The time and energy spent bringing perishables to a hot, sunny parking lot is less than ideal for many small and specialty farms. Thus the LFP program allows a small producer access to consumers without forcing them to be present at the market. LFP also ensures for a farmer the sale of his products. The farmer is only required to bring what has been already presold versus harvesting lettuce that may just wilt in the sun. The convenience of the program extends to the point of sale, which has been digitized so that all the purchases are consolidated for both the buyer and seller. The buyer only has to make one transaction to purchase from a multitude of farmers, while the farmers receive a single check from all the buyers of the week. This reduction in paperwork and administration can be substantial in tasks ranging from making exact change to tracking inventory. The LFP format also serves as a collective marketing tool for smaller specialty farms to produce sales with a minimal investment in marketing or time spent soliciting direct sales. The format allows farmers to charge retail prices with only a small percentage loss to overhead costs, thus improving the margins of profitability.

Following the second summer LFP expanded to offer year-round sales. In the off-season, orders are delivered to Main Street, where distribution space is donated by the UPS Store. This partnership provides a highly visible and accessible location for the food distribution.

While the LFP program has been a tremendous success overall for farmers and the general public, there are some shortfalls with the program. For farmers who sell only a small quantity of products during any given week, the trip to town for a minimal drop-off can be an irritation. Likewise, buyers can be inconvenienced by the specific pickup hours and the technological expertise required to shop online. While these nuisances are being addressed with cooperation, teamwork, and public education, the difficulty that remains is financing the administration of the program. Grant funding and initial enthusiasm can catalyze a project start-up, but enduring programs require a predictable income to cover costs. To derive this income, revenue must be generated by advertising or by dedicating a percentage of sales to program administration.

Collaboration with local community organizations is both rewarding and challenging. By bringing in collaborators, fresh energy, skills, and ideas are infused into a program. Defining commitment is crucial in the planning phases to anticipate the degree of responsibility of each collaborator through the completion of the project. It is also important to formulate a decision-making process and parameters for conflict resolution within the group if success is to be insured. Finding community partner organizations with the character to follow through with the hard work of project implementation can be a process of trial and error. Through this experimentation all of the partnering organizations learn, and enduring relationships can form.

Local Goods Guide

Following our success with the Local Foods Plymouth program, we sought another outlet for connecting farmers directly with the public. While LFP had a substantial core of committed consumers we realized the limitations of this Internet-based structure. Our goal was to broaden the possibilities for producers and consumers to interact by providing an open access through free distribution of printed materials.

During the first year (2006) we solicited eight local farms to provide their information for a trifold-style pamphlet. The pamphlet exhorted the public to buy directly from local producers and offered the farms listed as alternatives to supermarket chains. In addition the pamphlet listed farmers' markets and LFP possibilities for buying local. From our desktop publishing operation we produced and distributed about five hundred copies of this flyer. The public responded positively to the pamphlet, which we distributed at the farmers' market, the local health food store, and various tabling events.

The next year we grew the Local Goods Guide by accepting sponsorships and developing advertisements to offset the costs of printing. By the third year we decided to outsource the printing. At that point the increased quality provided by local printers and their willingness to support the project through discounts and sponsorships made the price of off-site printing reasonable. As the Guide expanded, the need for a graphic design specialist who could meet the technical challenges and bring fresh creative energy became apparent. To address this issue we approached a local artist and entrepreneur named Stacey Lucas as a collaborator. As a visual artist and gallery owner she had a vested interest in the local economy, while her expertise in desktop publishing and graphic design elevated the professional polish of the Guide. We also recruited another farmer employed in the computer mapping industry to help refine the geographical Guide elements.

With the addition of qualified community partners the Guide began to grow in terms of content and circulation. By the sixth year the Guide included nearly fifty farms, as well as locally owned galleries and secondhand stores. We also expanded the agricultural categories to include hay production, forest industries, and other cottage industries. The community support expressed in terms of advertising has provided the revenue to increase the circulation. In addition, through consistent oversight and planning we have also refined the circulation system so that distribution is ensured in a timely and effective manner.

In 2009 we coordinated the first Local Goods Guide launch event in downtown Plymouth on the Common. It was a day of farm-fresh samplings, a short presentation on local economies, and live music.

Undertaking this effort requires community support. As the Guide has grown we have realized the limits of our organizational capacity to annually produce such an extensive publication. While the winter months allow time for the necessary administrative tasks the immense job of soliciting advertising and updating the information for a spring deadline does not provide ample opportunity for rest and hibernation. The organizational enthusiasm to meet the challenges of publishing the Guide on deadline diminished over time, and the pressures of covering the costs rose as the budget expanded. After six years of producing the Guide, we sought community partners who were willing to undertake the responsibility of overseeing the seventh edition. Fortunately Stacey Lucas was willing and able to

fulfill this challenging task. Although Regina continued to assist by providing farmers' information, we were able to shift the solicitation of advertising away from farm staff, thus reducing the stresses of springtime substantially.

While circulation can be challenging, we have developed several methodologies to enhance distribution of the Guide. It is essential to utilize the advertiser and farmer network to assist with distribution, so we inquire about the projected numbers each business anticipates it will be able to distribute to. Once this initial distribution list is established, printing quantities can be estimated and a launch event can be planned. Generally we have launched the Guide in conjunction with the first Plymouth farmers' market of the season. This enables distribution to participating farms and provides a high-visibility location for pickups. If we are unable to coordinate a pickup, Guides are then sent out in a timely fashion via either U.S. mail or personal delivery. If additional Guides are available we solicit high-traffic locations such as the chamber of commerce and visitor centers, where they will be distributed.

The Guide has now become a fixture in our region. People recognize and look forward to the annual edition. It has become a tool that demonstrates continued creation of localized resiliency. In the future our organizational goal is to continue to foster the Guide as a community collaboration without being solely responsible for its production. By nurturing the Guide to this point, we have provided the foundation for the future to perpetuate and refine the Guide.

Multi-Farm Community-Supported Agriculture

A traditional community-supported agriculture (CSA) program offers buyers a share of seasonal farm products. This is effective for farmers because they can presell produce prior to planting, and that contract provides the cash flow necessary to pay for four seasons of farming. But as small farmers on a limited land base we are reluctant to engage in the CSA model. At this juncture we simply do not have the capacity to generate a worthwhile variety of farm products throughout a growing season for a number of families and also meet the needs of our on-site operations, which are based not only on our land base but also our interests and objectives.

In spite of our reluctance to accept the responsibility of producing a full array of CSA products, we have still been enthusiastic to experiment with the model. In the fall of 2011 we began planning for a multiple-farm approach to the CSA model. In creating this option our intention was to provide staples and to highlight locally produced specialty items. We contacted various farmers to discuss product options for a ten-week winter program. We were confident that D Acres could supply garlic, potatoes, and bread for the program, in addition to adding specialty products such as tea and sauerkraut. Other local farms were eager to supply milk, cheese, eggs, and meat. We designed the price point for shares of the program to allot a fair price for each unit to the farmers, which ended up being three hundred dollars for the ten-week program.

During that first winter we chose to limit the program to thirty buyers, and the shares were quickly sold. The public was glad to have the pickup arranged in a central location at the local public library. Every Thursday we would arrive with the products from the various farms and wait for the pickup between 3:30 and 6:30 p.m., which complemented the Local Foods Plymouth program and our existing town-run schedule.

The program was a success on many levels. The producers were pleased to receive a fair price for sales that were predefined both in terms of timing and quantity. The public enjoyed an array of fresh products delivered to a central location from diverse farmers. Our organization received public exposure and support for investing in and arranging this venture. Following the success of the first ten-week program we organized subsequent spring, fall, and winter programs.

While the initial success of the program was exciting for the organization, over time the external costs of organizing the program began to undermine the viability of the operation. The large amounts of time spent preparing and administering the program were not adequately accounted for within the cost structure. These issues coincided with staff turnover; rather than continue a program running in deficit, we chose to divest from the program and negotiated for the Multi-Farm CSA to be absorbed by the existing LFP program. As a subsidiary of the LFP program it helped build the diversity and income-generation base without expending resources from their existing overhead structure.

Our model of germinating programs and then passing along the administrative duties to the broader community reflects our capacity as an organization with transitory personnel. In circumstances where enthusiastic and creative personnel have been willing to initiate programs, D Acres has supported those efforts, though the organization may be unable to maintain the burden of that program if those individuals move on. Hence, it is important to consider how to shift program responsibilities beyond transitory individuals and develop strategies for enduring programs with broad bases of support. Many of our efforts have focused on inoculating models and ideas that can be transferred to other communities and taken on by the broader regional community.

On-Site Educational Programs

Educational programs provide a portal to the public for D Acres services. We invite the public to participate as visitors for a variety of activities. Each program is evaluated for its own efficacy, from the public's arrival to their departure, as well as part of a holistic outreach to the community. It is crucial to be able to gauge community awareness of each activity and how it correlates to attendance, which is essentially the effectiveness of our public relations. This helps us in the future to identify target audiences, then design experiences that both meet the goals of the organization and are positive for the public.

Community Events

Hosting events at the farm requires several crucial elements of infrastructure. We use desktop publishing and a printer to make advertisements, workshop handouts, registration forms, and menus as necessary. We can provide food for large numbers of people in the commercial kitchen area. We also must maintain adequate sanitary facilities and parking, which should give older, less mobile clientele preferential treatment geographically. For larger outdoor events tables, chairs, and tents help provide comfort from the hot sun and rain.

Making the connections that build a community network for outreach requires intention. While word of mouth may ultimately be the most effective

information dissemination tool, direct communication and solicitation is a proactive way to interact with the community. Spreadsheets that retain contact information and prior correspondence provide a historical record and are a baseline for continued communication.

To host successful larger events and gatherings requires significant planning and administration. Prior to the event attention must be paid to scheduling the activities and speakers for the program. Once the lineup is solidified the promotional campaign begins to reach out to potential attendees. To plan adequately it is important to ascertain the number of participants, particularly in regard to food. Generally we require a preregistration so we can anticipate the numbers. Other factors such as parking and the weather must be addressed and planned accordingly. One-day events can be simpler to organize, because of the complexities involving overnight accommodation. Volunteers can be especially helpful with parking, setup, food, and cleanup duties.

Farmer's Dinner

Through the years we have hosted a farmer's dinner during the month of February to provide a forum for agricultural networking and planning. This dinner provides an opportunity for an informal meeting to make acquaintances and renew friendships among farmers in the area. The purpose of the meeting is to diffuse competitive instincts and infuse a collaborative atmosphere in an attempt to build our group potential.

Buying Club

Food buying clubs reach out to the community and build resilience. By ordering large quantities together from a food distributor, we save on packaging and distribution costs. The food club dropoff is at D Acres, and we pay for the delivery when it arrives. The food club members then pay for their products when they pick them up. Members can share bulk purchases between households to obtain a lower cost per unit. This cooperative food-buying approach allows costs to be shared among households and increases the diversity and quality of food and household products that can be obtained.

In our area there are two operations that supply this service. One is a smaller local business that specializes in New England products. The other distributor is owned by a large multinational corporate conglomerate and offers lower prices and automated computer service. In general we attempt to buy from the food club distributor that supplies local food. Once or twice a month depending on volume, food club participants buy online or place an order via phone, and the drop-off is made at D Acres within three to five days. We then sort the packaged material so there is convenient pickup within a couple of days. We charge 3 percent of the total ordered to help offset administration and handling time.

At the moment the food club is operating as a rudimentary streamlined service. This is an example of a program that the organization is maintaining with a minimum of investment. That said, with specialized personnel on-site or attention from participants the food club could be enhanced into a more effective mechanism for mutual support and community building.

Community Food Events

The concept of the community food events is to provide a direct connection between people and their food. By preparing seasonally based meals and offering a relaxed casual atmosphere we are able to expose neighbors to the flavors of rural community life. Often rural neighbors who might only be recognized by their car or the house in which they live have become acquainted in person at these events.

The Trought family has traditionally held a celebration annually with a summer pig roast, and 1998 was the first of such New Hampshire–style events. Farm Day, as it has been known since, generally consists of tours, pig roasting, and live music. In the early years it was more of a Trought family event, though it has grown into a broad community celebration.

Volunteers help snip beans in preparation for Farm Day in 2007.

D Acres' residents and volunteers wash potatoes in preparation for potato salad to be served at the 2011 Farm Day.

Farm Day has gained a reputation for good eats, attracting people from miles away, just to eat some slow-roasted pork.

The Farm Day buffet line is a testament to homegrown and local. All the food served for the meal was grown at D Acres or sourced within a fifty-mile radius.

The basis of our food events began with the traditional potluck. The idea of sharing food with neighbors in an informal and nonfinancially biased format appeals to our inclinations of an organic community-based democracy. The community potluck was our initial foray into utilizing our commercial kitchen space as a tool for community interaction. For the first couple of years we would host sporadic potlucks and call a list of invitees. These potlucks became extravagant affairs that provided familiarity of the structure to the public but strained the capacity of the building and was an overwhelming experience for the residents. Through the years we began to schedule the potlucks with the full moons to provide a predictable natural basis for the rhythm of the event. To minimize the confusion further we now host the events on the last Friday of each

Once the food has been set out, the line is formed—folks don't mind waiting; they get to socialize and anticipate their plateful.

month. At this point we are combining the potluck with an open mic event so the community can enjoy music into the evening throughout the year.

For all our food events we generally ask the public to contribute on a sliding scale of five to fifteen dollars. While we do not require a payment, we do display a cash pouch prominently so our capacity to receive revenue is obvious. We generally receive about $7.50 per participant, and we do account for the numbers of people served as well as the actual servings so that we can maintain a sense of the costs incurred for each event.

Farm Feast Breakfast

On the first Sunday of every month we serve breakfast to order for from 25 to 150 people between 10:00 a.m. and 1:00 p.m. We offer a choice of potatoes, sausage, greens, pancakes, and eggs cooked to order. The food is prepared in open view with ingredients derived primarily from the farm or from local sources. At 1:00 p.m. we offer a deluxe tour of the farm, which lasts up to three hours if participants have the stamina.

In a sense the Farm Feast Breakfast has become a prototype for our organizational program. By our supplying the food directly to the public, they receive the highest quality product and connect with its source. This direct exposure to the real flavor of food is the crux of our local food challenge. We are reestablishing links between health and local agricultural vitality that fast food fostered by clowns has erased from collective memory. By

A local group of friends are able to spend time together at the Farm Feast Breakfast.

closing the gap between consumer and farmer we offer an alternative to the homogeneity of the modern food system.

The breakfast attracts diverse participants. While many local neighbors attend the event, people also drive for several hours to share the breakfast experience. The participants range from college students to churchgoers to families. We serve as a convenient meeting spot for friends and family to gather socially. The road bike club plans their Sunday rides so they can ride up and enjoy the event. This informal atmosphere inspires community. We have chosen to offer shared seating and self-serve beverages so people are forced to interact with the other people in the room. While this event may serve as an introduction to the farm for newcomers, there are many long-term participants who regularly attend the event as well.

Pizza and a Movie

Pizza and a Movie Night provides a monthly excuse to indulge in the popular, prototypical American convention of dining and entertainment. The commonplace nature of the food choice and entertainment medium provides a portal to introduce the public to subtle yet important distinctions in our event. Pizza Night food is prepared with farm-fresh toppings that highlight seasonal garden ingredients. Thus in the spring we celebrate pizzas with fresh greens and asparagus, while in the summer zucchini and tomatoes are likely toppings, and in the wintertime we delve into the stores of squash and potatoes. The pizzas are

Regina takes a hot pizza straight out of the cob oven, ready for serving.

prepared in the wood-fired cob oven to illustrate fossil fuel alternatives in cooking. As entertainment the movies we screen are films that offer information or social commentary beyond what is typically exposed in the mainstream media. The format of sharing food and thought-provoking films allows folks to digest and ruminate on ideas in a safe space for further discussion. This informal gathering process serves as a platform of information and face-to-face communication, which can ferment and catalyze cooperation and innovation within our community.

Seasonal Soup Night

We have initiated Seasonal Soup Night at different times in our history. The focus of the night is to provide soup made with ingredients that are seasonally available in New Hampshire during that time of year, either from the field or the root cellar. By focusing on seasonal ingredients we hope that the participants will realize the potential of a seasonal diet in our region, both for flavor and socioeconomics.

We have also operated this Seasonal Soup Night theme in collaboration with local restaurants. During 2011 we held monthly soup and culture events at Mark's Speakeasy Café in Plymouth. These events featured local musicians and performers on stage while the soup was served. Though we were pleased by the turnout for these events, the venue is no longer open.

Youth Education

Imparting practical skills and knowledge to youth is a primary goal of the organization. At this time we have begun to request payment for our services as organizers of work and tours. We have a standing policy to request payment on a sliding scale of from two to ten dollars per participant for one- to four-hour work projects and an additional two to ten dollars per participant for tours. We will consider certain groups with whom we are familiar for less, but this policy helps offset the tremendous time and effort necessary to work with learners.

Working with off-site learners should never be viewed as an opportunity to accomplish work. While accomplishing tasks may result it should never be expected or anticipated with any degree of probability. Tasks on the farm, if they are indeed a priority, should not be delayed with the expectation that a school or youth group will accomplish the tasks.

While the value of work is an important concept, seldom do visiting volunteer participants qualify as the same caliber as an experienced laborer. Sometimes, volunteers can do more harm than good in certain situations; inferior work may often cost more than free labor. The education activity in itself is the goal. The goal is to expose participants to the theories and principles, but there is not time to produce calluses. On a day at the farm the best you can hope for is blisters and sore muscles. It is essential to provide projects that impart education without overextending the attention span.

While focus can diminish toward the end of the school year as summer break approaches, the quality of instruction and supervision affects the success of youth education endeavors. Summer camps for middle and elementary school kids with teenage counselors will not provide adequate oversight; likewise, visits arranged by a school district or other administrative personnel (conveniently absent on the day of the visit) for whole classrooms led by disinterested teachers are often unproductive. Summer camps set up for overnight are generally more organized and funded than their day camp counterparts.

If participants arrive enthusiastic they will generally leave with greater excitement. If they are

This group of high school students from Hollis, New Hampshire, were on an overnight retreat at the farm and volunteered a few hours each day during their stay. With a staff member that can effectively coordinate the flow of the project, a work effort like planting potatoes can happen in just a few hours.

A small group of kids from Sant Bani School work with Josh to load the trailer with firewood. While not the most efficient way of getting it done, it gives the young kids an opportunity to use their muscles and accomplish a simple task as a group.

Tyler Durham (resident in 2007–08) served well in a position as youth educator. He visited schools in the area and engaged tours and visits from all age school groups.

hesitant, distracted or resistant, distant and morose, their mood can be improved. While they may not be converted into Michael Pollan overnight, they may still be gleaning knowledge to help inform decisions into the future. While large groups can be difficult to engage, with practice and planning the process of working with youth groups plants the seeds for an abundant future.

Generally with youth groups we offer a tour based on their interests and aptitudes, then provide work-related activities. Depending on the season and the projects at hand we attempt to find activities that are organizational priorities and manageable with the groups at hand. Evaluating the character of the participants is a challenge. Without prior

knowledge of the individuals it is difficult to predict their responses or aptitudes when faced with certain tasks. Shoveling manure could be deplorable and despicable or quite enjoyable, depending on the individuals or their moods. I have had wonderful experiences with at-risk youth and horrible interactions with affluent private school students, and vice versa.

Our success with these groups has been difficult to predict. It can be a challenge to engage youth that have been forced to visit a farm against their will. The stigma of dirt, manure, and flies sends shivers through many people who wish to keep clean sneakers. While some youth are enthusiastic and open minded to all farm activities, many are reluctant to even step foot off the bus. This resistance can be

Morgan Casella gives a tour to a group of day camp kids from Ashland, New Hampshire.

accentuated further if the youth anticipate the most feared word in their age group: work. Especially volunteer work. It is difficult to ensure a perfect experience, though it is important to provide the opportunity.

There are several keys to improving the probability of successful interactions with youth groups. High teacher-to-student ratios can be helpful depending on the activities and participants' aptitudes. By developing learning objectives and a relative idea of the participants' current learning level prior to their arrival, the participants' teachers as well as the staff can adequately prepare an educational session. This requires an investment in time communicating shared objectives and planning.

Activities should be well thought through beforehand. For typical work events questions revolve around having sufficient numbers of tools or how to organize the work among a group. Consider how many people can participate in any specific task to avoid bottlenecks, promote fluidity of the activity, and offer high-quality education for the group. Time in which people are standing waiting for instruction may be counterproductive. Planning work activities is complicated by many factors and nimble adjustment must continually be made during activities based on monitoring the evolving situation.

Because many tasks require breaking into smaller groups, it is often more functional to focus on large-scale projects and chain efforts. School groups are great for harvesting garlic, for example, but less so for planting, weeding, or bed prep on the basis of skill and reliability. In addition to heavy work such as shoveling manure, there are also tasks such as shelling beans or peeling garlic that evoke opportunity for group dialogue and conversation.

Weather and conditions such as bug season can severely affect the activities that can be undertaken in any given day. Rainy conditions and biting insects undermine morale and can distract from the focus of the group. It is important to set up multiple possibilities for groups, considering weather and other variables such as temperament and physical aptitude, which are difficult to determine until the group arrives and begins the activities.

We have dabbled with on-site summer camps and after-school programs. It would be ideal if the school bus dropped off a cadre of latchkey farmers to help with evening chores and meal preparation. Because of the lack of staff dedicated to this purpose, these activities have not been fully realized as potential community outreach mechanisms. Developing the connections to establish programs such as these

requires dedication. For the program to endure and blossom, coordination by personnel with education training or life experience is fundamental.

While we would like to assist with every request that we receive, we must be selective at this time because of the amount of requests and the need to prioritize organizational time. In the near term we would like to consider financing for adequate staffing that would expand our youth education potential. By partnering with schools to provide classroom limits and aggressively pursuing school-age children during the summer months, we hope to further the capacity for empowering the youth with knowledge.

Adult Education

In addition to the residential education component, we offer workshops and educational events to adult learners throughout the year. We take advantage of our on-site demonstrations and facilities to offer sessions on gardening, woodworking, construction, and food preparation. Within the Community Building the kitchen, root cellar, woodshop, and art

Mark Fulford leading a group through the gardens, talking soil biology and health

Local artist Barbara MacAlister lovingly demonstrates and teaches wood carving. We hosted this particular workshop during an already scheduled community food event to draw attention and encourage participation.

Modern Times Theater (Rose Friedman and Justin Lander) have performed at several events at the farm, including this performance at a Farm Feast Breakfast.

studio are spaces that can be dedicated to practical public workshop space. In the open-air learning environment we can focus on hands-on gardening, construction, and forestry activities. The yoga room can host a multitude of events, including Reiki shares, formal presentations, puppet shows, theatrical performances, and films, as well as yoga sessions.

Often we will augment a regularly scheduled event by including an extra educational or entertainment bonus. This helps ensure an audience for the presenters and adds to the value of the regularly scheduled event for the public.

The content of our workshops varies, depending on the interests of the community and the expertise we can provide to present information. In offering workshops taught by on-site personnel, we limit the

diversity of skilled instruction and expertise that can be offered. Consequently, we match the potential workshops provided by on-site staff to the activities with which we are most familiar and often schedule them to coincide with the seasonality of the work. Hence, pruning workshops are scheduled for the spring and a garlic planting workshop is planned for late September. To enhance our on-site knowledge and that of the public, we also regularly invite speakers and presenters to offer their expertise. These sessions provide the public opportunities to interact directly with experts in the field. To fund these sessions and provide affordable access to the public we will seek community funding sponsors, apply for grants, or internalize the financial costs. For example, when hosting Michael Phillips for two separate presentations during the spring, we must

Organizing events like this evening presentation given by Dave Jacke in 2012 at Plymouth State University takes time and energy in advertising, promotion, and sponsorships. But the end result reaches a wide and captive audience.

allocate funds to pay for his services. To recoup costs for the first session, which was free and open to the public, we sought cosponsorship from the university and local businesses. The following daylong session was offset by the registration fees and the tuition from the permaculture course.

Public Art, Theater, and General Fun

In an attempt to create events that foster a relaxing and fun social atmosphere, we have hosted traditional dances, theme parties, bands, and dance parties. While these events are often arranged at the impetus of the on-site residents to fulfill their good time quota, we have also honored requests by local musicians to appropriate the space for public performances.

Community dances are a tradition in New England, and the dances are similar to square dances or contra dances in that particular songs have particular dance steps. These steps are orchestrated by a caller, who tells the dancers how to proceed. Such dances are great fun for young and old and provide a time in which people can interact without politics. While these events can be held in a mowed grass field, a community building or a barn with an expansive wood floor is more suitable. Attention should be paid to temperature control and adequate refreshments. Musicians can be expensive, and marketing is crucial; the success of the event depends on attaining a critical mass of dancers.

Through the years we have hosted Bread and Puppet Theater and the Beehive Design Collective

Themed parties make residential life a good time! This circus party was a good one.

Farm residents sometimes need to create a way to bond, have fun, and socialize outside of working time. An organized gathering can do that.

D Acres has cosponsored several barn dances; this particular one was held at the Dorchester Town Hall with fiddlers Jacqueline and Dudley Laufman.

for public presentations. We have held workshops and collaborated with art school students to produce sculptural installations throughout the property. These collaborations enhance the aesthetics of the property while allowing a site for experiential group processes to manifest.

In addition to the monthly open mic event, we also have chosen to host musicians and bands for the public. These performances provide exposure and experience for local musicians. The dance parties and theme parties enhance the social atmosphere and help to shake the winter doldrums with ecstatic movement.

It must be said that social music parties of this nature always raise liabilities to consider; In particular, the presence of smoking and drinking: Though the organization does not regulate the choices and acts of others, it is imperative that the organizational mission and community-driven purpose is kept in the forefront.

Permaculture through the Seasons

The Permaculture through the Seasons course is an opportunity for community outreach. By basing the course on the curriculum developed by Bill Mollison we combine that learning experience immersion at D Acres over the course of seven weekends. This quality of time spent over the growing season allows participants to both grasp the permaculture

The PDC class participates in a hands-on session, transplanting annuals in the garden.

PDC students work in groups to compile a design charrette for a practice presentation.

Guest presenter Dave Jacke joins the PDC class to explore the North Orchard forest garden.

Dave Wichland from Keene, New Hampshire, has been giving workshops at D Acres for several years. He serves as a guest presenter for the PDC, sharing his vast knowledge and passion for mycelium.

design process and become a part of the D Acres community. By spending these weekends in an educational format and sharing the fruits of the garden we imbue a community spirit and engender a cadre of permaculturalists to disseminate sustainable design processes.

Overnight Accommodations

The overnight accommodations program is an opportunity for people to be immersed for multiple days at the farm. By hosting the public in the facility overnight we have the capacity to broaden the experience that can be attained through a tour or workshop. By sharing the space on the farm for longer periods of time the public receives more intimate exposure to the working of the operation.

Our overnight accommodations allow guests to participate in the day-to-day workings of the farm. They are invited to participate in chores and other farm activities. In addition they have exposure to fundamental aspects of sustainable systems, such as the compost toilet, and renewable energy systems through actual usage. While touring the grounds they have the opportunity to grasp an understanding of the gardening, animal husbandry, construction, and forestry work that is ongoing on the farm.

On-Site Visitors

People arrive at D Acres with many different expectations. Some arrive from the highway looking for fresh vegetables or farm store products, while others visit to walk the grounds and view the animals. For us, we hope to meet the expectations for the visitors while also ensuring that our organizational activities are not disturbed by unscheduled arrivals.

While the safety of visitors should be protected, we also have concerns about the welfare of our domestic animals and cultivated landscape. It is important to allow the public to interact and connect with the animals with minimal stress. Consequently, it is important to maintain adequate signage and

verbally assert to guests the parameters of their visit. While chickens and ducks might appear innocuous, the dangers of roosters and open bodies of water should be considered before allowing children free reign with these animals. Territorial critters such as geese or sows with a litter may react aggressively to invasions from the public. While electric fences do provide some security, guests unfamiliar with them may endanger themselves as well as the farmscape. Our plants also require boundaries to prevent the public from stomping through the gardens.

The trails and woods at D Acres provide a multitude of opportunities for the general public. The varied terrain and landscape is an ideal outdoor classroom for the study of the natural world. The variations in the landscape are also traversed by trails that offer different degrees of difficulty. The Loop Trail provides a brisk thirty-minute outing or an hour and a half of leisurely strolling depending on your pace. The Property Line Trail is a solid half-day hike with steep ledge sections where the terrain is difficult to negotiate.

The overall goal is to bring people on-site and provide a forum in which others without land can feel at home. The grounds are open to the public, and while we state that visitors should feel free to follow any path to where it leads, our policy is to prohibit the public from entering personal living spaces without prior agreement.

Conferences

Through the years we have been involved with several conferences. Conferences have the potential to be transformative events that can manifest exponentially positive results. They can take months to plan and be undermined by foul weather. For conference event organizers there is a high potential for professional burnout, as months of preparation culminates in an anxious, nerve-racking event, which leaves the organizer exhausted and unfulfilled. The high stakes of undertaking a conference provide a risky proposition for seasonal farm plans. Time and effort involved in the planning and implementation is a process that requires months of preparation.

Our first organizational experience administering a conference occurred in 2004. The Small and Beginner Farmer organization sent out a job description seeking a coordinator for their winter conference, which was to feature Joel Salatin. Having been introduced to the charismatic and outspoken Mr. Salatin at a University of New Hampshire conference the prior November, I urged Abby to utilize her organizational skills and become involved. Abby was hired and organized two successive years of winter conferences for the organization. While the conferences provided publicity for our efforts at D Acres, the off-site activity drained energy from our farm system.

In 2008 and 2009 we decided to organize on-site conferences. The themes of these conferences were based on the seasonal staff's interests and aptitudes during those summers. In 2008 the theme was Health and Wellness and consisted of diverse informational and participatory workshops. Participants arrived on the morning of the event and were welcomed at an opening introduction circle. The morning and afternoon each offered two sessions with several different workshop options, including herbal health and nutrition, beekeeping, and soap making. There were also ongoing skills sharing of Reiki and massage throughout the day. During the lunch hour the attendees gathered for a farm-fresh feast and potluck dessert, then enjoyed a keynote presentation by author Jim Merkel.

The following year we hosted a fall festival known as the Traditional Arts Fair. For this event we invited regional artists to bring materials and demonstrate their techniques throughout the day. This allowed attendees to tour the grounds and see a wide range of specific techniques being practiced. During the day we also scheduled planned activities for children and adults that provided the opportunity to create art. Specific projects included fabric dyeing and a group sculpture project from found materials. The grounds were divided into several subcategories including metalworking, ceramics, and fiber. There were also workshops offered in dance, painting, bookmaking, and basketry. We also hosted a performance of Howard Zinn's one man play *Marx in Soho* and a Punch and Judy skit by Modern Times Theater and closed the night with traditional barn dance callers Jacqueline and Dudley Laufman.

Signage and Accessibility

Accessibility is a major concern for our on-site public access program. While snow may be viewed as posing a dilemma, it is the muddy season following the snow that wreaks particular havoc on our grounds. We have attempted to provide a route to tour the immediate vicinity of the Community Building. In designing this route we have considered the groundwork, educational vistas, and geographical progression to enable large- and small-tour access.

Tour routes are often dependent on the capacity, inclination, and clothing of the particular group. Inappropriate footwear and clothing may limit the duration and course of the tour. Certain groups may be interested in specific features and structures on the landscape. The tour route may also be tailored to include seniors groups, as well as those with crutches or wheeled vehicles who must avoid stairs or extremely muddy terrain.

To improve the grounds and enhance the tour experience, we have adjusted our routes and improved the paths of passage. By improving the drainage with water bars and ditch work, we find the roadways dry faster. We have also added ledge pack materials onto heavily traveled areas, strengthening the paths by providing a firm, resilient, well-drained surface.

Our signage is an attempt to familiarize guests with the location and provide for fluid knowledgeable discussion of the on-site geography and infrastructure. The signage helps identify essentials such as the outhouses as well as the sense of place regarding the surrounding mountains and swimming holes. By standardizing the names of certain buildings with

A tall signpost stands beside our welcome kiosk to invite visitors to explore the farm.

signage, such as the Ox Hovel, we feel that agreed terminology enhances the possibilities for constructive dialogue and information sharing.

The artwork on the grounds also reflects our desire to provide sensory pleasure for residents and visitors alike. Stone sculptures and metalwork have been installed throughout the property. It is our goal to provide an experience rich in culture and aesthetics, as well as hands-on nitty-gritty practical.

The playground is a fantastic component of the farm system for kids young and old. This location provides a space to relax and enjoy the outdoors. Features include a picnic bench, a sandpile, a big rock pile, a seesaw, and some tire swings. In addition, we have begun construction of a musical zone with chimes and percussion instrumentation. In the future we hope to add outdoor horns and stringed instruments, as well as slides, more swings, and play equipment.

We have worked to enable self-directed education and enjoyment of the farm to improve the efficacy of the program. The self-direction of visitors enhances their experience by allowing them to discover the rewards of absorbing the natural elements of the farm in their own time and pace. If the route of the self-tour is visible and easily negotiated it functions well for families. Small children may run or bike ahead from their parents pushing strollers.

To enhance the experience of the self-tour we have built two kiosk-type structures. The structures provide weather protection for panels with maps and other information that introduces the project and guides a visitor's stay. We have also designed several pamphlets to introduce and explain the grounds. One map offers a marked tour of perennial plants, while another focuses on the built infrastructure and animals. There is also a trail map that denotes the extensive hiking, skiing, and mountain biking opportunities.

There are also visitors who arrive at the farm intent on arranging a personal tour of the grounds with senior D Acres personnel. When people contact us we attempt to channel them toward the Farm Feast Breakfast tour once a month, but alas, often people arrive with the assumption that we have the time and inclination to offer them a personal experience at their convenience. While we are obligated to greet these folks and welcome them, we must maintain organizational priorities, so we try to continue with the project at hand and point these visitors toward a self-tour or invite them to the next meal or future event.

At times enthusiastic visitors may arrive unplanned and attempt to be involved with the work. It is important to consider the safety of the individual and the farm crew and address potential liabilities to avoid accidents. While we certainly encourage help in the work at the farm, planning suitable activities for specific volunteers' capacity is crucial.

In general, while we invite the public to grounds during daylight hours, we are not readily available to offer personal interaction other than a brief greeting, introduction, and orientation to the grounds. While some may arrive with the expectation of an afternoon of touring the grounds and enjoying tea and crumpets as the clouds roll by, we attempt to be forthright in the limited personable exchanges we can offer outside of planned public events. While we are dedicated to welcoming the public, we must also retain focus to actualize the working farm activities.

Opening a farm as a public business reduces personal privacy and adds responsibilities. Community projects that transpire over seasons take planning and budgetary energy that could be spent on other priorities. Through the years we have learned to channel visitors to the self-tour or our community events rather than be frustrated by an intrusion. The goal is to fulfill a prioritized task list by channeling the potential distractions of community outreach into positive outcomes through creative people management.

In terms of our programs and organizational capacity, we are flexible in developing further programs based on staff and community interests, limited only by the constraint of meeting organizational budgetary baselines. The goal is to creatively finance programs without creating a dependency on specific personnel or deprioritizing organizational necessities.

Catering to the Community

While we have offered a soup night on the farm, we have also utilized the mobile nature of the entrée to provide the portions off-site in collaboration with other organizations. In 2010 we collaborated with the Common Man's Flying Monkey Movie House and Plymouth Area Democrats to provide a soup and movie showing of *Fresh*. This soup and movie were followed by a panel discussion of local farmers and elected officials, which addressed the local food crisis. By combining the farm-fresh food with an engaging documentary and dialogue, we presented a rich format for food issue immersion in a popular Main Street location.

We have also contributed to food-focused events with the New Hampshire Master Chorale and the annual Souper Bowl fund-raiser for Artistic Roots. While this role is similar to a catering contract and can provide revenue, these avenues also allow us to interact with other constituents to raise awareness and broaden our educational efforts.

Weddings

We have had one experience thus far with hosting a wedding on-site as a part of the D Acres operations, and it is unlikely we will host another event. The complexities of meeting the bride and her mother's expectations proved difficult for the D Acres wedding coordination team. Unfortunately, while the couple was initially interested in D Acres because of our educational capacity based on agriculture and food, I think our on-site agricultural focus was undermined by the intensity of a wedding preparation.

While other events such as conferences can also diminish our agricultural intensity, a wedding is an intense life event that, while potentially lucrative, can be otherwise challenging. The high-pressure atmosphere and emotional nature of such proceedings may be best handled by a professional. While the farm setting may present an ideal to the wedding participants, the rigor of meeting standards anticipated by convention is substantial. Clear dialogue, resulting in a written agreement regarding the expectations for the event, should happen months in advance of the event so that mutual expectations are aligned toward a successful event, from planning to fruition.

Collaboration with Local Organizations and Businesses

Over the course of many years we have developed a working relationship with many Main Street businesses in Plymouth, New Hampshire. The Peppercorn Natural Foods store and the Chase Street Market/ Biederman's Deli have been particularly helpful as collaborators through the years. The Peppercorn has been cooperatively owned and managed by three women since the 1980s. The ownership has always been supportive of our efforts through advertising and sponsorships for public events and programs as well as buying our surplus produce. The Biederman's Deli operation is a family-run restaurant and grocer and is one of the bedrocks of the Plymouth Main Street businesses. The Chase Street Market also is the location of the local coffee roaster shop, Cafe Monte Alto. These businesses provide access to a large segment of the Plymouth community, so we have worked hard to maintain close and cordial relations. These local business owners and their families are the flesh and blood of building strong

rooted relations with locations of commerce in the Plymouth region.

In addition to the event sponsorships and donations we solicit from the Main Street businesses we also organize three weekly pig food pickups. This is a fine opportunity for face time and conversations that build connections on Main Street. On occasion we also have utilized sidewalk space in front of these businesses to table and inform the public of special programs such as Local Foods Plymouth or the Multi-Farm CSA. This valuable high-profile opportunity has also been utilized for multifunction public food events. When we have a surplus harvest of a commodity such as kale or garlic scapes we will offer small-scale cooking demonstrations, provide samples, and share food information with the public. Bringing seasonal surpluses into the public domain in this manner provides tangible results. By preparing somewhat obscure though nutritionally dense food such as kale, garlic, shiitake, and nettle stir-fry to the public, we provide inspirational evidence to the Main Street audience via the aromas and tastes. These cooking events stimulate sales of seasonal surplus produce and educate the public regarding nutritional, readily available cooking options.

During the summer of 2012 we collaborated on a more formal event with similar intentions. Board member Bryan Felice arranged for food activist and author Didi Emmons to prepare a meal with Regina at Biederman's, to be served as a buffet-style revelation on the potential of local food menus. We sold tickets for thirty-five dollars per person and booked a local musical group. We also invited several local food microenterprises, such as a brewer and a fruit jam canner, to provide tastings and share their wares. Regina and Didi prepared a tremendous food ensemble, and Didi shared her thoughts on the food crisis and its impact on urban youth with the assembled audience.

During the summer of 2009 we organized a community event that focused on the shared natural resources of the area. The Baker River Appreciation Day combined the concept of place and shared responsibility with a definitive goal of improving water quality. We organized with the Baker River Watershed Association as partners to build the integrity and public awareness of the event. We contacted town officials so that the trash picked up could be brought to the town dump. On the day of the event we embarked with a flotilla of teams on kayaks and canoes that scoured the river for trash and recyclables. After pulling tires, gas grills, beer cans, and plastic bags from the river in the morning, we enjoyed an afternoon and evening of music at a local outdoor restaurant. Restaurants donated food to nourish the volunteers, which subsidized the costs of the event. Grant funding helped offset the costs of entertainment, which had been generously offered for reduced rates. Artists set up booths to demonstrate and sell their wares, and there was a raffle of donated local art as a door prize. The day was enriched by a bicycle tour of local renewable energy sites, as well as several teams sweeping the roadsides searching for debris.

The following year there was less interest in the river cleanup. More of the focus and intention was invested in the music festival–type event. While this was disappointing to our organization, we chose to focus on other projects. In 2012 we were contacted by a revitalized Baker River Watershed Association president who was interested in organizing another river cleanup. We reconnected the association with the organization that had come to be known as BRAMF (Baker River Arts and Music Festival) and have recommitted as partners to organize a day for cleanup and culture in the Baker River Valley.

Plymouth State University has an environmental and social justice student organization. Over the years we have helped the group organize their meetings and increase their organizational capacity. We support them by participating in such club events as Earth Jam and Apple Fest, along with providing our facilities for Common Ground organizational retreats. By collaborating with this organization through the years we have developed, by extension, a familiarity with the student body.

We have also contributed to the physical infrastructure by providing materials and labor for the facilities at the university EcoHouse. By offering assistance in construction of a compost facility and garden beds at the EcoHouse we are able to share our skills and resources. We have also offered materials and assistance with construction of shelving and furniture for the EcoHouse. These projects allow the students to contribute to the design and construction process.

In seeking collaborations with the local community it is important to consider existing organizations or pools of the populations to be served. Churches and schools, for example, provide an existing network for information sharing. To engage the student populations we approach faculty in the arts and sciences departments who know of students interested in our work. Pastors and parole officers provide possible additional routes for engaging with the public as a community organization. There are also senior enrichment programs that are designed to retrain and reengage the elderly in the workforce. These programs are additional opportunities to educate and be immersed in the broad fabric of the community.

Experimentation with Academic Partners

Through the years we have considered the concept of partnering with academic institutions. There has been discussion of teaching a college class and incorporating students in an experience that combined campus and on-site sessions into a semester-long format. College students have also arranged independent study credit for their experiences at D Acres, and we have considered offering an on-site intensive that would give credit equivalent to a full-time academic course load.

We have also experimented to develop research projects with institutional and community partners. On one occasion our facility was designated as a weather data collection site by a meteorologist at Plymouth State University. In addition to sensors that collected high and low temperatures we also had access to wind and solar radiation data. By placing the sensors strategically throughout the property, we were able to accurately monitor greenhouse temperatures and our wind power potential.

Once the weather data was accessible we conjured up an experiment that combined this data with the performance of our solar hot water systems, as discussed in chapter eleven. We positioned two different domestic hot water heating systems on the roof of the Community Building. By monitoring the energy produced by each panel in correspondence with the weather conditions, we were able to accurately compare the efficacy of the competing technologies.

While the information obtained from these academic collaborations was notable and the weather sensors on the farm provided insight into our environment, the experiment is still unresolved. Some of the data is skewed because our woodstove operations fluctuate the temperature of the input temperature differential and thus the energy generation necessary from the panels. Since the equipment is still in operation and performance could degrade with time, we hope to continue this analysis into the future.

Regarding the collaborative process, it is important to define the objectives and establish responsibilities at the onset of the partnership. Often the process of grant writing, with an examination of the goals, implementation strategy, and budget, provides insight that can assist with addressing the necessary concerns of a group project. Also it is important to properly match the goals of the academicians with the priorities of the farm operations. There is an intellectual disconnect between our purposes in implementing an evolving system and experimentation in academia. At times we felt manipulated—as pawns being used to meet the tenure or résumé goals of others.

Art Cooperatives

As a part of our cottage industry endeavors, and as supporters of the arts and cooperatives, we have participated in several art cooperative efforts in the area. When Artistic Roots was founded in 2004 we were part of the opening day roster of artists. Ten years later D Acres is the only member that has been active throughout the entire period. I have served the cooperative by directing marketing and organizing and facilitating meetings for the membership.

In Canaan I served as president of the Cardigan Mountain Art Association. This artisan co-op gallery formed in a burst of creative enthusiasm. Unfortunately this robust and ambitious start was short-lived, as the initial leadership dissolved in acrimony, leaving a void in organizational oversight. At that juncture, in desperation, the co-op members solicited the general public, seeking someone to help them organize meetings and develop structures to meet their objectives. My principle role was to facilitate meetings to ensure clear and concise process and to maintain fiduciary responsibility. Under my leadership the organization attained nonprofit status, grew the board of directors, and provided educational services to the region.

These art co-ops provide an excellent opportunity to reach horizontally into the community and pool creative talent and ability. In addition to the actual sales that we have made through these retail venues, we have also profited from the exposure. As the public passes through these storefronts they have the opportunity to view our display and learn about the organization from the gallery location. We also have gained from our experience as participants of a cooperative with a membership diverse in life experience and artistic training. While balancing the variations of profit and service motivations among the members has been challenging, we have benefited from this immersion in cooperative business and service efforts.

Speaking Engagements and Presentations

One of my first forays into public speaking was as a presenter at the first Northeast Organic Farming Association of New Hampshire (NOFA-NH) Winter Conference. While questions from the audience allowed for improvisational moments, it was important for me to come prepared with talking points on the subject material so that pertinent information was provided in a sequential and rational format. Prepared notes reliably forced the presentation to deal with the topic at hand instead of rambling around the topic.

Since that first presentation, I and other D Acres staff members have offered over one hundred such talks throughout the region on topics of agriculture, sustainability, health, and community. By attending speaking engagements in the community we offer the public opportunities to interact with our staff and ask questions; in turn, we are able to assess the interests and knowledge base of the community. We have spoken at garden clubs, civic organizations, and college classrooms, recording our presentations to broaden their potential exposure via the Internet. Recently I was invited to be the commencement speaker at Sterling College, a progressive leader in experiential higher education. I am honored to have been asked to address the graduating class of this institution.

Internet and Video

Using the Internet, we are able to reach out to the general public with information pertaining to our activities and operations from 218 Streeter Road. Documents, including the Organizational Manual and Project & Goals, include broad statements of vision and purpose, as well as descriptions of the minutiae of day-to-day life and are freely accessed through the website. The site serves as a repository of information so that other organizations may use it as models for their own organizational methodology. In addition we have archived articles and workshop handouts that have been produced over the years on a range of topics, including construction, forestry, gardening, and community living.

We have expanded our outreach capacity through the years by experimenting with two technological phenomena: desktop publishing and social media. Desktop publishing has allowed us to readily produce, in-house, such materials as annual appeals

and reports, as well as professional workshop handouts. Social media provides another avenue to organize and promote events and share educational articles and videos.

To a degree our capacity to serve the public has been enhanced by the Internet and advances in video production. For this technology to function for the benefit of the organization requires a design team with aesthetic and creative instincts as well as computer savvy.

Our website has provided a valuable service as a portal to the local and global constituency. The website was initially devised by Monika Chas during a winter residency in the winter of 2000. Her vision of a clear format with an appealing color scheme provided the basis from which we could regularly update the site. In 2012 resident Matt Palo upgraded the website and transferred the archived material to a new platform.

Our on-site video production has advanced with assistance from the local cable access channel. The cable channel lent equipment for the purpose of recording and producing cultural and educational videos for the general public. Our first efforts were mostly documentation of large events, such as conferences or public presentations. Over time we began recording the Sunday Farm Feast tours at D Acres. With the patient guidance of Jamie Capach at PB-Cam, the tour footage became the basis for a short film entitled *What Is Permaculture?*, which portrayed the food preservation, animal husbandry, gardening, construction, and cottage industry techniques practiced on-site. The movie was shown on the local public access channel, and I also traveled to eleven locations, including college campuses, public libraries, agricultural conferences, and fairs, throughout Maine, Vermont, and New Hampshire to show the film. After each showing we allocated time for question-and-answer sessions intended to inspire a solutions-based dialogue.

During initial phases of working with PB-Cam we opened Vimeo and YouTube accounts. These video-sharing platforms allowed our video productions to be accessed through the Internet by the general public. The purpose is to provide practical, hands-on content to a global audience. Over the last year we have begun adding video projects to the YouTube account on a consistent basis. A couple of times a month we will produce videos documenting farm activities, a tour, or a guest presentation. By regularly producing these videos we capture seasonal activities that can be viewed throughout the year. On occasion we have also taped particular events such as lectures or dances to be submitted as programming on the public access channel. This provides a platform for interaction with a local as well as a global audience. In-house production helps us retain control of the substance and character of the message portrayed through the media.

We have also been documented by visiting filmmakers. The first film produced was part of a college class project by a summer intern. Jessie Richardson made a feature length film that delved into the roots of our motivation to seek this lifestyle. Several years later George Packard of Parrot Creek Productions arrived to shoot over eighty hours of footage for a documentary titled *The Root of Food* (currently in production). These projects have increased our familiarity with the filmmaking process as we continue to evaluate the best methodology to reach out to the public.

CHAPTER EIGHTEEN

COTTAGE INDUSTRY

The term "cottage industry" evokes images of the feudal era in Europe and the beginnings of the industrial age. Before production was centralized into factories during the industrial revolution, individual homesteads would manufacture and assemble various parts of a product produced for retail sales. This home-scale process was termed cottage industry.

In the modern era the term cottage industry has many of the same connotations, referring to goods and services produced from home with locally available resources and qualified personnel. To produce a functional product of value with reasonable pricing requires skill and expertise. Especially in our era of cheap, disposable, factory-made products, cottage industry products must stand out in quality, as assessed by price, durability, functionality, and factors such as aesthetics and local appeal. Value can also be assessed based on availability and the degree of expertise it took to produce.

To fully promote cottage industry products consumers must be educated to distinguish cottage industry merchandise from its mass-produced competition. While many people are aware of the problems created by the cheap, disposable products ubiquitous in the global market, alternatives need to be adequately marketed. Educating the public regarding the methodology of a home-scale operation provides an understanding of the process and invites appreciation. Highlighting the people involved and

the place of production allows the consumer to feel a profound connection to the merchandise.

The appeal of cottage industry products is the result not only of the craftsmanship and the characters involved in production but also the sense of place that they evoke. This sense of place is appealing to local area residents as well as visitors who become nostalgic about their time in the area, so products that contain familiar flora, fauna, vernacular, and landscapes can be particularly sentimental for the purchaser or recipient of gifts. Often products in our region are embellished with the likeness of moose, bear, or the "Old Man in the Mountain," a scenic rock formation and immortal image of the Granite State.

Why Cottage Industry?

To generate revenue from home-crafted goods and services derived from natural resources and community skills requires an element of specialization. This differentiates a cottage industry from corporate competition. It is not the price that determines value but rather qualities such as durability, aesthetics, and the environmental footprint of the product.

Within the natural system there are surpluses and shortages. Cottage industry is a means to store surpluses in intellectual, physical, and natural resources through investment in time. This initial investment has the potential to provide security in times of shortages through trade or consumption. Success requires addressing the economics. Ultimately we must provide a margin of dividends that outweigh the cost on a balance sheet. In real terms a low-cost raw ingredient can be value-added with ingenious creativity. With patience and perseverance a systematic strategy of implementation can be constructed to add value to available resources.

Value is determined by the purchaser. Goods must be marketed appropriately to increase the appeal to the consumer with an emphasis on local character, integrity, and handmade value as well as the principles of a localized economy. There is no viable product without a viable market. Exposure is important, although the time spent marketing and promoting must also be considered as a cost of production.

Ultimately, as a farm and an organization, dividends from the work, materials, and time investment must be constantly evaluated. What is accomplished through the production of a given cottage industry? What value does it bring to the farm organism? There is a broad range of metrics utilized to evaluate the cottage industry production. Is this product valued as a commodity, a necessity, or a luxury or for the quality of life it brings? The goal is to provide the greatest dividends, based on organizational priority, with the least amount of investment.

Cottage industries provide a platform to grasp the complexity of a functioning natural system. There is ample evidence of natural cycles and feedback loops that can be identified and whose study refines the process. From conception to point of sale and beyond, the process is complex and interrelated. There are also exponential benefits attributed to this farm's industry. The various programs in combination build a momentum of interest and enthusiasm that is compelling and inspirational to the public, thus accomplishing the mission of cross-pollination through various product lines and services. This, in turn, provides multiple avenues to educate and collect revenue and ultimately for organizational perpetuation.

As you can see, traditional farming is not our objective. We do not wish to be a market farm. We are unsatisfied with the commodity market farm model and are seeking other viable opportunities for modern rural life. Our objective is to demonstrate a model for perpetuation on this rugged landscape. Rebuilding a handshake economy requires thoughtful conscientious choices. Our goal is to present models and offer alternatives to generate that type of localized economy.

Elements of Successful Cottage Industry

The three fundamental elements of cottage industry are the product, the producer, and the process. Within the D Acres organization an additional element of this process is the educational component, to inspire similar ventures within the region and offer the opportunity for individuals to practice this entrepreneurial experiment. By offering workshops and presentations we share actual, practical skills and ideas for motivated individuals. For instance, we have offered frame-making classes to the public so local artists could begin producing their own distinctive frames to enhance their paintings or photography.

Determining what goods and services to produce is often part of a natural evolution. A specific good or service is initially produced to meet the needs of the working farm. After successful experimentation with a prototype of a product for application on-site, the product is produced in sufficient quantities to develop an inventory and explore outlets for sale.

We encourage specialization by personnel in skills about which they are passionate. By developing expertise in a craft, workers devote positive energy and enthusiasm into production. This passion also drives the individual to continually refine his or her talents and exert creative energy in the process.

The implementation of a cottage industry program requires attention to various phases from conception to sale. Cottage industry builds skills and experience in the process of craftsmanship. To master the processes of creating value requires practice and dedication. In addition, marketing requires an implemented plan with associated costs and an unpredictable level of success.

In the production of cottage industry merchandise there must be a balance between the time and skills invested by an individual and a price point that can guarantee sales. The value of the raw materials utilized in the production process must also be reconciled. What can unskilled personnel produce with low-value raw material that will be desirable in the open market? Unsold inventory represents an investment without yielding a profit.

The perishability of the goods is also a concern at all phases of production. For example, wood that has been sawn and stored under cover from rain is usable for many years, while ripe raspberries must be processed within hours. The time and expense that is incurred to ensure the materials are safeguarded must also be considered.

To realize success economically in a cottage industry endeavor requires a combination of several factors. While the raw materials and personnel are essential for the actual production, marketing and points of sale are also crucial for consistent purchases, as is the quality of the product. Material selection, uniqueness, design, and manufacturing efficiencies all contribute toward building a viable process. In addition, cottage industry can be responsive to trends of popularity that determine what merchandise is produced.

Although branding is a common advertising ploy, we have used logos, product placement, and conventional advertising strategies to spread a concept rather than to sell a product. The organizational priority is not to sell more products but rather to spread our philosophies to a wider audience. While sharing information is the goal, we are using the conventional routes of production to market alternatives.

It is prudent to produce items with various price points to match the consumers' buying preferences.

Low-cost items priced less than ten dollars can satisfy impulse buyers as keepsakes. Ten to twenty-five dollar items are ideal for gifts for friends and acquaintances, while twenty-five to one hundred dollar items are gift material for family and loved ones. Items above a hundred dollars are more likely produced for collectors and to satisfy custom orders. Reproducing prints and cards from original artworks increases exposure while continuing to add value to the original production. These pricing scenarios are based on experience at the cooperative retail outlets in Canaan and Plymouth, where the customer base is split between local and seasonal tourist clientele.

While specific products may become mainstays and pillars of the production line, it is important not to burn out the staff's creativity and enthusiasm through overproduction of any specific product. Ideal products are those that build skills yet can be replicated simply and routinely, while also allowing for creative investment to combat the boredom of repetitive fabrication. The goal is to meet orders and production goals while still retaining positive energy for the work. The popularity of a single product can be negative for people forced into incessant monotonous fabrication. Also, dependence on a specific product may be foolhardy in the evolving marketplace.

The drawbacks of dependence on a specific product have been illuminated by observing other operations. When we first came to live in Dorchester, we met a woodworker who had developed a line of mail-order products. The pet bowl stands he produced became very popular, and he focused effort and energy on scaling up the production and hired an assistant. Unfortunately, imported, low-cost pet bowl stands eventually flooded the market, and the enterprise collapsed. On a larger scale hammock production at the Twin Oaks community in Virginia was a mainstay of their economic system for many years. The production was ideal for individuals to self-determine their income-generation needs and pace of production. The community enjoyed a bloom of activity and flourished in the midst of a strong economy. Unfortunately the narrow scope of this nondiversified economic system was unbalanced. When their national retailer abruptly chose not to renew the contract with the community and outsourced production to a foreign country, the community was shocked into facing a more balanced alternative economy.

Recognizing this lesson, we therefore maintain a varied and evolving income stream from multiple cottage industry enterprises. Based on our natural resources and the fluidity of skilled personnel and being responsive to both farm demands and community needs, the following cottage industries highlight the breadth of activities undertaken within the D Acres organization.

Woodwork

The woodworking program grew from tradition, personal interest, natural resource availability, and necessity. When we arrived at the farm we inherited a woodshop filled with Uncle Delbert's tools. He had spent considerable time refurbishing antiques and creating reproductions of Shaker furniture, as well as the cabinetry for the homestead. This legacy dovetailed with my personal interest in building and woodcrafting. The heavily wooded property and culture of the Northern Forest accentuated the possibilities for a value-added wood-based economy.

We spent our first fall at the farm doing inventory and cleaning Delbert's shop space in the two-car garage. Soon thereafter, because of the threat of rodents, it was necessary to begin constructing cabinetry for the rudimentary kitchen in the Red Barn. At that point we began to evaluate the limitations of Delbert's shop space. The heating system of the detached building consisted of a decrepit woodstove, which, while effective, presented a fire danger and was not convenient for intermittent usage. The shop was crowded, and in fact, access was so limited that Delbert had occasionally been forced to remove a portion of the wall to bring larger items in and out.

Dust was also a major health concern in his space, as there was neither a collection system nor adequate ventilation, particularly in winter.

Another factor that led to our wood-based cottage industry was my father's lifelong interest in woodworking. His vision was to retire from work in the hospital to spend his golden years as a woodworker. As the concept of our community multiplex began to manifest, a new woodworking space became a priority.

The woodshop was designed to serve the needs of the farm as well as the cottage industry enterprise. Two large doors that open into the garage space assure access for large and small projects. The garage houses the dust collector, which is piped to various portals throughout the shop. The inclusion of the shop in the Community Building also provides for consistent room temperatures and convenient access to the living space. The combination of accessibility and dust collection provides convenience, cleanliness, and air quality that were unobtainable in Delbert's shop. This upgrade in infrastructure propelled the possibilities for the woodworking program at the farm.

As soon as was feasible during the construction process of the structure we began to use the woodshop facility to complete the building as well as meet the farm's needs. We utilized the shop to construct cabinetry units and trim throughout. Once the house was occupied our activities in the shop began to shift more toward products for sale.

Simultaneously, our forestry program gained in expertise and capacity. We began harvesting wood with the oxen to be milled into planks by Jay Legg. We also began collecting character wood and unique roundwood for varied uses, including walking sticks, tool handles, and furniture. In this manner we were enabled with abundant material and infrastructure.

Our woodshop production, however, was limited without dedicated, skilled personnel. We were therefore particularly fortunate to host Sam Payton as the resident woodworker for several years. Sam had recently graduated with an undergraduate

Sam Payton hosting a workshop on rustic furniture making in the woodshop

degree in computer design and graphics but had spent his youth under the tutelage of his father, who is a master craftsman. Sam used his variety of skills to create unique furniture and woodcrafts during his stay. We marketed the woodwork at farmers' markets, on-site, and through the Artistic Roots co-op.

In addition to providing products for sale Sam led workshops for the community that focused on various aspects of the woodshop. These sessions offered hands-on experience to the public and on several occasions resulted in commissioned work for Sam. The organization benefited from the prestige of hosting a high-caliber woodworker along with the publicity and revenue that resulted.

The process of hosting Sam provided other benefits as well. Sam was given creative license to pursue a wide range of products and techniques. This open studio expanded Sam's knowledge and allowed him the space to realize imagined designs. In addition, this process was valuable for Sam because it helped solidify his choice of vocation. He chose to depart D Acres in pursuit of formal training as a violin maker.

Sam's presence here also extended our capacity to meet the learning goals for D Acres' on-site participants. Interested residents rotated through tasks to assist with production and attain hands-on experience in the shop.

Following Sam's departure I have been the primary woodworker at D Acres for several years.

During the winter months I spend time making spoons, bowls, furniture, and birch-bark frames. My woodcraft products have emerged as a result of my carpentry experience, sharing ideas with Sam, a lathe class, and, most importantly the time, materials and space to practice.

The first products I produced were kitchen utensils and birch-bark picture frames. I still have the first primitive spoon produced on-site. Without tutelage the process lacked time-saving and aesthetic refinements, and we spent eternal hours with laborious hand sanding. I began making the spoons because we needed some for the kitchen, and they seemed like a good fit as a commodity to begin producing for sale. We had the tools and abundant materials. While the spoons continue to be labor intensive, we have made several improvements in the fabrication process by building our skills and techniques and by acquiring specific woodworking tools. With seasons of experience we have dramatically improved the onerous time spent sanding the wood and improved the quality of the product as well as our potential profit margin.

Over the years we have produced thousands of spoons. As the quality and efficiency of production increased we began to market them to a wider audience. Today they are sold in an array of venues ranging from retail stores to farmers' markets and health food stores. The spoons are durable and should last for a lifetime, providing a continuous reminder of the spirit and philosophy at D Acres.

Spoons are an ideal cottage craft because they have a relatively high value in a small volume. They are easily transported and are not perishable, nor

An inventory of the unique variety of spoons for sale

do they require a conditioned space for storage. In an ideal fashion the spoons invoke nature through a handcrafted product that has purpose and utility in our daily lives.

Birch-bark frames are another signature item at D Acres. We originally began producing them for the Community Building to help improve the aesthetics of our mandatory postings, posters, artwork, and photos. The frames are relatively simple to produce but provide a unique and natural appearance. To achieve the effect birch bark is laminated onto a rough wooden frame. Once the bark is adhered the wood is covered so that the bark itself frames the image. We produce the frames to custom sizes for artwork and mirrors. They are a convenient retail product for us because they can be hung out of the way on the walls to complement the overall aesthetic of the Community Building until sold or simply serve as advertisement for custom orders.

While the birch is relatively easy for us to preserve when we are chopping wood, the material is novel and unique for many people visiting the region. The character of the material complements the rustic charm of the typical furnishings of log cabin–type decor. By highlighting the beauty of this material we hope to bring natural flair back within sight of people's daily lives.

Eating vessels are an essential part of our daily rhythm. The uniqueness and natural character of wooden bowls stand in vivid juxtaposition to the composite and plastic materials common in our lives today. Work on the lathe requires specific tools and training, though information is readily available regarding these ancient, fundamental techniques of shaping wood. To become comfortable on the lathe my father and I took a weekend course at a crafts school in Maine. When we returned home we tooled up and began bowl production. We turn the wood while still green to attain the rough shape, then store it to dry for nine to twelve months. The process of drying and twice turning the bowls inhibits the tendency of

the wood to crack as it dries. The bowls are then carved to final dimensions, sanded, finished, and marketed. There is merit to the delayed satisfaction of returning and finishing the project, a process that has not been expedited by technology and must be accomplished via patience and persistence.

Wooden hearts became a staple of our inventory through experimentation. During a period of inclement weather in which we had an abundance of personnel interested in woodworking, I attempted to conjure a product that we could assemble as a group. We had an excess of slab wood from milling, and I began to experiment with cutting out hearts and peace signs from these flat pieces. The peace signs are labor intensive, requiring careful cutting and shaping and tedious sanding to produce a saleable product. While the peace signs are an extremely popular item, the hearts are even more universally successful. The hearts can be cut with less precision and more artistic interpretation. In addition to the message they impart, the hearts highlight the character of the wood from which they were carved. While the process of shaping each heart can be monotonous, these wall hangings routinely fetch between twenty-five and seventy-five dollars, for far less than an hour's work.

It is interesting to note that on the day we started to fabricate hearts, there was a revolt in the woodshop. One of the participants who had some woodworking experience termed the work "whimsical" and compared producing the heart shapes to cartoonish buffoonery. He imagined a custom order of a dining room or bedroom set. Unfortunately, while those projects are cherished by woodworkers, at that juncture we did not have the personnel, the materials, the time frame, or the commission for beginning that type of project. While this individual was disappointed by the lack of high-end furniture production that epitomizes typical woodworking aspirations, the hearts served to build our collective skills by utilizing a waste product and also garnering revenue.

The display at Artistic Roots in Plymouth: hearts, spoons, frames, bookshelves, cards, and blacksmithed items

The wood products are an educational tool. From the forest where the wood is harvested, then transported for milling and drying, to the showroom floor where it is sold, the process involves many steps and corresponding skills and knowledge. This woodworking program provides hands-on experience in not only the cutting, sanding, and finishing techniques that will create an exceptional product but also the process of pricing and distribution necessary to get a product to the point of sale. The wood is educational for those that fabricate the products and for those who purchase the goods. The purchaser is rewarded with a fuller understanding of the natural resources and a conception of what it means to have a localized, people-powered economy.

In addition to Sam's matriculation as a violin maker, many other individuals have followed a similar journey in which their skills in woodworking have blossomed based in part on time spent at D Acres. My own growth as a woodworker comes from having the on-site resources available. By providing the training, tools, materials, and direction we have stoked the fire of a wood-based natural resource economy. When I arrived at the farm my experience was limited to rough framing techniques. Over time my work has progressed through experience using my hands and increased knowledge of materials and tools. As my skills grew I was able to produce a refined aesthetic in my work. Consequently, my capacity to earn revenue in the woodshop has grown.

Sanding a bookshelf before finishing and moving it into the house library

An additional component of the educational process in the woodshop is our electrical generation capacity. The woodshop tools are partially powered by photovoltaic panels. In the future we are hoping to further enhance the program by producing power directly for woodworking applications by integrating wind and water, as well as human and energy applications.

It is important to highlight the minimal display needs of these products. Wall space and simple shelving adequately market these goods to visitors. In contrast, the space required to transport and display furniture is restrictive. Damages such as scratches can be difficult to repair far from the woodshop. The economics of showrooms adds considerably to the retail cost of the product. Store owners must pass along commission costs to the purchasers, which inflates the retail price. Rather than work to find retail space in a furniture store and absorb the commission costs, we have chosen

Putting together the rustic bed frame for the Green Room

to try to market products on-site. While roundwood furniture in Manhattan has a greater saleable value than in Dorchester, our hope is to continue to inhabit the Community Building with furniture to serve as representatives of the style and attract custom orders.

Metalwork

Blacksmithing and other forms of metalworking fill several niches as an integral part of our community economy. The metalworker has a fundamental role as a fabricator and repair specialist for the many critical components constructed for our day-to-day lives in agricultural operations. In addition to on-site necessities the metalworker has traditionally

offered services to the community at large. The type of work can range from fixing a small component to creating a custom product.

At D Acres blacksmithing skills are indispensable assets for our remote, frugal existence. In many instances on-site repair of a manufactured product can save countless hours of transportation and repair costs.

There is also a significant history in the art of blacksmithing from which we have inherited tools and infrastructure. My uncle had built a blacksmith shop area in the lower half of the two-car garage. The space was ideal for small projects and farm-related repair work. He built many of the hinges, hooks, and hardware that decorate the homestead with functional, enduring art. As we considered the

Joe Vachon demonstrates the craft at the Lebanon Farmers' Market.

future, fire safety and limited space in this lower half of the building made the room unsuitable for classes or daily operation. To address this concern we utilized a portable forge and moved the shop first to a tepee and then into the spacious garage space in the upper section of the building.

Though Sam came to D Acres with metalworking skills and offered an occasional workshop to the public, our first full-time metalworker was Joe Vachon. Joe arrived fresh from college training in the arts, through which he had been exposed to the power of the forge. His passion for metalworking and primitive skills allowed him to focus on developing his craftsmanship. Over the course of several years Joe's skills progressed from basic to an advanced level of proficiency. While his first

products were predominately hooks, door handles, and bottle openers, his skills progressed to encompass ornate chandeliers, railings, and fireplace kits. Not only did the functionality improve but also the quality of the design as he added refinements in the shape and ornamental characteristics.

Joe not only progressed as an individual craftsman but also enhanced D Acres' educational component. He was able to offer regular workshops and private tutoring. In addition, he found the time and energy to offer public demonstrations at a plethora of summer events. By attending craft fairs and music festivals he could educate the public by demonstrating his craft and spreading the word about the D Acres project. His presence in the public eye helped attract and create an area guild that gathered

There is a long history of metal-working in in Delbert's Smithy.

regularly to share skills and ideas regarding their mutual interest in metalwork.

Since Joe's departure we have continued offering occasional workshops by partnering with area blacksmiths. While the workshop space has been maintained and is utilized for farm and community projects, we do not offer an inventory of products for sale or advertise work on commission at this point. This serves as evidence that, in addition to infrastructure, resources, and viable markets a functioning cottage enterprise requires trained personnel.

Herbal Products

Herbs grow prolifically on this landscape. In contrast to demands of typical annual crops, which require heavy inputs of compost and fertility, herbaceous species thrive in our soil.

Drying herbs for culinary and tea blends provides an excellent opportunity for a cottage industry. This industry dovetails with the needs of a typical homestead. Because herbs are a household necessity for cooking, daily beverages, and medicinal purposes, they are fundamental to health on the farm. By practicing the arts of harvest and preservation for our household purposes we have become proficient in expanding the production process to make the product available to the public. While our main focus is to meet the needs of the on-site residents, an abundant harvest is a valued commodity.

Often the teas and herbs that we dry are vigorous plants in which cultivation stacks functionality in the farm system. Plants such as thyme, mint, and oregano provide flavor along with flowers that feed our pollinator population and sustenance for birds and mammals. By cultivating fresh herbs on-site we enhance the quality of the food products that we offer the public, and the herbal tea serves as an invitation for conversation and relaxation for the many visitors to the farm. In addition to direct consumption, herbs have the potential to be harvested for tinctures and aromatic effects. There are also herbs such as raspberry and nettle that provide fresh food products, as well as foliage for drying.

The relative ease of the production of tea means that it does not compete with other high-priority seasonal tasks. Another advantage of the process is that all the facets of production can be accomplished by supervised groups of volunteers. The initial stages of cutting and hanging the herbs can be accomplished in short order. Herbs are generally cut at the base and hung upside down via bailing twine in the hayloft of the Red Barn. During sunny days even into the fall the metal roof of the barn heats the upstairs of the building to well over 90 degrees, and the porous envelope of the building allows air to ventilate the structure. After between five and fourteen days, depending on the type of herb, temperatures, humidity, and solar incidence, herbs are dried to leathery consistency. Evaluating the preservation method to achieve the highest quality with the longest shelf life is essential. If the herbs are overdried to the point they can be easily crushed into a fine powder, they may lose flavor and nutritional benefits. If the herbs are not dried sufficiently they will mold in their storage containers. After drying, the herbs are packed into airtight glass jars and stored in a location without sunlight exposure.

Packing the jars can be accomplished during rainy days and evening sessions during the growing season to maximize outdoor work time management. Once stored, the herbs can be blended fresh to meet orders as they come and supply our own needs throughout the year.

The recipe potential for herbal products is limited only by creativity. While a long shelf life, abundant raw materials, and demonstrable community need all create a viable market for a home-based herb business, there is significant competition from international and corporate producers who aim to control herbal access. To compete with the low price points of imported and mass-produced herbal products, small producers must emphasize their freshness and quality along with the true economics of localized herb production.

Our herbal tea utilizes cultivated and wildcrafted raw materials to be blended into a healthy and beautiful product.

There are many applications for herbal products to replace the synthesized chemicals polluting our daily lives. Herbs can be used in personal hygiene products as well as cosmetics, fragrances, and perfumes. Herbs are also valued for their medicinal and tonic attributes, which aid in human wellness, and can be utilized to produce cleaning products for the home and pest and weed inhibitors for the garden. By relying on natural herbal products instead of synthetic chemicals in our daily lives, we can avoid contributing to the crisis.

Baked Goods

For several seasons we experimented with a bakery as a value-adding operation. We initially designed the commercial kitchen with an eye toward cottage industry food production. The bakery concept was appealing for several reasons. While food regulations are typically onerous, they are less restrictive for this type of product when compared to those for canned or refrigerated items, so complying was simple. The production process, quality of results, and shelf life were also fairly predictable. Baked goodies could be prepared with a specified recipe as part of a routine within a predictable time frame. The products were made with all organic ingredients that we sourced through our food order. In addition to bread, we concocted several recipes for sweet breads as well as nutritional and energetically rich snack bars.

Once the recipes had been created we began experimenting with portion size and packaging options. We also began observing the time the product could be offered before its quality diminished so that we could determine the shelf life. We developed packaging and stickers that listed the ingredients and offered information about the nonprofit organization. Once the products, recipes, and packaging had been perfected we approached various Main Street businesses, soliciting them to retail the products.

We offered samples at the farmers' market and in front of various Main Street businesses and began deliveries on Mondays and Thursdays. This schedule allowed us to exchange any unsold products with freshly baked goods. The public was excited about the organic option that we provided, and the products generated substantial dialogue in the community. D Acres interns benefited from the commercial kitchen and small business operation experience. We all enjoyed the flavor of our "day-old" returns as a snack at the farm. Deliveries and pickups in the Plymouth area were combined with necessary errands and supplied the opportunity to connect with the community at large.

But there were several drawbacks from the program that eventually lead to its cessation. While we had grand intentions, our farm food system was unable to supply a majority of the ingredients, so we were forced to procure the ingredients from a regional supplier. This compromised our commitment to the localization of the food system. This process magnified the need to emphasize a seasonal diet and develop the preservation and storage capacity necessary to revitalize our region's food network. Though sales were promising, profits were marginal, and scaling up would have required additional investment and commitment. Without a specific individual with the passion and motivation to operate the bakery the weekly grind of fulfilling our commitments through four seasons became onerous.

In addition, production of value-added food products is complicated by several factors. The market is highly competitive, with cheap, mass-produced food products readily available and extensively promoted. Fresh food also has a limited shelf life, and the perishable nature of the merchandise makes production a risky venture. Food choices change with fads and diets, so producers must respond to evolving tastes and preferences while maintaining the consistency consumers expect.

As we closed the doors of the bakery business we realized the lessons of the endeavor and vowed to refine our efforts, focusing on connecting and strengthening the local food system with our organizational enterprise.

Plants

Plant nurseries are a common crossover market for traditional farm industries. Conventional operations import plant stock from distant locales via tractor-trailers. At D Acres we are interested in propagating and proliferating species and varieties that have been successful for us on-site, and in sharing with people in this climate the genetics that have been effective permaculture plants for us in this climate. To operate such a business requires planning and knowledgeable personnel to implement, but we have innumerable opportunities to propagate plants on-site. Be it seed saving, grafting, division, or layering, we have the mother plants!

We have been successful with several nursery plant ventures. When we have had an excess of a particular annual in the spring we have sold them at food events or the farmers' market. In addition, we have potted up various herbs and shrubs that needed division or were encroaching on others. At times we maintained a small nursery of potted plants, though weeding and watering were difficult to ensure without interested and dedicated personnel. These prolific plants have survived despite our consistent neglect. Maintaining soil moisture is critical, but the small volume of soil in pots makes this difficult. Our most successful method thus far has been to pile wood chips around pots to maintain moisture in the partial shade under the solar panel

Nursery stock waiting to be planted out or sold

by the driveway. This location provided high visibility for daily monitoring of moisture levels, access to water, and a convenient point of sale for visitors.

Design Consultation

We have experimented with a variety of cottage industries based on the skills and interests of our personnel, our on-farm needs, and marketability. There are several genres of products and services for which we have sufficient capacity in terms of experience, infrastructure, and raw materials. Food products are an obvious option, given our ability to produce raw ingredients and having the infrastructure of a commercial kitchen to craft a value-added product. Our experience with gardening and

designing the landscape in this region, along with our on-site plant inventory, makes it natural to offer those services to the public.

But while landscaping and design work could also offer the opportunity to share our knowledge, we have been reluctant to offer these services professionally. Personally I am philosophically challenged by the concept of designing systems for others. For success these systems require significant perpetual evaluation and adjustments as well as participants who develop their own profound understanding and ownership of the evolving design process. The purpose of the system itself requires an immersion into the landscape that requires physical investment and a wholehearted attachment to place and commitment to its perpetuation. It is important that the participants in a design are familiar with the food and flora they are cultivating in their landscape. Transformation of infrastructure and landscape is fruitless without a correlating spiritual and intellectual metamorphosis. In essence the design springs from the participants through a regime of study and hands-on practice in a place. The process of professional design and landscaping services manifests in climate-controlled, armchair environmentalism.

The dissonance created by a perpetuation of the design consultant paradigm is akin to the guru mentality common in our society. Our imaginations are attracted to absolute solutions, perfect scenarios, and miracle elixirs that answer all our problems. The common response to challenges in our society is to seek a specialist's expertise to resolve our dilemmas. The design work promotes the ethos of specialization while we desperately need generalist implementation. In the case of perpetuating culture there is a belief that a designer is paid to provide answers. For me permaculture is a tool for people to create their own solutions. To create a permaculture-based system in which the people are crucial components requires their intimate observation and examination of the system from conception through the process of realization. The choices and decisions affecting the landscape, flora, fauna, and socioeconomic system are obligatory for those who are a part of the design. Without that level of ownership in the process an essential element will be lost by the participants.

Fiber Arts

Another cottage industry with regional tradition and abundant resources is fiber arts. Since the colonial era people have utilized natural fibers and dyes to produce products for decorative and essential purposes. While much of this tradition has been diminished by synthetic and industrially produced products, handmade fiber art is a process worth reviving. Our capacity to produce value-added products has been realized through Regina's dedication and efforts.

Regina arrived at D Acres with experience and equipment. With her passion for traditional techniques we have the ability to process wool from cleaning and carding to a finished product. By maintaining the knowledge and equipment necessary to see this process through completion, we are preserving a skill that may be essential for our future economy.

We have been able to amass several looms and equipment to augment what Edith and Regina have accumulated. There is inexpensive and plentiful raw wool available from the many hobby farms raising sheep, which is not competitively marketable because of the abundance of cheap fiber. By creating a studio space in the basement of Betty and Bill's home we hope to preserve the craft through practice and skill sharing.

While our passion is for the localization of economics, our prioritization is tempered by the contemporary abundance of fiber and clothing that surpasses the current needs of the people in this region. The future of this cottage industry is dependent on investment of time and energy on-site, as well as evolving regional marketability of this craft.

Education

Equally important components of our cottage industry are educational offerings such as classes, private lessons, and events. The Early American decorative art painting classes that my aunt offered on-site, for example, not only provided homesteading skills but also a viable saleable product. To perpetuate that tradition we have offered classes that provide lessons applicable both at home and in commercial production. Once the resources and expertise have been obtained, sharing the information is crucial to strengthen the community at large.

Tourism

Recreation-based tourism is a vital component of our local economy. The winter snow brings snow sports enthusiasts to the region for activities such as snowmobiling, skiing, and hiking. During the summer the lakes fill with boaters, and traffic is steady on the mountainous hiking trails. The fall draws tourists to witness the glory of the foliage.

While we appreciate the spectacular natural resources every day, we also value the possibilities for sharing this resource with visitors to the region. In particular the recreational opportunity of rock climbing attracts a particular niche of the population. While sport rock climbers may be maligned as a subculture with its own unique terminology and dress code, they are a demographic into whose collective conscious we have inserted ourselves.

Our services are based on experience traveling and visioning an operation that would not diminish the work on the land and our community outreach programs. The farm-based accommodations with such amenities as the Community Building and Outdoor Kitchen offer an exciting alternative to the conventional hotel stay. The farm-fresh food is incomparable, and particularly attractive to the upwardly mobile, hip, health-conscious, recreationally obsessed urban dwellers. On their pilgrimage to the country they are greeted with star-filled nights and bird calls, in stark contrast to the electric vistas and human-created audio experience of the city. Aspects such as the solar panels and compost toilets are shared as educational tools for the limited attention span of their visit. Sharing a single fire pit, as well as community living and kitchen space, encourages the visitors to recognize a collective focus beyond conventional individualistic glory.

Regina awaits the arrival of guests for a farm-fresh dinner buffet.

The overnight accommodations program accomplishes multiple goals for the organization. Our guests represent a paying, captive audience for our message. This combination of revenue and education allows us to perpetuate and build our audience. The rock climbers serve as envoys and messengers to a larger segment of the population, disseminating information about the project geographically through their conversations in travels as well as through their careers and familial connections. Thus the ideals and philosophy of the project can be spread exponentially by being an iconic aspect of a subculture.

In addition, the dollars spent on-site provide a portion of sustenance for the farm project. Overnight visitors are prime candidates to enjoy food prepared on-site, which offers a premium for education and income generation. Farm-fresh food served for meals is a value-added commodity in itself. Meal times and quantities for reserved guests are predictable and can be scheduled in an organized manner. By serving fresh meat and vegetables to our guests, we offer a visceral connection to the land, the producers, and the quality of the product. This product fundamentally cannot be replicated and affirms the value of place-based meals.

Instead of the typical model of a commodity farm stand or CSA, we believe that the optimal use of our farm food products is on-site production. This allows our harvest to be predicated on the number of reservations rather than guessing how many people will attend the farmers' market. We are able to recoup a higher premium with less investment by offering prepared foods throughout the seasons from our farm products. The model garners the most value by taking the shortest, most direct route from seed to plate.

In making that connection we also are building support for the organization into the future. Visitors see the care taken to demonstrate and educate sustainability techniques. By clearly stating our nonprofit status and providing information about the opportunities to support D Acres we are planting the seeds of the future. Even if guests are unable to contribute financially to us as members or donors, they grow to see the value of such an organization. The educational and economic elements of the cottage industry program are woven together to create viable natural systems of perpetuation. Once a natural resource base is established to feed the system the ascension of product quality is based on skilled labor and infrastructure to deliver the highest-quality product with minimal cost.

CHAPTER NINETEEN

MARKETING AND PROMOTION

Marketing and promotion are a controversial undertaking. Self-promoting is construed as narcissistic propaganda. This exercise in egotism is expensive and time consuming. Compounding this situation is the difficulty of accurately assessing the efficacy of advertising. We have spent extensive hours crafting marketing materials for the organization, and while we have significant renown in the region there are many local people who still have yet to visit the farm. It is important to recognize that the investment and sacrifice of marketing and promotion endeavors are necessary to accentuate our impact in providing service. Throughout the process it is crucial to evaluate what benefits are reciprocated by each avenue we consider. Accomplishing our goals of community outreach requires multiple mechanisms for connecting and communicating with people.

Our intention is to utilize the channels of media to accomplish our goals of sharing knowledge and providing service. The focus of our efforts depends on the specific audience and the service we are offering. To attract college-aged interns we contacted career services departments in academia with program information. If we are seeking assistance in the kitchen, we contact culinary arts schools or food employment websites.

There is intense competition for people's attention in the media-rich world of North America. We hope to reach a large audience, introducing alternatives to convention that run counter to the prevalent message that encourages consumption. We would like to impact our audience with usable content that will inspire, enhance, and enrich their lives. Be it event promotion or information sharing, our goal is to do so with a manageable outlay of investment. Investments of time, effort, and dollars depend on their potential returns offered for connecting with segments of the population.

D Acres has several distinct yet overlapping audiences. For each of these audiences we develop a particular strategy for outreach. As an example, to encourage attendance at local events there are many avenues of outreach distinct from when we are attempting to reach a wider audience with general information. To promote events on the local level we want ubiquitous public notification that is both inexpensive and effective. The local paper has a section of community events in which space is provided for nonprofits to advertise. Community notice boards and storefront windows are ideal for canvasing with flyers that attract the public to events. Local cable public access channels have opportunities for community nonprofit promotion.

Advertising can be done in many creative ways. Using our truck as a moving billboard certainly caught attention.

Our style of marketing comes from studying the techniques of contemporary advertisements. From the orchestration of political campaigns to the marketing of oil companies in *National Geographic*, our society bears witness to a barrage of convincing messages. By assessing and evaluating the various strategies, and choosing to invest in implementation of our own, we are reluctant participants in this overwhelming assault on our senses. In addition to our own observations, we have benefited tremendously from consultation, assistance, and advice from acquaintances in the professional fields of advertising, design, and marketing.

Our marketing techniques are rooted in the music concert event-promotion styles of my college years, when impromptu, guerrilla-style marketing of events was achieved with minimal printing costs and foot or bike distribution to campus, in student housing, or at locations with heavy college-age foot traffic. Flyers were produced using color copiers at Kinko's with collage or hand-drawn artwork. While early D Acres flyers were similar to this model, we now generate most of our materials with Adobe InDesign and a standard printer. Occasionally woodcut prints augment the artistry of our flyer program.

Curbside promotion in the form of an organized spectacle is a practiced art. My experience with curbside promotion is based on participating with Rainbow Caravan and Bread and Puppet creating parades. By creating a spectacle in the streets we were able to attract an audience to attend a specific show or event. At D Acres we have given this an agricultural twist by offering cooking demonstrations and farm-fresh food samples to passersby.

The side panels of our vehicles have served as a canvas for our promotional undertakings. We have painted billboard-scale murals on the sides of our Fuso delivery vehicle. These murals have advertised specific events and overall messages on sustainable food and community. As we carry on with our work-related activities in the vehicle, this mobile platform provides unique exposure to a broad swath of the populace. On occasion we have also parked the vehicle in a high-traffic area as a temporary advertisement.

While our advertising methodology has similarities to other organizations' and existing models, the marketing of the organizational entity is distinct from most traditional schemes. We market the organizational entity of D Acres to promote sustainability awareness and provide information that introduces alternatives. We are marketing not only products and services but also concepts and philosophy. In our quest to market the concept our practice is to distribute the focus among the many contributors to the project. Instead of the traditional reliance on individual guru experts, our marketing is representative of a coalition of participants. We are marketing the capacity of an organization rather than the expertise of an individual.

Our marketing is designed for several purposes. We would like people to visit the farm and discover the inspirational work that has been accomplished on the grounds. After their initial visit to the property we hope to encourage them to return often for events and to utilize the public facilities. For those unable to join in the on-site opportunities we seek other routes to share information and inspiration. By offering these multiple layers and channels we are creating a recognizable voice within our region.

Logo

Our logo design developed as Monika and I discussed the ideal symbols to represent the organization. We were seeking something that could be easily

The D Acres sun

reproduced, minimized, or enlarged while still being identifiable. The symbol could not be too detailed yet must be distinctive. We spoke of using the image of the sun and incorporating a spiral to indicate the journey of life. She produced several images with subtle variations on the symbol we now use.

The D Acres sun now stands as a recognizable image of the organization. This logo serves as a wordless identifier of D Acres, our work, and our philosophy and transcends a name or use of the English language. Similar to a flag, the logo provides a tool to unify the many complex facets of the overall organization.

Desktop Publishing

The technological shifts of the last twenty-five years have had a definitive effect on our ability to reach a broad audience. The introduction of desktop publishing has allowed us to use on-site skills and infrastructure while incurring minimal costs. We have been able to produce aesthetically pleasing, legible promotional materials in-house. Prior to the introduction of desktop publishing, producing promotional materials of this quality required contracting with professional printing and production services. But while this has been liberating for creative exercise, it also adds complication to our lives. Dependence on the technology and proficient operation of it requires initial and ongoing costs in equipment, time, and skills acquisition.

The era of desktop publishing has alleviated some costs in production, but it means that the responsibility for production quality shifts on-site. A visual production that is boring or similar month to month rapidly loses its appeal and does not generate results. Incorrect information or errors in the text tend to detract from the purpose of the advertisement.

If necessary, it is possible to contract all phases of marketing and promotion. From concept creation to printing and distribution, professional services can be acquired for a price. While we occasionally utilize professional printing services for high volumes and poster-size reproductions, it is an economical and philosophical choice to create our own marketing materials.

Posters and Flyers

Each flyer carries a cost in terms of graphic design and printing. It is important to get the most exposure to the selected audience in a timely fashion

Katie Cristiano greets interested passersby under the agricultural tent at the Maine Organic Farmers and Gardeners Association Common Ground Country Fair.

with minimal cost. Consequently, we have developed lists of locations suitable for our promotions. This includes community boards at supermarkets and public kiosks and available wall space on college campuses. There are also locations with potential traffic at agriculture supply and hardware stores and health services locations. It is important to develop positive relations with the property managers of the locations you are exploiting for promotional purposes. If you utilize window space with downtown merchants, it is important to find ways to show your appreciation and maintain this collaborative relationship.

A typical farmers' market table display with a variety of goods such as woodcrafts, veggies, and baked goods

Flyers are most effective when they have clarity and aesthetic appeal. It is important to clearly state the purpose of the event along with details such as cost, time, location, and contact information. While art and graphics are important, that content should not detract from the overall purpose of the promotional effort. At times special events propel us to print full-size posters with dimensions of 11 by 17 inches. This is an expensive choice, as we do not have the capacity to print larger than 8.5 by 11. Rationalizing this decision requires an event where the return of public education and involvement justifies the cost.

Brochures

Utilizing desktop publishing, we have produced various brochures through the years as an introduction to the organization or various programs. The production content has differed, based on the nature of the subject. Brochures have been designed for informational purposes such as trail maps or as an overview of the organization. These materials provide an educational tool that can be used as another positive public interaction mechanism.

Rack Cards and Business Cards

Rack cards and business cards are ubiquitous marketing mechanisms in this era. We have produced rack cards for distribution through various contractors, using chamber of commerce and state highway rest area kiosks as distribution points. On one occasion we contracted with a distributor that offered a package geared to outdoor recreationists with kiosks in urban sporting goods stores and athletic clubs.

Rack cards are expensive to print and distribute, and our sense is that their importance as a mechanism for attracting participants is dwindling as electronic alternatives dominate the public's attention span. While tourists continue to utilize these sources occasionally, the digitally inclined are quick to seek online alternatives.

Business cards are relatively inexpensive tools for providing contact information. Their size facilitates mobility, though the space is adequate to provide only the essential information. They can be used in transit to encourage a connection during a brief encounter in a restaurant or on a bus ride. They can also be distributed to revive memories for trade show and on-site visitors. Care should be taken in the design, particularly regarding accurate information into the future. While high-volume printing is cheaper per unit, five thousand cards can take years to distribute, and changing the e-mail or phone info on each card is painstaking. In addition to ensuring viable information, make sure the graphic is reasonable. We had a difficult time distributing our first cards, which featured a leafless and seemingly lifeless tree.

Website

The Internet has provided another technologic avenue for sharing information and marketing to the public. During my college experience the World Wide Web that allowed e-mail and websites became a reality. The Internet era has allowed us to develop our rural enterprise, facilitating communication and marketing to a broad audience. The reach and minimal cost of this technology has transcended and surpassed what was possible through communications via phone, snail mail, and print advertisements. When we arrived we set up an office space in the Red Barn and relied on a dial-up modem to communicate via the Internet. As the web exploded with content we began to make plans to create our own D Acres site.

The arrival of freelance extraordinaire Monika Chas in the winter of 2001 initiated the construction of the first website. Her emphasis on clear functionality with appealing color combinations created a simple, elegant tool for communication. The website provided information about our organization along with educational materials. In addition, a calendar of events, as well as opportunities for living at D Acres, were clearly provided on this platform.

After using this original site for over a decade with only content updates, the website received a substantial facelift in 2012 through the diligence of Matt Palo. After years of various personnel attempting to manage the platform, files were organized in a confusing chaos that was difficult to decipher and manage. Matt converted the existing content and updated the graphics and photographic archives to new software so that we could return to regularly updating the site in-house.

E-Mail

The capacity of e-mail to connect with high volumes of highly desirable audiences at a low cost is enticing. Over the course of years we attempted to manage our promotional e-mails using standard office software. Unfortunately, over time the volume of addresses outgrew our ability to effectively send out information. Our clientele were being blocked from viewing our e-mails by their spam-detection software. Eventually we chose to hire a mass e-mail management company. This service provides reports to verify delivery and analytics to assess who opens the e-mail and who is responding to links. While expensive, these tools provide valuable feedback regarding the efficacy of our promotional efforts.

Social Media and Video

In addition to our website and e-mail capacity, we utilize social media and video-sharing platforms to communicate with the public. In 2006 a member of our board of directors created a Facebook page for the organization. Since that time we have provided regular content and created events using that social media tool.

As discussed in chapter seventeen, video sharing via the Internet also provides a mechanism for connecting with a global audience. At this point we have posted over two hundred videos on YouTube and Vimeo. The hours of footage range from lectures and classroom-style presentations to garden montages and recreational journeys captured digitally. The length of the videos varies from over three hours of lecture to minute-long panoramas of the garden. The purpose is to inform, inspire, and attract people by offering opportunities to consider sustainable alternatives and appreciate the natural resources. The video technology provides high-resolution sights and sounds from throughout the year in a convenient format for educational and cultural revelations.

I feel that, relative to print ads and rack card stands, the efficacy of dollars spent on Internet communication and advertising is superior. The technologic tools that allow strategic audience selection and mass e-mails have allowed us to track the campaigns as they unfold. Data is available on the number of visits to and views of our web portals. Our surveys and evaluations indicate that a substantial percentage of participants find us through the web. However, we do have ongoing concerns regarding the time spent deskbound and technologically dependent.

Additional Outreach Tactics

Contributing to other nonprofits is a part of our service mission and also provides an opportunity to promote the D Acres project. By donating goods and services to raffles and silent auctions held at charity events by other organizations, we generate a positive reputation and goodwill for the organization while raising revenue for the community. Instead of merely

Every space on the property is an opportunity to advertise or display our artistic endeavors.

being a cash expense the contribution provides an alternative form of advertising and promotion.

We participate as members of the League of New Hampshire Craftsmen and Artistic Roots not only to sell cottage crafts but also to gain exposure for the organizational service activities. By providing brochures and other informational materials in these retail venues we are utilizing the storefront to broaden our organizational outreach beyond the farm. By contributing a percentage of the operating costs in commission and member fees, we receive Main Street exposure while supporting the community's artistic endeavors.

Yearly Marketing Plan

While we continue to experiment with various annual approaches to marketing, consistent strategies are the basis of our promotional plan. One of the methods has been to devise a theme each year. This strategy gives focus to the organization by building an outreach program around a specific topic while also supplying the public with a distinctive context through which to understand the organizational objectives. This also grounds our grant-writing process by offering a new perspective on standard programs each year. Themes such as "Year of the Northern Forest," "Year of Traditional Arts," and "Year of Health and Wellness" have provided clarity for both the organization and the public in reaching our collective goals.

An annual calendar has been another simple and successful way to promote organizational events. This calendar is available online for printing and at times has been mailed with an annual solicitation for support. These calendars can provide an easy refrigerator reference to the events on the schedule. In addition quarterly and monthly calendars provide an avenue for updating the public and providing a promotional mechanism.

Face-to-Face Contact

In addition to the many print advertisements and Internet options that exist, face-to-face contact and word of mouth are still reliable forms of spreading a message. Attending events such as farmers' markets or tabling at exhibitions and conferences provide opportunities for direct interaction with a receptive audience. By participating at this level we are making credible connections that are memorable and impactful. Personal invitations often have a greater success rate than mass electronic transmissions.

Mailing Lists

Through the years we have developed many kinds of lists for direct marketing of our services. For instance, we have lists of career services and environmental departments at colleges and universities throughout the United States. When seeking college-age interns we have used this list to mail brochures and flyers to be posted within their offices and job resource binders. Another list we maintain promotes occupancy in our overnight accommodation, focusing on Northeastern bike and ski shops, hostels, outing clubs, and rock climbing gyms. We send these entities information about our services as a means to connect with particularly suitable clientele.

In addition, we have a general list to which we send regularly scheduled mail as well as our annual appeal and report. Over the years we have built this list from on-site sign-up sheets and tabling events, as well as a "join our mailing list" aspect of our website. This general list of five thousand–plus provides a broad spectrum of people with whom we have had direct contact through the years. This list provides a lifeline to the general constituency that we are perpetually growing. We also have developed other specific lists for promotional and outreach purposes. At times we have collected addresses of regional farmers as a means to communicate with the agricultural community. We also have an extensive list of media outlets for posting event information and connecting with journalists and editorial staff.

Over time we have shifted away from reliance on snail mail, using it only for specific promotional, fund-raising, and public support campaigns, and for people who have requested this as an option. Our regular correspondence is done primarily through digital media.

Public Participation

When we began as a nonprofit we adopted a membership structure for the organization, as discussed in chapter eight. We offered annual memberships of fifteen dollars for seniors, students, and low-income supporters; twenty-five dollars for individuals; and thirty-five dollars for family memberships. The structure also included a "sustainer" level for those who contributed more than a hundred dollars per year. As benefits we offered a 25 percent fee reduction on workshops offered by D Acres staff, 10 percent discounts on overnight accommodations, and an invitation to annual member dinners. We had no formal mechanism for inclusion in organizational decision making by members, although we did solicit them regularly for feedback and offered the annual member dinner as a forum for discussion. To me this construct represents the status quo, in which inclusion depends on financial contributions.

We are currently revising this structure. To reduce the administrative burden we are mailing letters soliciting support just once a year instead of mailing out annual renewals monthly based on initial signups. In addition to revising how the membership program is administered, we need to consider both the goals and consequences of the program. The construct of a club or other exclusive venue that is reserved for individuals based on their ability to pay must be dissolved. The goals are to build a mechanism of broad-based community support that is recognized and appreciated. We must find a way to market and promote local ownership as determined by the commitment to provide service to this land-based community project and create a construct that adequately provides a voice to a public ownership more expansive than our current board of directors model allows. This cadre would compose the core of our volunteer and community support.

As the recognition of and equity in the valuation of our participants has evolved we've searched for terminology sufficient to denote their wide range of investments and commitment. We have initiated the Many Hands program to replace the membership program as our primary mechanism for collaborating with the public. The Many Hands program includes people who contribute to the program as volunteers, monetary donors, and frequent attendees of events. By choosing this method of inclusion and recognition, we hope to invoke equity among the contributors to this project.

All of these various strategies provide a platform to maintain connections and share information with the public. By choosing a diversity of media to direct information, our outreach is holistic, providing multiple points of access for the community. This dispersal helps create a broad base of community participation and support, thus augmenting the overall goals of the project. The strength of our community equates with our organizational vitality.

CHAPTER TWENTY

IDEAS TO COME

D Acres began with a vague notion to provide service and education through actualization of alternatives in homesteading techniques. The mission of the project has been formulated to address the crisis of our humanity's current existence; however, the particular actions necessary to implement the mission are neither permanent nor specific. Our technique has been to evolve the program.

The evaluation of our evolution is based on ongoing observation and implementation. We are responsive to cultural and socioeconomic patterns on the local and global level. We are attempting to steer a course that will both benefit the on-site residents and offer alternatives for global dilemmas.

Taking Steps

Throughout the preceding chapters, steps to reduce our environmental footprint, conserve energy, ensure local food sovereignty, and utilize efficient infrastructure have been detailed, including the removal of the microwave, toaster, dishwasher, and clothes dryer; the use of bike power for simple machines; food preservation; and root cellaring.

In the future we are contemplating additional conservation measures, such as power-down days, during which we shut ourselves off completely from the electrical grid. These days would force us to unplug and reconnect with nonelectrical and grid-independent energy alternatives. By evaluating our consumption and seeking systematic strategies of conservation, we can agree to significant changes and reductions. Through this process we are making incremental steps in pursuit of the mission rather than attempting to address all the complexities at once.

The diet at D Acres, too, will continue to evolve based on individual choice and our relationship to the landscape. While we initially depended on soy imports to provide our protein, our current diet includes a mixture of chicken, eggs, pork, and beans. In the future we hope to increase the diversity of vegetable and tree nut protein produced on-site. As localized independence from animal protein becomes more feasible, we may continue to raise animals but with less focus on meat. Into the future we will continue our quest to minimize our dependence on energy and food imports.

In addition, we are working to evolve other organizational components. This requires refining the substance of our lives, from daily work and personal interactions to addressing equitable decision making and ownership criteria. To continue to evolve as a species we must as individuals seek a conscious shift of values.

This transition of spirit and culture will require social and economic changes. There will also be shifts in how we approach governance and civics.

Decorative D Acres logo integrated into the cob wall of the G-Animal

But these shifts are based in values rooted not in the religion of consumerism but in the glory of the natural world. Instead of the competitive, hostile environment that worships the individual, our mentality must shift to honor all of our kindred earthlings.

We must accomplish this task at different scales. D Acres is a microcosm of the complex world in which we live. The challenges we face as an organization mirror the broader society's. We are generally dissatisfied with government and authority, though we are unable to offer a fully articulated alternative. We often face the depression that comes with conscious awareness in this era. To overcome pessimism we continue to adapt.

People Power

The breadth of our endeavors is determined in large part by the human component of our operations. We depend on people who are able and willing to work diligently toward a sustaining, nourishing, collaborative vision.

We initially based our organization on the World Wide Opportunities on Organic Farms (WWOOF) model. This system, popularized in Australia, is an international mode of connecting farmers and workers. The general agreement is farm labor in exchange for room and board, but agreements may vary. Farms may only need assistance during certain points of the growing season, or the help they need may be with construction or child care. Farms provide descriptions in a listing that can be viewed either online or in a printed version. Willing workers can then peruse the catalog and contact farms directly.

I feel quite fortunate to have participated as a willing worker on other farm projects. The work for room and board model provides rewards at many levels. The experience enhanced my abilities by providing hands-on practice in alternative methods. Work at various farms provides insight into the vast range of possibilities and provides the tools to grapple with the unique challenges of a particular site.

Another dimension of the WWOOF experience is the ability to be a participant in the various roles of seasonal and short-term farm labor. By experiencing the reality of being a migrant worker and a landowner, you begin to understand a broader vision of the various perspectives of participants in a farm project.

At D Acres we have particularly enjoyed hosting international willing workers. While language translation issues and travel considerations can add significantly to the complexity of hosting, these short-term visitors invigorate the rural experience.

We are therefore shifting from a transitory education program that accepts large numbers of short-term workers to a more stable, perpetuating farm model. While our roots are based in the willing worker model of agriculture our dependence on transient labor creates a precarious farm economy. Though we have enjoyed many successful willing worker experiences, this is not always the case. Occasional ineptitude at farm labor or workers' unreliability sometimes creates more challenges than their work provides in benefits.

Transitory status can easily become the culture of an organization or institution. We are struggling with the reality of reaping what we sowed. By offering a training center for beginner farmers we have catered to their needs without focusing on nurturing the on-site community. This cycle has negatively influenced our ability to retain longer-term staff. Staff consistently have become exhausted by the rapid turnover of residents. The stimulation provided by transition wears on one's sense of security and comfort. Short-term visitors can require a tremendous investment to orient and meet their daily needs without their making enough contributions to return that investment to the organization. Then upon their departure, the organization must adjust to that absence of trained labor. This perpetual lack of continuity prevents potential for advanced development of the projects on-site. To reduce the unsettled chaos of successive staff turnovers we are now prioritizing our energy toward the needs of a few, dedicated, long-term residents.

Funding staff is an investment. In addition to occupancy costs, potential salary, taxes, or benefits, staff necessitate the time, energy, and experience of oversight and management. Will funding a staff member provide returns to the organization? What is the proper time frame and evaluative tool for assessing these returns? People must come with reasonable experience and serious intention to compensate the organization for their occupancy. This process demands long-term investment and continuity of interest and effort. How to budget for an increase in personnel? Personnel must be prepared to recoup costs of their occupancy. Through their endeavors they must provide an agreed portion of operation costs plus any salary or personal expenses incurred.

While some of the following positions are currently being undertaken by existing staff, we are hoping that focused, skilled, and passionate personnel in these areas will accentuate and amplify the organizational capacity. While many of these roles may seem fantastical, they are all currently accomplished in some fashion today by the existing staff. We are seeking to assemble a team of willing individuals to determine a route forward for this land-based service organization. Their positions are likely to be hybridized and integrated as personnel with varied interests arrive and depart through the years. The idea is to encourage program development in conjunction with long-term continuity while providing organizational resilience to inevitable changes in personnel composition. While we may seek expertise in certain fields, this specialization does not mean exclusion from communal tasks. The strength of our system is the collective acceptance of responsibility.

Within the garden an herbalist would be an excellent addition. By maintaining and enhancing our herb gardens this staff member would be investing in our living apothecary. Products could be developed for the health and well-being of the residents, as well as being designated for sale to the general public. In addition, an herbalist could plug into the community by offering classes or coordinating presentations that focus on plants for health.

A wellness specialist would utilize a passion for health and wellness to provide the basis for an outreach program to the broader community. There is a wide range of workshops, classes, presentations, and shares that could be offered in this area, including exercise and diet, as well as mental health and treating illness. As an organizer or instructor of classes for fitness and recreation, such as yoga, biking, skiing, snowshoeing, and canoeing, the wellness specialist would provide a valuable service to the community. Other programs such as Reiki share and instruction or nutrition classes also need direction and continuity to become a reality.

A resident woodworker would help us realize the value of our immense forest resource. By offering woodworking classes to the public, we serve to rekindle the skills of the Northern Forest people. Retail and custom-order products could be produced from the forest to the showroom on-site. A woodworker could collaborate with a forester to coordinate the outdoor wood activities. In addition to timber for milling harvest, the forester could be responsible for providing firewood and wood chips. Beyond logging, a specialist could assist with trails and wildlife interpretation.

A metal fabrication specialist would find abundant opportunities to pursue his or her craft. Welding, cutting, and shaping skills are helpful for equipment repair, design and building of structures,

Codonopsis root can be used medicinally to enhance vitality and boost energy—acting similarly to ginseng.

and farm devices. We would also like to encourage metal art for use on-site and to sell to the public.

Our goal in recruiting specialists for the garden and on the grounds is to encourage enhancements for the community. By focusing on the following particular areas and projects, our work can be refined and made suitable for broadcast as inspiration to a global audience. A specialist could also enhance our community capacity by serving as a designated point person to coordinate educational opportunities for the broader region.

Our waterworks could be improved with the focused attention of an aquaculture specialist, coordinating power generation, wastewater treatment, and food production efforts. The landscaped systems of ponds, ditches, and swales continue to evolve, and subsequent development of the plant and animal species necessitates attention. A specialist could also improve the irrigation systems currently in place, including water pumping, catchment, and delivery systems. Another possible direction of focus would be the development of a wastewater treatment or living machine structure.

An alternative construction specialist could find extensive opportunities to apply that craft on the property. The materials necessary are readily sourced in the area. On-site there are ample round-wood and stone for projects, and clay is available in the valley. The existing earthen structures would benefit from a focused plaster effort. Projects such as dwellings, greenhouses, herb dehydrators, and a sauna have been tabled until adequate personnel become available.

Seed saving requires the time commitment and focus of a specialist at D Acres. This four-season task requires concentrated effort to meet the community's needs. While replication and multiplication of our existing seed inventory is desirable, this role could also serve the wider community by offering workshops and educational events dedicated to this subject. There is also the potential for cooperative seed production and sales to the general public. In addition, a seed specialist could work toward acquiring equipment for large-scale grain and bean processing. This equipment would be designated for collective use, which would add a necessary element of food security to the region.

Some roles could be designated to develop our biological farm assets. We have an extraordinary inventory of edible fruit, nut, tuber, and berry species, as well as perennial herbs. Additional focused personnel could prune, manage pests, and harvest. We can also divide and multiply our existing plant stock to increase production as we move toward a perennial food system. There is a tremendous need to disseminate these biological resources in the region. Toward this end a perennial nursery person would add a significant element to the operations at D Acres. By propagating the adapted species on-site we have a potential to produce a diversity of valued plant species for sale. This role would require a three-season commitment that focused on multiplying the existing stock and making it available to the public. An entomologist of sorts could manage honeybees and increase the native pollinator population. By utilizing on-site and procured ingredients,

a director of compost could manage this operation and help increase the fertility of the land. Within this role could be a focus on growing green manures and cover crops.

While Regina capably wears the title of the kitchen manager, she is open to collaboration in the kitchen. Additional participants able to contribute with food preservation and cooking for guests would be helpful. As in the past the role of baker could provide an abundance of goods for the residents, guests, and farm stand.

While mechanic Kevin Maass is incredibly reliable and knowledgeable, his years of effort have not been compensated financially, and he deserves a rest from the ceaseless repairs and maintenance necessary to keep our array of mechanized implements operable. While major repairs might always be beyond our on-site capacities, our needs would make an ideal vocation for someone passionate about farming and motorized equipment.

For the organization to function fluidly and effectively we need personnel willing and able to fulfill the role of household services. Handyman duties such as painting, heavy equipment operations, plumbing, and electrical must be accomplished to expand and maintain services on-site. Either as a supervisor for willing workers or as an individual capable of self-directed, skilled labor, we need a coordinator to maintain and enhance our on-site infrastructure.

There are a variety of projects that could be undertaken by a renewable energy specialist at D Acres. We have an assortment of existing hot water and photovoltaic systems in operation on the property. These systems require monitoring, maintenance, and possibly expansion. In the past we have utilized several college-age interns to construct specific renewable energy projects. During their stay we have built bike-powered appliances and a small-scale biogas project. While such projects provide learning experiences, they seldom endure as functional components of our infrastructure. A consistent project specialist could not only construct but also maintain and refine installations,

Louie Holland drives Auguste and Henri through deep snow to create a ski path for residents and visitors.

along with providing the basis for researching and documenting for public dissemination. We would also consider expansion of the renewable energy systems to include wind and microhydro applications. In addition, an interested specialist with skills in design and construction could help us with such projects as an icehouse or living machine.

To accomplish the fundamental tasks of animal care and gardening we are looking for collaborators. While we retain responsibility for our existing domestic animals, we are considering a gradual change in the species composition on the farm. We are interested in transitioning from a pig-intensive operation to one with a goat dairy component. The rapid expansion in field space made possible with the pigs has outgrown our ability to manage with limited personnel. We would like to work toward

maintaining open spaces and concentrating our animal operations on sloping, forested land that is not as suitable for crops as field space. By switching our efforts we hope to attain the bounty from shepherding the goats without compromising our most desirable plant production zones on the landscape.

The future will also require decisions regarding our oxen team. After ten years the boys have begun showing signs of their mortality, with Auguste developing severe arthritis in the knees. As we have reached our goals of opening up field space within the forest, our reliance on the team for intensive logging efforts has diminished. With the acquisition of mechanization there are also alternatives readily available. Soon we must decide if the numerous benefits outweigh the costs of maintaining an on-site team.

An animal specialist could be extremely helpful with implementing an effective rotation and grazing plan. While we have experimented with chicken tractors and raising forage crops, a specialist could focus time and energy on refining the productivity and effectiveness of these systems. From breeding to butchering there are many facets to the operations that require careful attention. In addition, fiber and dairy both require specialization and focus.

Organizational development could also use the practiced attention of a specialist. Personnel focusing on grant writing, fund-raising, outreach, and event coordination would amplify our limited experience and reduce the workload on other staff. Particular media-related skills such as graphic design and videography would be helpful to augment our in-house publicity capacity. We have also considered roles for personnel linked with public relations and services. An individual interested in marketing, publicity, and media relations and correspondence could expedite the mission of the organization. Someone capable of coordinating our overnight accommodations as a hospitality manager would free the farm team for more time in the field. It would be helpful if an individual could actively engage with the youth education opportunities. By serving to coordinate with local schools and summer camps we could provide presentations and assist with gardening and sustainability programs. In addition, this individual could help plan and implement tours of the on-site facilities. In developing child-age education we need a credible and committed representative who will invest in the community and the content of the programs. Particular events and workshops geared toward youth provide a connection in a designed format. After-school and D Acres summer camp activities are other ways to provide service and education to the community.

Service to the Community

In addition to work dedicated toward developing a model of land stewardship and community living, we continue to serve the general public with specific programs, inviting experts in their fields for workshops and public presentations as well as arts and cultural events. To support the agricultural development of the region we hope to provide infrastructure to enhance our regional capacity. Through the acquisition of a silo for grain storage we will be able to buy grain in bulk for distribution to our neighboring small farms, significantly reducing feed costs. We also hope to acquire agricultural equipment to increase the region's food processing capacity. Capital-intensive equipment to automate tasks such as sifting compost or threshing beans could be shared or lent to other operations, empowering all to produce more local food. It is an organizational goal to find significant points of leverage to actualize regional sustainability.

Infrastructure

With regard to future infrastructure development, the following are the essential projects for which we have the highest need. Turning these plans into reality relies on the correct combination of materials, personnel, skills, time, economics, weather, and commitment.

Waterworks are a crucial component of our perpetuation on the landscape. By buffering the zone from extreme inundation or drought we are providing resilience. While the importance of ponds and waterworks is elaborated elsewhere in the book, this priority cannot be overstated in our plans for the future. It is crucial to install waterworks before climate severity wreaks unmitigated havoc.

A living machine component would serve both a utilitarian and an educational purpose as an alternative to the conventional septic and leach field setup. Because we installed an approved septic system as a precursor to our kitchen licensing, we have not needed an alternative wastewater system for the Community Building. Therefore the project has been deprioritized until targeted funding or personnel become available.

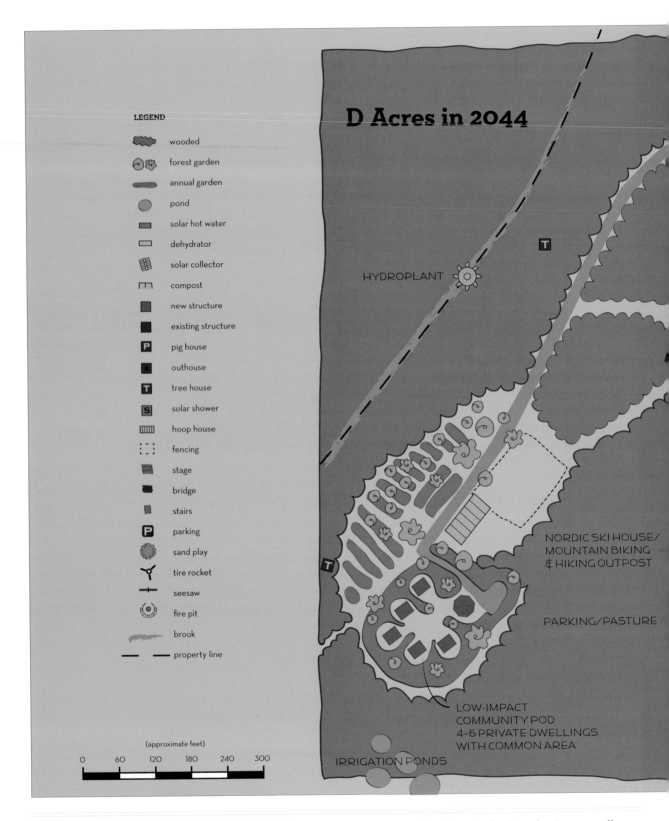

LEGEND

- wooded
- forest garden
- annual garden
- pond
- solar hot water
- dehydrator
- solar collector
- compost
- new structure
- existing structure
- **P** pig house
- outhouse
- **T** tree house
- **S** solar shower
- hoop house
- fencing
- stage
- bridge
- stairs
- **P** parking
- sand play
- tire rocket
- seesaw
- fire pit
- brook
- property line

(approximate feet)

0   60   120   180   240   300

D Acres in 2044

HYDROPLANT

NORDIC SKI HOUSE/
MOUNTAIN BIKING
& HIKING OUTPOST

PARKING/PASTURE

LOW-IMPACT
COMMUNITY POD
4–6 PRIVATE DWELLINGS
WITH COMMON AREA

IRRIGATION PONDS

The 2044 map illustrates development aimed to address our future needs in housing and food production, as well as
Illustration by Marylena Sevigney.

RICE TERRACE FIELD

LIVING MACHINE

AMPITHEATRE

GOAT PASTURE

ICE HOUSE

FOREST GARDEN

COMMUNITY GRAIN SUPPLY

VISITOR CENTER ART/FARM STORE

PASTURE

EQUIPMENT STORAGE

economic and recreational goals. These investments are intended to perpetuate the vibrancy and viability of the project.

After the construction of the outhouse, we had originally planned to build a sauna. Since then the sauna has consistently been mentioned as a possibility; we have gathered materials and made plans, but we have not yet broken ground. Lack of confidence in our design and building strategy has perpetually delayed the project. In recent years the conceptual plans have been expanded to include a hot tub as well as the functionality of an herb dehydrator.

We are hoping to provide security and sense of place to residents by offering personalized structures to meet their long-term needs. Our first candidate to be the recipient of a new construction project is Regina. Over the next couple of years we are hoping to build a structure that offers more permanent, personal space, including a sleeping and studio room for four-season usage. At this time we are determining its location and design. We are considering additional units for the future and developing a plan to share resources and access to provide proximity yet privacy, as well as reduce costs of redundancy in road and facilities.

This type of infrastructure adds complexity to the operations at D Acres. The organization must decide how to approach constructing this type of personal space. While we are versed in construction techniques we must address new questions regarding tenure and ownership, including who will pay the construction costs and whether there will be a lifelong lease of sorts.

A Framework for the Future

As a youth my frame of reference for discussing community living was limited to preexisting stereotypes. The term "commune" has many connotations. Drugs, free love, and laziness are often the images of American commune life in the 1960s. When discussing groups of unrelated people collaborating, the term is frequently utilized as derogatory. In order to move beyond these presumptions we must provide new terminology to the American lexicon that describes land-based service projects. While "cohousing" explains part of the concept and "workers' collective" describes another facet, perhaps the closest existing terminology is "ecovillage."

We would like to develop personnel continuity and provide the possibility for longer-term on-site relationships. We would also like the property to remain in use as a farm into the future. We fear possible subdivision that would cut the property into pieces. We would like to provide a home for residents and meet peoples' needs collaboratively without defeating the concept of collectivization through subdivision. Our goal is to maintain the contiguous property by sharing in public ownership.

People crave the security of a home, not only a place to keep their possessions but also a place where someone cannot tell them to leave without due cause. For many, a lifelong lease could satisfy their need to declare a space "home." Security can also be provided by granting decision-making status or legal ownership. We intend to provide such security through transparent management processes, as outlined in documents such as the Organizational Manual. While the current governance structure of the organization allows open access to voice concerns, it does follow the corporate structure of a 501(c)(3). This board-directive governance must be responsive to the needs of the on-site personnel and the constituents.

To feel secure people need to have a voice in decisions by which they are affected. The status or weight of an individual voice within the community is a constant point of discussion. Some people can be more influential in a decision-making process through intellect, personality, or powers of persuasion. As we move forward we must continue to be open to compromise, define community expectations and meet them, and ensure fairness.

Is it ideal to create a flat governance structure without hierarchy ruled by consensus or sociocracy? Or is it inevitable that a hierarchy is established based on commitment, investment, and experience? If this inevitably occurs, how do we address and define the structure of decision making to provide fairness to all parties involved?

We are intent on setting limits on our expansion and growth. The goal of the land-based project is to maintain a rural population density for the benefit of the ecological balance, including the human element. On the two-hundred-acre property, our maximum population should be fewer than thirty people, including children and the elderly. To achieve this density will require the time and investment of many and may or may not occur. At this juncture we are keeping the doors of possibility open regarding our plans for the future.

Public Ownership

Our goal is not to continue the downward spiral of rural development, whereby subdividing our farmlands and woodlands is destroying resources to sell the illusive American Dream. Such suburban and rural housing does not invest in agricultural capacity; by allowing this development we are losing real security by destroying the peoples' ability to create food from the land.

There are too few alternatives for legally devising a way to perpetuate shared responsibilities of land tenure, stewardship, and ownership. Land trusts provide a method for public ownership; however, the operations of a trust are ultimately dependent on the people who serve as the administrators of the trust's mission.

We do not have a current strategy to resolve the question of public ownership. The property is owned by the Trought family. We have considered gifting the property to the nonprofit under the provision that the organization would pay a reasonable share of property taxes. We have not proceeded with that option because of two primary concerns. The first concern is the viability of the organization. Are we certain we have the ability to maintain a positive cash flow in perpetuity? What would happen if D Acres ceased to exist? The second concern regards concentrating control of the assets with the D Acres

board. We have investigated the possibility of forming a separate trust charged with maintaining a conservation mission. The primary concern of the trust would be land management, while the service and education operations would remain the domain of the D Acres board.

Cultural Shift

Because our society has shifted so far from our agricultural heritage, it is an extreme shock to be reimmersed in seasonal endeavors and physically demanding, time-sensitive activities. While mainstream society romanticizes farm life, the consistent effort to accomplish farm work requires inspiration sourced from idealism and necessity. We must find ways to encourage the populace to embrace the challenges and rewards of perpetuating a sustainable socioecology.

Attitudes toward work depend on the individual. For collectives such as D Acres to endure, positive flows of energy embodied in people, in the ecosystem, and on the ledger sheet are necessary. Our survival depends on meeting goals for both our physical and spiritual existence. But sharing an investment in people and place is daunting in the face of our contemporary culture of individualism. The challenge will only be met when people are willing to sacrifice instant gratification for a long-term vision. Freedom from the perils of a society of depletion can be found in regenerative living. One of the sources of this regenerative energy must be within our physical bodies. We must undertake to perform physical tasks for our own survival. Without the crutch of industrial-scale energy and resource consumption our mechanism for survival will be reduced to our physical, intellectual, and spiritual capabilities.

The work of creating food subsistence and shelter through sustainable practices is necessary for our survival and has tremendous benefits; nevertheless, it is difficult. Long hours of muscle exertion can be tiring. Working through the challenges of living collectively can be exhausting. While we all realize the urgency of our circumstance, the burden of the immensity of work that needs to be done can quickly become a heavy, onerous weight, leading to negativity. But instead of being overcome with despair, we must manifest the joy of this labor. We must demonstrate the fruition of investing in people and place. By evaluating the bounty of results and abundance produced by our efforts we see its value in comparison to the destruction of the exploitative, capitalist model.

In addition to revaluing the essential nature of labor, we must invest in a connection to the food we grow. By valuing food produced by its consumer, we take personal responsibility to follow a noble, responsible path to sustainability.

D Acres—the organization, the community, the model—continues to evolve. As this progression occurs we return to the doctrine of localization. Localization of our economy recognizes the limits to growth innate in every system. By defining a limit, we can prioritize building the strength of the localized system. We receive inspiration from our belief that, while our efforts primarily occur in the Baker River Valley, we are immersed in the global challenge of designing human perpetuity.

Conclusion

"Fall in love with hard and patient work . . . we are the beginning, not the end."

SLAVOJ ŽIŽEK

In August of 2013 the Cuban permaculturalist Roberto Perez Rivero visited Keene, New Hampshire, and I had the opportunity to ask him what advice he would give North Americans as we approached the precipice of our energy consumption pattern. He voiced his concern that we would react violently to shortages and the rationing of resources, pointing out that a percentage of Cubans were unwilling to ration, choosing to risk their lives in boats as refugees and flee to Miami rather than face the reality of a subsistent, sustainable life. He

concluded by explaining that there are three phases to creating community: first we begin as individuals, then we form collectives, until finally we can become a community.

Through observation and rational assessment of humanity's current trajectory, we need to recognize the urgency of action. I am not suggesting that D Acres is the solution nor preaching or advocating any singular solution for the situation at hand. But any response must involve community. My hope is that D Acres' work demonstrates feasible concepts as alternatives to the conventions of today. Rather than prescribe a particular course of action, it is important for us to develop a design based on the circumstances and goals specific to the participants.

Our motivation should not be to act out of fear, which breeds panic and poor, selfish behavior and decisions. We need to calculate the contingencies in a calm manner, making assertive moves to address and remediate our plight. We must become fortified by the knowledge that we can transcend this shallow, egotistical state and perpetuate a healthier existence into the future. To reach this goal we must abandon the doctrine of greed and exploitation and replace that creed with norms of collaboration and common good.

While some may prefer the status quo, our contemporary circumstance is one in which luxuries and conveniences are appropriated to a few at the expense of the masses. I embrace a future that offers fairness and equity. The challenge is to deliver this doctrine while nurturing a reality in which we can face our obstacles collectively, relying on our instinct for survival while our consciousness helps us transcend beyond our competitive primal urges.

It is important to recognize our competitive behavior and to channel this compulsive energy toward more productive purposes. Instead of focusing on sports or fashion, perhaps we can shift beyond the contemporary hyperglorification of an ego-driven self. By diverting our attention from the limited scope of the individual, we can direct our efforts toward the primal challenge of survival for our species. Seeking to join with an existing group or evolving our current living situation is as essential as founding a community anew.

It's not uncommon for visitors at D Acres to state that they "want to start something just like this." I certainly do not discourage other farm-based service organizations, but the process of founding a similar organization no longer appeals to me. While the notion of starting your own instead of joining an existing organization is pervasive, we must also recognize the value of uniting with and supporting existing efforts. Part of our current societal mythology involves the value placed on the individual and

his or her accomplishments above all else. While every circumstance is unique, it is vital that individuals join others to gain experience and strengthen the movement. Instead of "what I have done" and "what is mine to claim," we need to focus on what has been collaboratively built. Individual possessions provide an illusion of control and security. Give freely, and invest in community and the earth.

What must not be done is to stand aside and be a spectator. People often say that they will come to live with us after the shit hits the fan; frankly, that's too late. We must shift from a life in which materialism is the sole stimulus that engenders responsibility

Frog thinking outside the box

and commitment. We must change our attitude and share in mutually beneficial behaviors instead of the divisive individual ownership model.

Instead of seeking grass in distant greener pastures, take a stand. Get somewhere and start doing; plant roots. Persistence pays: The point is to make the grass green where you are. Be present and accountable. There are immense rewards for working within an ecological system. A primary benefit is seasonal fresh food, which serves as medicine to promote nutrition and exercise. Health, provided by this lifestyle, is the most valued commodity on Earth. We are hoping to realize a paradigm that is built with wealth expressed in terms of health, and quantified by the strength of the ecosystem's resilience.

Another benefit is that of staying rooted in one place for thousands of days. It is important to love a place, wherever you may be geographically. This gift is earned through time and daily systematic observation. The soul of a place is a combination of many things, including people, but also the bricks and mortar, soil, scents, sights, birds, and bees. We must also address the challenges—social, economic, geographic, climatic—that place presents. Above all we must take a stand . . . the ground beneath your feet is yours to share for the common good.

Instead of everyone needing his or her own dog, kitchen, bathroom, boat, vehicle, house, second home, second vehicle, and personal vacations, we need to share these conveniences and luxuries. We also need to share responsibility and commitment.

A Plymouth State University environmental science class visits D Acres for an Earth Week tour.

Rather than a culture of mobility without an enduring social fabric we need a stable fertile field for community relations.

We live in an era that values recreation, but what are we lacking in our lives that we need to recreate? Excitement, a sense of accomplishment, physical exertion? Risk-taking behavior? These can all be generated through realizing daily tasks of doing. Let us change our vocabulary from "them" to "us," and from "me" to "we." Instead of taking, we must recognize the value of sharing. Share what we claim to own: ride share, share space, share property, rent a room, share food, host potlucks. We have enough.

We are blessed by our existence. And while I am thankful for that gift, I have questioned, "What is the purpose of humanity?" My belief is that our purpose is to find our place on this earth, to cultivate our health through food and community. To achieve this level of being, alternatives to destructive reality must move from theory to practice. There is no time for more handwringing, aspirational chatter. We must value each precious day.

The corporate mentality, which demands immense profit margins to pay large dividends to shareholders, is unsustainable. The time has come to make conscious decisions about our future. The solution is to regain our independence. Localize our economy. If unable to buy your own farm, join an existing one to learn the ropes—don't reinvent the wheel. To cope with the travails of living with the land you have to believe it is worth the effort to do it. Our success is defined by our action. The choice to face challenges and follow an alternate path is a success in itself. While we may seek measurable terms for analysis, realization and first steps are the movement that we need. Break the inertia of the conventional.

This is the basis of D Acres and its personnel, to collectively steer our course and perpetually refine our operations, to react to our environment and adapt. I do not believe in a fairy-tale future full of bliss and leisure. I do not have faith in technology to save mankind, nor do I wish to continue with our exploitative contemporary culture. Meeting the challenges of the future requires sacrifice, but it also yields profound rewards. To find our purpose and truly own up to our destiny, we must define and pursue our goals with intention.

I am inspired by visitors who bring their children and grandchildren to D Acres. It invokes a feeling that our work is valued and that we can share it with the next generation. While we are intent on striving for mutual well-being, we must also seek an internal serenity. We must take the time to stand in the sun and realize the contentment of life, the joy and fulfillment of sharing these moments on the planet. Enjoy your time here doing good work!

ACKNOWLEDGMENTS

I would like to thank all of the D Acres participants through the years, as well as the supporters near and far of the project.

Much gratitude to Beth Weick for her help editing the manuscript.

I am thankful to the many teachers of permaculture who have dedicated their lives to the propagation of a perpetuating culture. I am especially thankful to have shared ideas with Micki Visten, Dave Jacke, Keith Morris, Lauren Chase-Rowell, and Steve Whitman.

Thanks to Peter Bane, publisher of the *Permaculture Activist*, and the Intentional Communities Network for keeping their lights on.

Thanks to Bree for being the right person at the right time.

Thanks to Chelsea Green for taking the contents and making a book, and especially for the brilliance to bring Michael Metivier onto the project.

I mention many other people, businesses, and organizations by name throughout the book. My goal in specifying by name is more than an acknowledgment. I prefer to recognize that it is actual individuals and businesses who have chosen to contribute as the glue of the community. People often ask how we have accomplished as much as we have. While the short-term visitors have provided a portion of the labor, it is the local community partners who have provided the continuity, stability, expertise, and reliability that has ensured organizational support. Through this process our mission in becoming a community organization has been self-fulfilling.

RESOURCES

Intentional Communities and Ecovillages

Fellowship for Intentional Community—www.ic.org
The ultimate source of information about living in community. Produced a printed catalog for many years which is now online. Also produces *Communities* magazine.

Dancing Rabbit Ecovillage—www.dancingrabbit.org

Sandhill Farm—www.sandhillfarm.org

Red Earth Farms—www.redearthfarms.org
Three unique and successful ecovillages and sister communities located in Missouri, they all function without the restrictions of land zoning and are mutually supportive, though each has distinctive goals and interests.

Caravana Arcoiris por la Paz—www.caravana arcoiris.blogspot.com
A mobile ecovillage. I was a member for a brief part of their journey, which originated in the ecovillage Huehuecoyotl in Mexico and traveled throughout Central and South America for over a decade.

Lost Valley Education and Event Center—www.lost valley.org
Similar mission and focus to D Acres with educational events and a permaculture focus.

Occidental Arts & Ecology Center—www.oaec.org
Well-established project in Sonoma County, California, that has evolved through several permutations since 1974.

The Farm—www.thefarm.org
Well-known survivor of the community movement of the '60s, which is now home to second-generation residents. Houses several successful microenterprises.

Sirius Ecovillage—www.siriuscommunity.org
Massachusetts pioneer of sustainable living, combining education and ecovillage concepts.

Agua Blanca—www.machalillanationalpark.com /agua_blanca_community.shtml
An indigenous community on the coast of Ecuador that served as an exemplar of minimalist cooperative living and forest gardening during my travels with La Caravana.

Twin Oaks—www.twinoakscommunity.org
A well-established intentional community in Virginia that relied heavily on a single stream of income—making hammocks—until their major buyer outsourced production to overseas manufacturers. This lesson enforced the need for diverse streams of income at D Acres.

Acorn—www.acorncommunity.org
Offspring of the Twin Oaks community. Has grown through the success of its microenterprise as a seed purveyor.

Pendle Hill—www.pendlehill.org
Serves as a retreat and education center that focuses on the Quaker traditions of conflict resolution and peaceful coexistence.

Earthaven—www.earthaven.org
A North Carolina ecovillage based on permaculture principles.

Permaculture

Finger Lakes Permaculture Institute—www.finger lakespermaculture.org
Serves as a regional training center for permaculture principles and education.

Central Rocky Mountain Permaculture Institute—
www.crmpi.org
>Founded by Jerome Osentowski, CRMPI was
a tremendous influence during my time at
Solar Energy International. My sister also
served as a resident as the D Acres concept
was being formulated.

Prospect Rock Permaculture—www.prospectrock
permaculture.wordpress.com
>Founded by Keith Morris, who also created
the Northeastern Permaculture Network
listserve and helped found the Permaculture
Institute of the North East, this site specializes in education, design, and permaculture
plant nurseries.

Bullock's Permaculture Homestead—www.perma
cultureportal.com
>The Bullock's permaculture design course
trained two of D Acres' original designers,
Charles Dalton and Brenna Nies, for the D
Acres project on Orcas Island.

Portland Permaculture Resilience Hub—www
.resiliencehub.wordpress.com/
>Pioneers of urban permaculture, with a store-
front community center in Portland, Maine.

Organizations

Maine Organic Farmers and Growers Association—
www.mofga.org
>The oldest statewide organic organization in
the country. Hosts the premier agricultural
fair in North America in Unity, Maine, every
third weekend in September.

The Good Life Center—www.goodlife.org
>The site of Helen and Scott Nearing's final
homestead, on the coast of Maine.

Aprovecho Research Center—www.aprovecho.org
/lab/home
>For many years Aprovecho offered a sustain-
ability immersion program similar to that of
D Acres; they have evolved to focus primarily
on their principal interest in supplying rocket
stove technology.

Beehive Design Collective—www.beehivecollective
.org/en/
>Founded during the same period as D Acres,
this collective disseminates its message of
environmentalism and politics throughout the
globe from their base in Machias, Maine.

The Bread and Puppet Theater—www.bread
andpuppet.org
>Founded by Peter Schumann and his wife,
Elka, granddaughter of Scott Nearing, in
1963. As a groupie I have been influenced by
their theater, dance, printmaking, parades,
and puppets, as well as their focus on food
and the power of demonstrative activism.

Artistic Roots—www.artisticroots.com
>I am the last of the original founding
members of this local art and education organization, which demonstrates the difficulties
as well as the perseverance of the cooperative
model of business ownership.

Plymouth Area Renewable Energy Initiative—
www.plymouthenergy.org
>Local organization founded to provide
mutual support within the region as we face
the peak of oil production. PAREI continues
to innovate and evolve with both changing
technology and needs of the people.

Hostels and Camping/Recreation

The Hostel in the Forest—www.foresthostel.com
>A tree house–based tourism business that
served as a model for D Acres.

Deer Isle Hostel—www.deerislehostel.com
>Combines traditional homesteading with
unique overnight accommodations.

Doe Bay Resort & Retreat—www.doebay.com
>An inspiration for D Acres during our time on
Orcas Island, for its recreation opportunities
and local food sources.

Websites and Magazines

Permaculture Activist magazine—www.permaculture
activist.net

The premier source of information on permaculture in North America.

Adbusters—www.adbusters.org

Originators of the Occupy movement, this magazine is the vanguard of the antiglobalist alternative. Provocative and influential magazine for activism.

Edible Forest Gardens—www.edibleforestgardens.com

Dave Jacke's web presence offers information on forest gardening.

Local Goods Guide—www.facebook.com/LocalGoodsGuide

The website for this guide originated at D Acres and has morphed into a private enterprise managed by our long-time collaborator, Stacey Lucas.

Market and Supplies

Fedco Organic Growers Supply—www.fedcoseeds.com

Major source of D Acres seeds and rootstock of the on-site orchard, the catalog provides helpful plant and political information.

Local Foods Plymouth—www.localfoodsplymouth.org

While the concept originated at D Acres, the program was initiated through collaboration with PAREI and has been maintained since 2006 based on a broad base of community support and perpetual innovation.

Education

The Farm School—www.farmschool.org/

A Massachusetts-based school that provides a variety of farming education programs for all ages.

Turtle Island Preserve—www.turtleislandpreserve.com

Founded by Eustace Conway (made famous in Elizabeth Gilbert's *The Last American Man*), teaches traditional skills to all ages.

INDEX

Note: page numbers followed with p refer to photographs

root zone, 257–58

rose hips, 213

Roy, Rob, 123

rubble trench foundations, 118, 142, 144

rubus species, 273, 281

Ruby, K., 60

Rumberger, Emily, 46p

runoff, 182, 184

R-value, 118

rye, 236

Sachs, Paul, 236, 267

Sadhana Forest, 16

safety, 46–47

sage, 258

salaries, 48–49

Salatin, Joel, 323

sales and services, 75–76

Sanctu tree house, 156

Sandhill Farm, 377

Sant Bani School, 313p

sauerkraut, 220–21

The Sauna (Roy), 123

Schumacher, E. F., 14

Scott, 62p

scythes, 278

Seasonal Soup Night, 312, 325

seaweed, 234

seed calendar, 270–71

seeding, 266, 269–71, 273. *See also* potting soil

seed saving, 50, 268–69, 273, 363

self-sufficiency, 7

Senior Center, 302

shade, 128

shear strength, 115

sheep, 28, 36

sheet mulching, 231, 237–42

shiitake mushrooms, 276, 296

shrub layer, 259

Siberian peashrub, 262

signage, 323–25

Silent Spring (Carson), 10–11

silo, 113, 130, 160–61

Sirius Ecovillage, 14, 377

site selection and work, 115–16

skid steer, 247p

skid trails, 284, 290, 292–93

Skinny Shack tree house, 27p, 156, 157

slash and burn techniques, 230

slaves, 13

slugs, 296

Small and Beginner Farm organization, 323

Small is Beautiful (Schumacher), 14

smoking meats, 219–20

snakes, 206

snow, 25, 114–16, 120, 126, 154, 173, 182

social activities, 318–20

social media, 354–55

Society for the Protection of New Hampshire Forests (SPNHF), 82

soil amendments, 236–37, 268

soil fertility and management, 34, 37, 206, 227–31. *See also* garden development

solar cookers, 176–77

solar dehydrator, 174–76, 210–11

solar design, 126, 127–28, 139–40, 170–71. *See also* photovoltaics; renewables

solar energy, 265–66

solarium, 136, 272

solar shower, 169p, 172p, 185

Solheimar, 14

Solviva: How to Grow $500,000 on One Acre & Peace on Earth (Edey), 15

Solviva Solar Dynamic, Bio-Benign Design, 15

sourdough, 220

southern microwatershed, 185–86

speaking engagements and presentations, 328–29

specialists, 49–50

spicebush, 262, 263p

spinach, 269

SPNHF (Society for the Protection of New Hampshire Forests), 82

spoons, 336–37

spray foam insulation, 119

squash, 219, 258

squirrels, 27

ABOUT THE AUTHOR

Josh Trought helped to found D Acres of New Hampshire in 1997, and today serves as its director. He is a Phi Beta Kappa graduate of the University of Colorado Boulder with a degree in environmental conservation, and has worked for organizations as diverse as the National Oceanic and Atmospheric Administration and Costa Rica's Tapantí National Park. Currently he is a member of the Artistic Roots co-op in Plymouth, New Hampshire, and serves as treasurer of the Pemi-Baker Solid Waste District. His proudest accomplishments are as a participant at the Bread and Puppet Theater; at La Caravana Arcoiris por la Paz, a mobile ecovillage that traveled across Central and South America; and at D Acres, where he has specialized in forestry, construction, gardening, and administration. The primary motivation behind his life's work is the healthy perpetuation of our species within its various, diverse ecologies.